Proceedings of the Second International Oats Conference

WORLD CROPS:

PRODUCTION, UTILIZATION, DESCRIPTION

Volume 12

1. Stanton WR, Flach M, eds: SAGO. The equatorial swamp as a natural resource. 1980. ISBN 90-247-2470-8
2. Pollmer WG, Phipps RH, eds: Improvement of quality traits of maize for grain and silage use. 1980. ISBN 90-247-2289-6
3. Bond DA, ed: *Vicia faba:* Feeding value, processing and viruses. 1980. ISBN 90-247-2362-0
4. Thompson R, ed: *Vicia faba:* Physiology and breeding. 1981. ISBN 90-247-2496-1
5. Bunting ES, ed: Production and utilization of protein and oilseed crops. 1981. ISBN 90-247-2532-1
6. Hawtin G, Webb C, eds: Faba bean improvement. 1982. ISBN 90-247-2593-3
7. Margaris N, Koedam A, Vokou D, eds: Aromatic plants: Basic and applied aspects. 1982. ISBN 90-247-2720-0
8. Thompson R, Casey R, eds: Perspectives for peas and lupins as protein crops. 1983. ISBN 90-247-2792-8
9. Saxena MC, Singh KB, eds: Ascochyta blight and winter sowing of chickpeas. 1984. ISBN 90-247-2875-4
10. Hebblethwaite PD, Dawkins TCK, Heath MC, Lockwood G, eds: *Vicia faba*: Agronomy, physiology and breeding. 1985. ISBN 90-247-2964-5
11. Sørensen H, ed: Advances in the production and utilization of cruciferous crops. 1985. ISBN 90-247-3196-8
12. Lawes DA, Thomas H, eds: Proceedings of the Second International Oats Conference. 1986. ISBN 90-247-3335-9

Proceedings of the Second International Oats Conference

The University College of Wales, Welsh Plant Breeding Station, Aberystwyth, July 15–18, 1985

edited by

DUDLEY A. LAWES and HUGH THOMAS
Welsh Plant Breeding Station
Aberystwyth
U.K.

1986 **MARTINUS NIJHOFF PUBLISHERS**
a member of the KLUWER ACADEMIC PUBLISHERS GROUP
DORDRECHT / BOSTON / LANCASTER

IV

Distributors

for the United States and Canada: Kluwer Academic Publishers, 190 Old Derby Street, Hingham, MA 02043, USA

for the UK and Ireland: Kluwer Academic Publishers, MTP Press Limited, Falcon House, Queen Square, Lancaster LA1 1RN, UK

for all other countries: Kluwer Academic Publishers Group, Distribution Center, P.O. Box 322, 3300 AH Dordrecht, The Netherlands

Library of Congress Cataloging in Publication Data

```
International Oats Conference (2nd : 1985 : Univer-
    sity College of Wales)
    Proceedings of the Second International Oats
Conference, The University College of Wales, Welsh
Plant Breeding Station, Aberystwyth, July 15-18, 1985.

    (World crops ; v. 12)
    Includes bibliographies and index.
    1. Oats--Congresses.  I. Lawes, Dudley A.
II. Thomas, Hugh, Ph. D.  III. Title.  IV. Series:
World crops (Hague, Netherlands) ; v. 12.
SB191.02I57 1985      633.1'3         86-8396
```

ISBN-13: 978-94-010-8461-1   e-ISBN-13: 978-94-009-4408-4
DOI: 10.1007/978-94-009-4408-4

Softcover reprint of the hardcover 1st edition 1986

Copyright

PREFACE

The Second International Oat Conference, of which these are the Proceedings, was held at Aberystwyth, 15-18 July, 1985.

In the business meeting held at the close of the First International Oat Workshop that was held at Pennsylvania State University in 1982, an International Organising Committee under the chairmanship of Dr KJ Frey was elected to organise the next conference, and Aberystwyth was proposed and agreed as the venue. The final date of the Second Conference and the outline of the programme, including selection of the main speakers, was agreed by the International Committee but local arrangements and finalisation of the programme were delegated to a local sub-committee centred at the Welsh Plant Breeding Station. We wish to record our appreciation of the work done by this local committee in assuming various organisational responsibilities.

From the outset there was a firm committment to make the Conference as international as possible and every effort was made both to provide a programme that would attract participants from all parts of the world and to keep costs to a minimum so as to increase the chances of potential delegates being able to attend. A list of participants is included but it is worthy of note that twenty-three countries were represented.

The Conference was organised into seven half-day sessions. Six of these covered different topics relevant to the improvement of oats and the seventh was a visit to the Welsh Plant Breeding Station where laboratory and field work on the oat crop was demonstrated.

A considerable amount of practical and scientific information concerning oats and the improvement of the oat crop was presented in papers and posters and arose in formal and informal discussions. This exchange has gone a long way to rationalise and disseminate our knowledge of the crop and it is gratifying that there is now a permanent record of the Conference. This was made possible largely by financial assistance, towards the cost of producing the Proceedings, generously donated by the British Oat and Barley Millers' Association. For this we wish to record our gratitude. We would also like to thank Mrs Margaret S. Mack for typing and her endless patience.

The next International Oat Conference will be held at Svalöf in Sweden in July 1988. It is then proposed that, thereafter, this Conference should alternate with the American Oat Workers Conference and be held every four years.

D A Lawes
Hugh Thomas
Joint Local Organisers of the Conference
and Editors of the Proceedings

INTERNATIONAL ORGANISING COMMITTEE

Dr K.J. Frey, USA (Chairman)
Dr J.B. Brouwer, Australia
Dr R.I.H. McKenzie, Canada
Dr T. Morikawa, Japan
Dr H.L. Shands, USA
Dr H. Thomas, UK

Second International Oat Conference

ACKNOWLEDGEMENTS

The Organising Committeee are grateful to the following who contributed towards the staging of the Conference:

The University College of Wales, Aberystwyth

The Welsh Plant Breeding Station

Aberystwyth Town Council

The British Oat and Barley Millers' Association

Morning Foods Ltd

Nickersons RPB Ltd

Duraweld Plant Breeding Supplies

United Kingdom Agricultural Supply Trade Association Ltd

These Proceedings

are Dedicated to the Memory of

PROFESSOR E.T. JONES

Pioneer Oat Breeder and Research Worker

CONTENTS

OPENING OF CONFERENCE

SESSION I: GENETIC RESOURCES AND THEIR USE IN THE BREEDING OF OATS

SESSION II: PESTS AND DISEASES OF OATS

SESSION III: EFFICIENCIES OF BREEDING METHODS FOR OATS

SESSION IV: CROP PHYSIOLOGY AND PRODUCTION METHODS OF OATS

OPENING OF CONFERENCE

3

OPENING OF CONFERENCE AND WELCOME TO DELEGATES

Dr E. L. BREESE
Deputy Director, Welsh Plant Breeding Station

Welcome to the 2nd International Oat Conference and to Aberystwyth. I hope this University town hedged in by mountain and sea will live up to its reputation for a sunny clime and I trust that the relative calm of this College campus in vacation time, will provide a suitable environment for your deliberations.

It is a privilege for the Welsh Plant Breeding Station and its parent body, the University College of Wales, to host this conference and to welcome so many distinguished visitors from so many countries.

Your presence here acknowledges the long involvement of the P.B.Station with oat breeding. Although the fame of our founder Director, Sir George Stapledon, is associated with grassland improvement, he was well aware of the importance of cereal concentrates in the economy of livestock systems in Britain. Accordingly when the Welsh Plant Breeding Station was set up in 1919 to study grassland improvement and the breeding of grasses and clovers it also included an oat breeding department, - oats being the important cereal at that time in the west of Britain. Perhaps you will allow me to pay tribute to the name that will always be associated with pioneering improvements to the oat crop during those early days at Aberystwyth, - E.T.Jones. He took charge of the programme in 1922 and over the next decades made major advances in developing a series of varieties suited to mechanised farming, with an emphasis on resistance to pests and disease, and adapted to high and low fertilty situations for winter and spring sowing.

Professor E.T.Jones went on to make his mark in the wider area of improved crops for livestock farming as Director of the Station. But his first love and commitment was oat breeding and on his retirement in 1958 he maintained a close contact with the breeding programme in the Station. It was Professor Jones' final ambition to attend and participate in this International Conference - and we had hoped that that ambition would be realised. But it was not to be, - he died in April of this year at the age of 92.

On Wednesday you will have the opportunity to visit the WPBS and we look forward to showing you our oat breeding programme in the context of the Station's overall responsibilities.

This International Conference on crop breeding and production, with all the opportunities it affords for discussion, debate and exchange of information is perhaps particularly timely when the disparity in food supply throughout the world has never been more obvious (overt), ranging from costly overproduction in many developed countries, to stark-famine in those less fortunate. Few governments have the perspicacity to recognise the major part that plant breeding can play in contributing to a solution to both problems. A reaction to overproduction so often is to reduce funding for that very plant breeding research and development which is the surest way to reduced production costs on expensive fertilisers and agrochemicals, while putting a greater emphasis on dietary quality. Again the greatest aid that can be offered to less privileged countries is the scientific knowledge and skills to develop their own adapted food crops. I am sure there are lessons to be learned from a study of the oat crop which have relevance far beyond the crop itself, - and this conference will no doubt spell them out.

I am sure you would wish me to pay tribute to your Organising Committee for assembling a most attractive programme and ensuring that you were all able to meet here. I would like particularly to bear witness to the enormous amount of work put in at the local level by Dr Dudley Lawes and Dr Hugh Thomas and their team of willing helpers, - 'to make and orchestrate the arrangements, and to bring together the synopses of papers, - which is no mean feat given the vagaries of international communications.

The Local Organising Committee have asked me to record their grateful acknowledgement of support and financial assistance from a number of official and commercial organisations which have greatly facilitated arrangements. We shall make other opportunities during the course of the Conference to be more specific in these acknowledgements.

Ladies and Gentlemen, may I finally wish you a successful conference and a pleasant visit to Aberystwyth.

SESSION I

Genetic resources and their use
in the breeding of oats

Chairman: G. Ladizinsky

GENETIC RESOURCES AND THEIR USE IN OAT BREEDING

K.J. FREY
C.F. Curtiss Distinguished Professor in Agriculture,
Iowa State University, Ames, Iowa 50011, USA

1. SPECIES INTERRELATIONSHIPS

The USA World Oat Collection contains nearly 20,000 accessions, and 45% of these belong to wild and weedy species. However, more than 95% of the accessions of the collection are hexaploid.

The basic chromosome number for _Avena_ is seven. Four major genomic modifications, A, B, C and D have been postulated (Table 1) on the basis of cytogenetic studies, and minor modifications have occurred in the A (A_c, A_d, A_e, A_p and A_s) and C genomes (C_p and C_v). All hexaploid species have the genomic composition AACCDD, and they are interfertile with the F_1 hybrids showing normal bivalent pairing and recombination. Ladizinsky[1]

TABLE 1. Di, tetra and hexaploid _Avena_ species and their genomic compositions(2)

Ploidy	Species	Genome
Diploid (2n=14)	clauda pilosa	C_pC_p
	ventricosa	C_vC_v
	prostrata	A_pA_p
	longiglumis	A_eA_e
	damascena	A_dA_d
	canariensis wiestii hirtula strigosa	A_cA_c A_sA_s
Tetraploid (2n=28)	barbata vaviloviana abyssinica	AABB
	maroccana murphyi	AACC
Hexaploid (2n=42)	fatua sterilis sativa byzantina	AACCDD

Lawes, DA, Thomas, H (eds), Proceedings of the Second International Oat Conference.
ISBN 90-247-3335-9
© _1986, Martinus Nijhoff Publishers, Dordrecht. Printed in The Netherlands._

grouped the Avena species into six 'biological species' on the basis of morphological, genetical and ecological evidence, and his A.sativa biological species included all four of the classical hexaploid species. Understanding the genomic relationships among oat species and ploidy levels is critical to constructing a programme of interspecific matings for breeding superior cultivars.

2. CULTIVAR IMPROVEMENT TO 1985

USA oat breeding from 1900 to 1930 was restricted to pure-line selection from land cultivars introduced from Europe(3). Hybridization was begun in the late 1920s between A.sativa and A.byzantina to transfer crown-rust (Puccinia coronata avenae) resistance to A.sativa cultivars(4). Famous among those matings were Bond x D69 and Victoria x Richland, and cultivars derived from them with excellent crown-rust resistance were released in the 1940s. Concurrently, lodging resistance and test weight were improved immensely(5,6). Grain yielding ability was not increased very much, however, as shown in studies by Browning et al.(5) and Langer et al.(7) conducted in Iowa. They estimated that cultivars developed from hybridization up to 1970 yielded only 14% and 9%, respectively, more than did pure line cultivars from the 1900-20 era. More recently, Wych and Stuthman(6) showed that oat cultivars released in Minnesota in 1980 were 40% better yielding than pure-line cultivars selected a half century earlier. Rodgers et al.(8), using yield data from Uniform Nurseries, estimated that early and midseason oat cultivars released during 1971-75 in the midwestern USA yielded only 116 and 111%, respectively, of pure-line cultivars selected 60 years earlier. For cultivars released during 1976-80, however, the yields were 134 and 130%, respectively.

3. GERMPLASM USE

Most oat cultivars developed in the USA(9) until 1970 originated from seven land cultivars introduced from Europe. A.byzantina cultivars grown in the southern USA originated almost exclusively from Red Rustproof and Winter Turf. Most spring oat cultivars trace to Kherson, Green Russian, Victory, Markton and White Russian. Until 1970, the number of cultivars released in the USA and Canada that traced to these five introductions were: Kherson - 80, Green Russian - 50, Victory - 40, Markton - 37 and White Russian - 15(9). Obviously, the germplasm base from Europe used for oat breeding in the USA and Canada was very narrow.

It also is probable that the oat germplasm used to develop this crop in Europe was a very narrow sample of the genetic variability available in A.sterilis, the wild progenitor of cultivated oats (9,10). Oats were domesticated in Europe less than 4,000 years ago(11). The crop probably was carried from the Middle East to Europe as a weed in lots of cultivated wheat and barley. One can imagine that only a few non-shattering plants were selected from barley and wheat fields to initiate oats as a cultivated crop. Thus, a narrow germplasm base also would have been used to initiate this crop in Europe.

Langer et al.(7) suggest that the 9 to 14% increase in grain yield that occurred in oat cultivars in the 1940s probably resulted from recombination among genes from A.sativa and A.byzantina. Langer et al.(7) and Rodgers et al.(8) have suggested that further progress in breeding for grain yield would require the introgression of genes from exotic ecotypes into spring oat breeding populations, perhaps even from wild and weedy species.

4. WEEDY AND WILD RELATIVES

The recent history of use of exotic lines of cultivated oats in breeding programmes would be difficult to document and impossible to record in the space of this manuscript, so my discussion will be devoted to the use of genes from weedy and wild relative species of cultivated oats. Much research on interspecific matings of oats has not been recorded in the literature, so much of this report will be based on 'personal communications' from fellow oat researchers.

Over the last 30 years, several research groups have studied matings between Avena species from different ploidy levels. The Japanese(12,13), British(14,15) and Canadian groups(2) have used interploidy matings of Avena to study the origin of and genomic relationships among oat species. A tetraploid line derived by Zillinsky et al.(16), that contained a crown rust resistance gene from A.strigosa, was used by Frey et al.(17) to develop a putative hexaploid M isoline. In reality, this line, CI 9192, was an addition line with 44 chromosomes(18). The addition chromosome caused a 10% yield reduction(19). Thomas and Lawes(20) produced a fertile hexaploid addition line with an A.hirtula chromosome that carried mildew resistance, but this gene has not been utilized in a cultivar. Researchers in Wisconsin(21,22,23) have transferred the Pc-15 gene from A.strigosa to a true breeding hexaploid, and Brown (unpubl. data, Agron. Dept., Univ. Wis., Madison, Wis.) has identified a hexaploid line with the Pg-16 gene from A.barbata.

By and large, attempts to transfer genes from di and tetraploid oats to A.sativa or A.byzantina have been difficult to accomplish, and if accomplished, the gene tended to occur in an addition line. Exceptions are the transfer of Pc-15 and Pg-16 as translocation stocks. According to Burrows (pers. comm. VD Burrows, Ottawa Res. Stn, Ottawa, Ont., Can.), Gemini and Hinoat cultivars were selected from the matings A.strigosa^2 x Victory and A.strigosa x Abegweit. These are the only cultivars reported to have originated from di x hexaploid matings.

The use of A.fatua and A.sterilis as donors of genes to cultivated gene pools has been quite routine and rapid because they and A.sativa share nearly complete chromosome homology(24). Rines et al.(25) tested 2,200 accessions of A.fatua from North Central USA and found some with resistance to barley yellow dwarf virus (BYDV), tolerance to cereal leaf beetle, or high grain protein content, but none was resistant to stem rust (Puccinia graminis avenae) race 94. Suneson(26,27,28) developed 'Rapida', 'Sierra' and 'Montezuma' cultivars from matings involving A.fatua collected in California. The attributes they inherited from A.fatua were extreme earliness, extreme shattering resistance, and very large seeds, respectively.

Burrows (pers.comm.) developed 'dormoats' by transferring the seed dormancy genes from A.fatua to A.sativa. He forsees the benefits of dormoats to be "... improved grain yield (10-25%), disease escaping, early maturity, and improved grain quality ...". Suneson and Marshall(29) made A.fatua x A.sativa matings to increase the cold hardiness of cultivated oats. Bulk populations from these matings have been fall-sown in Pennsylvania for a number of generations, and as a result of natural selection, they have winter hardiness similar to elite winter cultivars (pers. comm. HG Marshall, Dept. Agron., Pa. State Univ., University Park, Pa.). Brinkman (pers. comm. MA Brinkman, Agron. Dept., Univ. Wis., Madison, Wis.) has used A.fatua extensively as a gene source,and some of his derived lines are as productive as standard cultivars. Since

A.fatua is either a secondary weed that evolved from or coevolved with cultivated oats(30), it is likely that its gene pool and that of A.sativa are quite similar. Thus, A.fatua probably does not represent a very divergent source of genes for improving the A.sativa gene pool.

A.sterilis is the progenitor of cultivated oats, and the USA World Oat Collection has 8,300 accessions of this species from 15 or more countries in the Mediterranean area and Middle East. The value of this species as an important source of crown rust resistance genes was first recognised by Simons et al.(31) a quarter century ago, and the introgression of resistance genes from A.sterilis into cultivated oats has been pursued vigorously. Seven of 10 crown rust resistance genes used in 'Multiline E77'(32) were derived from A.sterilis, and all nine genes used in 'Webster', recently released in Iowa, are from A.sterilis. Canadian researchers have discovered and studied ca. 30 genes for crown rust resistance from this species(33 and pers. comm., MD Simons, Dept. Plant Pathol., Seed and Weed Sci., Iowa State Univ., Ames, Iowa). Crown rust resistance in two A.sterilis accessions (CAV 1358 and CAV 1376) have a complex additive-type inheritance(34). McKenzie et al.(35,36) released 'Fidler' cultivar with Pc-38 and 'Dumont' with Pc-38 and Pc-39, both of which were found in A.sterilis. Also, crown rust resistance genes from A.sterilis have been used in 'TAM 0-301' and 'TAM 0-312' cultivars in Texas and in seven cultivars developed by Coker's Seed Company. These are 'Coker 227', 'Coker 234', 'Four twenty-two', 'Big Mac', 'Mesquite', 'H833' and 'Citation' (pers, comm., ME McDaniel, Dept. Crops and Soils, Tex. A & M Univ., College Station, Tex.). In contrast to the plethora of genes for crown rust resistance in A.sterilis, only two genes for resistance to stem rust, Pg-13(37) and Pg-15(38), have been found in this species. A gene for resistance to mildew (Erysiphe graminis avenae), Eg-3, was found in Cc 4347 accession of A.sterilis by Hayes and Jones(39). It is not clear whether a gene for mildew resistance from Cc 4346, reported by Lawes and Hayes(40) as being present in 'Mostyn' cultivar, is Eg-3.

Comeau(41) tested 20 Avena species and found tolerance to BYDV in nine of them. In another study, Comeau(42) tested 1,718 A.sterilis lines and found that 46% were resistant or tolerant to BYDV. Accessions from Algeria, Tunisia, Lebanon and Turkey had especially high percentages of resistance. Several accessions were resistant or very resistant across three years of testing. According to Landry et al.(43), genetic variability for BYDV resistance in seven accessions was due to additive gene action, and heritability of the trait was greater than 90%. Four or fewer factor pairs accounted for the inheritance patterns in matings of these accessions. Comeau(42) states "... many of the recent oat cultivars in the USA and Canada derive some BYDV resistance factors from Avena sterilis ..." and McDaniel (pers. comm.) says - "It is likely that both McDaniel and Harrison also have gained some BYDV resistance from A.sterilis lines utilised as sources of resistance to crown rust and/or stem rust". Frey (unpubl. data) found that P.I. 469112, a derivative from Holden x A.sterilis 2x Garland, has high BYDV resistance probably derived from the A.sterilis parent.

Certain accessions of A.sterilis have high groat contents of protein or oil. Groat-protein contents for A.sterilis collections have been reported up to 27.3% by Ohm and Patterson(44), to 28.0% by Campbell and Frey(45), and to 35% by Frey et al.(46). Campbell and Frey(45) studied 10 A.sativa x A.sterilis matings and concluded that groat-protein percentage was simply inherited in interspecific matings. Sraon et al.(47) estimated the number of effective factor pairs segregating for groat-protein percentage in A.sativa x A.sterilis matings to be from 1 to 25. Campbell and Frey(45) found additive gene action for groat-protein content in interspecific matings whereas Ohm and Patterson(48) found dominance for low groat-protein content. Heritability of groat-protein content in A.sativa x A.sterilis matings ranges from 30 to 76%(49).

Frey(50) has isolated A.sativa lines with high groat-protein percentage genes from A.sterilis, and these lines have protein contents as high as 21.2%. Cox and Frey(51) mated these derived lines to cultivars with high groat-protein genes from A.sativa, and on the basis of segregation, they concluded that the two species had complementary alleles for high groat-protein percentage. This led McFerson (pers. comm., J McFerson, Dept. Agron., Iowa State Univ., Ames, Iowa) to adopt recurrent selection to increase groat-protein content of oats.

Groat-oil content varies from 2.0 to 11.6% among accessions of A.sterilis(52,53), a range similar to the 4.2 to 11.3% groat-oil found for A.sativa(54). Matings of A.sativa x A.sterilis studied by Frey et al.(46) and Thro and Frey(55) showed that high groat-oil content was partially dominant and inherited polygenically. Thro and Frey(55) have shown that genes for high groat-oil content from A.sativa and A.sterilis are complementary. In a nobelised oat population with high oil genes from both species, Branson (pers. comm. CV Branson, Dept. Agron., Iowa State Univ., Ames, Iowa), after two cycles of recurrent selection, has recovered plant progenies with 12.5% groat oil.

A.sterilis has been used extensively in Iowa and Minnesota, USA, as a donor of genes for general oat cultivar development. One line developed and being increased in Minnesota, Minn. 80116, has A.sterilis in its parentage, and many advanced lines in the Minnesota breeding programme are expected to possess from 3 to 12% of genes from A.sterilis (pers. comm., D Stuthman, Dept. Agron. Genet., Univ. of Minn., St. Paul, Minn.).

The Iowa oat breeding programme has made extensive study and use of genes from A.sterilis that increase the vigour and productivity of cultivated oats. That A.sterilis possessed 'yield genes' was reported by Frey and Browning(56) and Frey(57), who showed that a crown rust resistance gene from CI 8079 caused an associated grain yield increase of 3.5 to 7.5% in disease-free conditions. Later, Lawrence and Frey(58) showed that more than a quarter of the lines segregating from the BC_2 to BC_4 of A.sativa x A.sterilis matings were significantly higher in grain yield than the recurrent parent. Frey(59,60) found that nine of these superior yielding lines tested over years and sites had agronomic traits similar to the recurrent parents and grain yields from 10-29% higher (Table 2).

Because grain yield = growth duration (days) x growth rate (q/ha/day) x harvest index (%), and growth duration and harvest index for the recurrent parent and the derived lines were similar, it was reasoned that growth rate was the trait that the introgressed genes were affecting. Indeed, Takeda and Frey(61) found that a quarter of the BC_0 and BC_1 segregates and 12.5% of the BC_4 segregates had significantly greater growth rates than the recurrent parent. Brinkman and Frey(62) and Bloethe-Helsel and Frey(63) showed that the derived lines with superior yielding ability had leaf-area duration (LAD) greater than that of the recurrent parent. In the later study, the LAD superiority of 21 to 47% was caused by the introgression lines having 10% larger leaf blades and delayed leaf senescence. The derived lines had an increase in spikelets

TABLE 2. Performance of recurrent parent (Clintford) and BC$_2$ to BC$_4$ lines of oats from A.sativa x A.sterilis matings(60)

	Grain yield (%)	Test weight (lbs/bu)	Straw score (1-5)	Harvest index (%)	Heading date (June)	Plant height (in)
Clintford	100	35.2	2.0	43	18	31
CI 9267	120	35.6	2.4	42	17	35
CI 9268	129	34.6	2.3	46	21	33
CI 9269	124	32.8	2.7	42	21	34
CI 9270	115	34.2	1.7	44	19	31
CI 9271	125	33.5	2.8	44	19	32
CI 9272	110	34.4	1.6	43	17	32
CI 9273	121	34.4	2.6	42	18	34
CI 9274	124	33.5	3.0	41	20	35
CI 9275	115	34.5	2.5	43	17	33

per panicle of 8% due to the availability of more photosynthate from the greater leaf area preceding anthesis. A 12% increase in seed weight was due to the leaf area remaining photosynthetically active for a longer period.

In A.sativa x A.sterilis matings and matings among A.sativa cultivars studied by Cox and Frey(64), the interspecific matings gave three and two times as many significantly positive segregates for biomass and vegetative growth index, respectively, as did the intraspecific matings. They conclude, "Thus, A.sterilis parents tend to be a good source of genes for high biomass and high vegetative growth index".

Robertson and Frey(65) found that A.sterilis cytoplasm increased grain yield, harvest index and heading date significantly (Table 3).

TABLE 3. Performance of BC$_2$F$_2$ derived oat lines with similar nuclear gene samples interacting with A.sativa and A.sterilis cytoplasms(65)

	Cytoplasm		
	A.sativa		A.sterilis
Grain yield (kg/ha)	2860	**	3005
Straw yield (kg/ha)	3725		3750
Harvest index (%)	43.6	**	44.7
Heading date (days)	54	**	55
Plant height (cm)	91.4		91.6
Vegetative growth rate (kg/ha/day)	75		74

**Cytoplasm effects significant at 1% level

One BC$_2$F$_2$-derived line from this study, D623-15 with A.sterilis cytoplasm, is being increased for possible release in Iowa. The gene pool being used for oat breeding in Iowa probably contains 10% of A.sterilis genes.

Marshall (pers. comm.) has used a unique method for developing A.sterilis introgressed BC$_3$ populations with a freezing-susceptible A.sativa line as the recurrent parent. This assures that freezing-resistance genes in the population have come from A.sterilis. His best introgressed populations are equivalent to 'Wintok' cultivar in freezing resistance.

5. SUMMARY

Because oats were domesticated in Europe from weeds in barley and wheat fields, the European gene pool for this crop probably is quite narrow. And in as much as seven European land cultivars were the primary contributors to the USA gene pool for oat breeding, the North American gene pool probably is even more narrow. Some researchers propose that the homogeneity of the USA gene pool was responsible for breeders making little genetic increase in oat yields from 1940 to 1970.

The gene pools of A.sativa and A.byzantina were introgressed more than 50 years ago, and segregates from these interspecific matings gave a 10% increased in yield. During the past 15 years, the use of exotic parents, including A.sterilis, in the breeding programmes in Illinois, Indiana, Iowa, Missouri and Minnesota have resulted in sizeable genetic gains for oat yields.

Diploid and tetraploid oat species, even though they carry disease resistance and grain composition genes that would be beneficial in breeding cultivated oats, have not been very useful. Difficulty in making crosses, nonhomology of genomes, and undesirable linkages are responsible for the difficulty in using genes from di and tetraploid species to improve cultivated oats. A.fatua, a weedy hexaploid, has contributed genes for dormancy, earliness, shattering resistance, and large seeds to the gene pool of cultivated oats. A.sterilis, the progenitor of cultivated oats, has contributed alleles for crown and stem rust and mildew resistance, for tolerance to barley yellow dwarf disease, for high content of grain oil and protein, and for increased biomass and grain yields.

The gene pool of cultivated oats needs to be expanded if breeders are to develop cultivars with greater genetic potential. When all traits are considered, A.sterilis probably is the most promising source for new alleles.

REFERENCES

1. Ladizinsky G: Israel J. Bot. 20, 133-151, 1971.
2. Rajhathy T and H Thomas: Cytogenetics of oats (Avena L.). Genet. Soc. Can. Misc. Publ. 2, Ottawa, Ontario, Canada, 1974.
3. Stanton TR: Superior germplasm in oats. In, US Dept. Agric. Yearbook. Govt. Printing Office, Washington, DC, pp.347-414, 1936.
4. Simons MD: Crown rust of oats and grasses. Monograph No. 5. Am. Phytopath. Soc. and Hifferman Press, Inc., Worcester, Mass., 1970.
5. Browning JA, KJ Frey and R Grindeland: Iowa Farm Sci. 18(8), 5-8, 1964.
6. Wych RD and DD Stuthman: Crop Sci. 23, 879-881, 1983.
7. Langer I, KJ Frey and TB Bailey: Crop Sci., 18, 938-942, 1978.
8. Rodgers DM, JP Murphy and KJ Frey: Crop Sci. 23, 737-740, 1983.
9. Coffman A: Oat history, identification and classification. US Dept. Agric. Tech. Bull. No. 1516, 1977.
10. Coffman A: J. Am. Soc. Agron. 38, 983-1002, 1946.
11. Coffman A: In, FA Coffman (ed.). Oats and oat improvement. Am. Soc. Agron., Madison, Wis., pp.15-40, 1961.
12. Nishiyama I: Jap. J. Genet. 5, 1-48, 1929.
13. Nishiyama I and M Tabata: Jap. J. Genet. 38, 311-316, 1964.
14. Thomas H: Can. J. Genet. Cytol. 12, 36-43, 1970.
15. Thomas H and ML Jones: Chromosoma 15, 132-139, 1964.
16. Zillinsky FJ, K Sadanaga, MD Simons and HC Murphy: Agron. J. 51, 343-345, 1959.

14

17. Frey KJ, JA Browning and RL Grindeland: Crop Sci. 13, p.290, 1973.
18. Dherawattana A and K Sadanaga: Crop Sci. 13, 591-594, 1973.
19. Brinkman MA, KJ Frey and JA Browning: Crop Sci, 18, 147-148, 1978.
20. Thomas H and DA Lawes: Euphytica 17, 404-413, 1968.
21. Forsberg RA and HL Shands: Crop Sci. 9, 64-67, 1969.
22. Sharma DC and RA Forsberg: Crop Sci. 17, 855-860, 1974.
23. Forsberg RA and S Wang: Can. J. Genet. Cytol. 13, 292-297, 1971.
24. McMullen MS, RL Phillips and DD Stuthman: Crop Sci. 22, 890-897, 1982.
25. Rines HW, DD Stuthman, LE Briggle, VL Youngs, H Jedlinski, DH Smith,
    JA Webster and PG Rothman: Crop Sci. 20, 63-68, 1980.
26. Suneson CA: Registration of Rapida oats. Crop Sci. 7, p.168, 1967.
27. Suneson CA: Registration of Sierra oats. Crop Sci. 7, p.168, 1967.
28. Suneson CA: Crop Sci. 9, 848-849, 1969.
29. Suneson CA and HG Marshall: Crop Sci. 7, 667-668, 1967.
30. Harlan JR: Crops and man. Am. Soc. Agron. and Crop Sci. Soc. Am.,
    Madison, Wis., 1975.
31. Simons MD, I Wahl and AR de Silva: Phytopathology 52, 585-586, 1962.
32. Frey KJ, JA Browning and MD Simons: Crop Sci. 25 (in press, 1985).
33. Simons MD, JW Martens, RIH McKenzie, I Nishiyama, K Sadanaga, J
    Sebesta and H Thomas: Oats: A standard system of nomenclature for
    genes and chromosomes and catalog of genes governing character.
    Agric. Handb. 509. US Dept. Agric., 1978.
34. Harder DE and RIH McKenzie: Can. J. Plant Pathol. 6, 135-138, 1984.
35. McKenzie RIH, JW Martens, PD Brown, DE Harder, J. Nielsen and GR
    Boughten: Crop Sci. 21, 632-633, 1981.
36. McKenzie RIH, PD Brown, JW Martens, DE Harder, J. Nielsen, CC Gill and
    GR Boughton: Crop Sci. 24, p.207, 1984.
37. McKenzie RIH, JW Martens and T Rajhathy: Can. J. Genet. Cytol. 12,
    501-505, 1970.
38. Martens JW, RIH McKenzie and DE Harder: Can. J. Genet. Cytol. 22,
    641-649, 1980.
39. Hayes JD and IT Jones: Euphytica 15, 80-86, 1966.
40. Lawes DA and JD Hayes: Plant Pathol. 14, 125-128, 1965.
41. Comeau A: Euphytica 33, 49-55, 1984.
42. Comeau A: Can. J. Pl. Pathol. 4, 147-151, 1982.
43. Landry B, A Comeau, F Minivielle and CA St Pierre: Crop Sci. 24,
    337-340, 1984.
44. Ohm HW and FL Patterson: Crop Sci. 13, 27-30, 1973.
45. Campbell AR and KJ Frey: Can. J. Pl. Sci. 52, 735-742, 1972.
46. Frey KJ, EG Hammond and PK Lawrence: Crop Sci. 15, 94-95, 1975.
47. Sraon HS, DL Reeves and MD Rumbaugh: Crop Sci. 15, 668-670, 1975.
48. Ohm HW and FL Patterson: Crop Sci. 13, 55-58. 1973.
49. Frey KJ: Crop Sci. 15, 277-278, 1975.
50. Frey KJ: Z. Pflanzenzucht 78, 185-215, 1977.
51. Cox TS and KJ Frey: Crop Sci. 25, 106-109, 1985.
52. Frey KJ and EG Hammond: J. Am. Oil Chem. Soc. 52, 358-362, 1975.
53. Rezai A: Variation for some agronomic traits in the world collection
    of wild oats (Avena sterilis L.) PhD Diss. Iowa State Univ., Ames,
    Iowa, 1977.
54. Sahasrabudhe MR: J. Am. Oil Chem. Soc. 56, 80-81, 1979.
55. Thro AM and KJ Frey: Inheritance of groat oil content and high-oil
    selection in oats. Euphytica 34 (in press, 1985).
56. Frey KJ and JA Browning: Crop Sci. 11, 757-760, 1971.
57. Frey KJ: Crop Sci. 12, 809-813, 1972.

58. Lawrence PK and KJ Frey: Euphytica 24, 77-85, 1976.
59. Frey KJ: Egypt. J. Genet. Cytol. 5, 460-482, 1976.
60. Frey KJ: In, TC Yap, KM Graham and BS Jalani (eds). Crop improvement research. Proc. 4th Int. SABRAO Cong., Dept. Genet. Univ. Kebangsaan Malaysia, Bangi, Selangor, Malaysia, pp.1-20, 1983.
61. Takeda K and KJ Frey: Crop Sci. 16, 817-821, 1976.
62. Brinkman MA and KJ Frey: Crop Sci. 17, 426-430, 1977.
63. Bloethe-Helsel D and KJ Frey: Crop Sci. 18, 765-769, 1978.
64. Cox DJ and KJ Frey: Theor. Appl. Genet. 68, 239-245, 1984.
65. Robertson LD and KJ Frey: Crop Sci. 24, 200-204, 1984.

This paper, presented at the Second International Oat Conference, Aberystwyth, is Journal Paper No. J-11905 of the Iowa Agriculture and Home Economics Experiment Station, Ames, Iowa, Project 2447

THE USE OF DISOMIC ALIEN ADDITION LINES IN THE TRANSFER OF OAT STEM RUST RESISTANCE TO HEXAPLOID OATS

P.D. BROWN*, R.A. FORSBERG**, R.I.H. McKENZIE* and J.W. MARTENS*

*Agriculture Canada Research Station, Winnipeg, Manitoba, R3T 2M9, Canada
**Department of Agronomy, University of Wisconsin, Madison, WI 53706, USA

1. INTRODUCTION

Oat stem rust (Puccinia graminis Pers. f.sp. avenae Eriks. and E. Henn) is an important disease in the north-central plains of North America causing reductions in yield and grain quality.

Dinoor and Wahl[1], in a study of the reaction of noncultivated oats from Israel to Canadian races of stem rust, found that Avena barbata Pott. collection D203 was resistant at the seedling and adult stage to all stem rust races used. A 44 chromosome disomic addition line with stem rust resistance gene, Pg-16, from the cross D203/5* Rodney O has been used in the stem rust differential set since 1977[2]. Since then, only one North American stem rust race has been found that will attack Pg-16[3]. This wide spectrum resistance makes Pg-16 extremely valuable and it would be beneficial if it could be transferred into stable hexaploid (2n=42) A.sativa L. lines. A lack of chromosome homology and pairing makes gene transfer between A.barbata and A.sativa difficult.

Radiation induced chromosome breakage has been used to transfer alien genes. Because the transfer is a rare random occurrence, large plant populations must be used to obtain the transfer. Techniques have been developed to identify and isolate plants with the desired gene transfer from within a large population. Sears[4] was the first to use radiation to induce interspecific gene transfer. In wheat plants with an added alien isochromosome for the leaf rust (Puccinia recondita f.sp. tritici Rob ex. Desm.) resistance-carrying arm, the alien chromosome was transmitted through male gametes at a greatly reduced rate. Resistant plants were X-rayed just prior to meiosis and then used as pollen parents in 'screening' crosses to a susceptible female parent. Any resistant F_1 plants contained a translocation between a host chromosome and an alien chromosome. To minimize cytological work and to make positive selections to avoid deleterious translocations, Driscoll and Jensen[5] and Aung, Thomas and Jones[6] irradiated seed of disomic addition lines in wheat and oats, respectively, in attempts to transfer disease resistance. Potential translocation lines were identified by the presence of monogenic segregation ratios.

2. TRANSFER OF Pg-16 TO HEXAPLOID OATS

In this study, seeds of 44-chromosome, resistant, disomic alien addition (DAA) lines were exposed to gamma radiation in an attempt to induce a translocation of stem rust resistance gene Pg-16 to A.sativa. In the first generation post irradiation, all plants were homozygous resistant. Any translocation between the alien chromosome and the recipient parent chromosome would result in the formation of a quadrivalent or bivalents between incompletely-homologous chromosomes. The result of this would be the formation of gametes lacking the gene for resistance with concomitant segregation in head rows of plants in the second generation post irradiation (R_2). Because of the lack of segregation in many R_2 rows, most of the material can be discarded at this stage and attention can be focussed on the critical lines which may potentially be carrying the translocation. By selfing and selecting for monogenic ratios in subsequent generations, lines possessing the desired translocation were identified. By selecting for near monogenic ratios, translocation chromosomes which did not pass through the gametes at normal rates were avoided. Also, by selecting for near monogenic ratios, monosomic alien substitution lines produced by the irradiation treatment could be eliminated. The method used in this investigation has several advantages. Because there are two chromosomes carrying the gene for resistance, the chances of achieving the desired random chromosome break are increased. No selection need be done during the first generation post irradiation and much of the material can be quickly discarded during the next generation thus permitting the exploitation of a large population which increases the chances of identifying a line with the desired gene transfer. Positive selection pressure for a close fit to a 3:1 monogenic ratio results in the identification of lines in which the translocation does not impede normal transmission through either the male or female gametes.

2.1 Induction and identification of the transfer

This attempt to transfer stem rust resistance to hexaploid oats involved three major processes with each process consisting of several steps. The first process was to induce and identify the transfer of Pg-16. Before irradiation, all seeds from DAA plants should have 21 pairs of sativa chromosomes plus one barbata chromosome pair carrying the gene for stem rust resistance. A radiation-induced break involving the portion of the barbata chromosome carrying Pg-16 might result in this chromosome segment being transferred to a sativa chromosome. Bulk samples of approximately 25,000 seeds from DAA homozygous resistant plants (2n=44) of the pedigree D203/5*Rodney O were exposed to one of four doses (15, 20, 25 or 30 kilorads) of gamma radiation and were planted in a field nursery. Rodney O is a backcrossed derived line of the cultivar Rodney in which all genes for stem rust resistance have been eliminated. A heavy natural infection of oat stem rust showed that of approximately 100,000 R_1 plants, all but four were resistant. Selection in the R_1 generation was unnecessary and individual R_1 tillers were threshed at maturity.

Plants in 14,000 1 m R_2 head rows were exposed to an artificially induced stem rust epdemic. R_2 head rows segregating for rust reaction were indicative of potential translocations of Pg-16 from the barbata chromosome source to any one of the 42 sativa chromosomes. Of the 14,000 R_2 lines examined, 222 were segregating for rust resistance. The higher levels of radiation produced segregating lines at a higher frequency. The segregation ratios within the 222 segregating lines were variable and ranged from 23R:1S to 1R:22S. At maturity 1,182 resistant R_2 plants in all rows segregating for rust reaction were pulled and threshed.

In the R_3 generation, potential segregation ratios in the presence of a translocation would be 3R:1S, 15R:1S, or homogeneous resistant. The 3R:1S ratio would be seen in lines in which Pg-16 was carried on either one sativa or one barbata chromosome. The 15R:1S ratio would be seen in R_3 lines in which Pg-16 was carried on both a sativa and a barbata chromosome. The homogeneous resistant reaction would be seen in lines homozygous for Pg-16 on either the sativa and/or barbata chromosome pair. An artificially induced oat stem rust epidemic was used to screen 30 to 50 R_3 oat plants in each of 1,182 R_3 lines. Each line was derived from a single resistant R_2 plant. Rust reactions of plants in these lines were read. Because of ease and assurance in identifying the 3:1 ratio, lines segregating close to this ratio were selected as potential carriers of the translocation for Pg-16. Of the 1,182 R3 lines screened for rust reaction, 517 were homogeneous resistant, 365 were segregating, and 300 were homogeneous susceptible. Within segregating lines, ratios ranged from 30R:1S to 1R:20S. Segregation in 41 R_3 lines was close enough to the desired 3R:1S ratio to warrant saving the resistant plants.

Up to 10 lines for each of 41 R_3-derived R_4 families were screened. Twenty to 50 seedlings per R_4 line were rust tested. Within-line and within-family segregation readings were made. In the presence of a translocation for Pg-16, R_4 families should segregate with a ratio of 1 homogeneous resistant line: 2 segregating lines, and the segregating lines should segregate 3R:1S. As an aid to selection, the chi-square statistic for goodness of fit to these two ratios was calculated for each family. It indicated that segregation within a family fit the ratio of 1 homogeneous resistant line: 2 segregating lines for 29 families probability > 0.05). There were 17 families which fit the 3R:1S ratio within segregating lines (probability > 0.05). In an attempt to obtain a better prediction of which families might possess the translocation, the probabilites of acceptance were increased to a probability > 0.25 for the within-family ratio of 1 homogeneous resistant:2 segregating lines and to a probability > 0.10 for the 3R:1S ratio within-segregating lines. On the basis of these higher probabilities of acceptance it was concluded that 11 families (R_3 Nos 281, 283, 288, 323, 954, 1002, 1087, 1089, 1129, 1132 and 1135) showed segregation that was close enough to both ratios expected in the presence of a translocation to warrant further study. These 11 R_4 families were derived from resistant plants in six segregating R_2 rows. Ten of these families were derived from seeds that had been exposed to 30 kilorads of gamma radiation and one (R_3 No. 954) was derived from a seed that had been exposed to 20 kilorads. Because of insufficient seeds, two families (R_3 Nos 281 and 1132) were not tested further.

2.2 Elimination of the alien chromosome pair

At this stage of the attempt to transfer Pg-16 to hexaploid oats, the second process, to eliminate the barbata pair of chromosomes in potential translocation lines, was initiated. The male gametic screen was used in this process.

Resistant R_4 plants from two homogeneous resistant lines and one segregating line from each of nine families that appeared to have the translocation for Pg-16 were used as males in crosses with Rodney 0. At least 12 F_1 seeds per critical family (four per resistant plant) were produced. Of the 108 Rodney 0/2/D203/5*Rodney 0 F_1 seeds planted, 99 germinated and produced seedlings that could be rust tested. In most cases the rust reaction of the F_1 plants followed the expected pattern, that is, F_1s arising from males of homogeneous resistant lines were all resistant and F_1s arising from males of segregating lines were either resistant or susceptible. The surviving 99 F_1 plants were the result of crosses between the female Rodney 0, which has 42 chromosomes, and R_4 males, derived from irradiated seed of DAA (2n=44) plants. Chromosome counts were made from three of 12 F_1 seedlings planted per critical family. Twenty-two of the 27 F_1s sampled had 43 chromosomes, and included both resistant and susceptible plants. This count of 43 and the independence of rust reaction and chromosome number were expected. There were, however, five plants, two resistant and three susceptible, from five different males which had 42 chromosomes.

Using six resistant F_1 plants per family as males in backcrosses with Rodney O, 567 BC_1F_1 Rodney O/2/D203/5*Rodney O seeds were produced with a minimum of 49 seeds per family. This backcross, using the resistant parent as the male, eliminated the barbata univalent chromosome by using the male gamete as a screening mechanism. To maximize seed production, the seedlings were space planted and were not rust tested. Selfed seeds from all 567 BC_1F_1 plants were harvested at maturity.

2.3 Verification of the translocation

The third process in this study of the transfer of Pg-16 to hexaploid oats was to verify the presence of translocation lines by rust testing plants in several generations and comparing observed segregation ratios with the expected ratios.

Thirty to 55 BC_1F_1 populations from each of the nine families that were potential carriers of the translocated gene, Pg-16, were planted in the glasshouse. Each population of 25-50 seedlings was rust tested. Of the 335 BC_1F_2 populations screened for rust reaction, 280 BC_1F_2 populations derived from eight of the nine potentially critical R_4 families were completely susceptible to rust or showed segregation ratios of 1 resistant:many susceptible. Based on previous work with Pg-16, in which we observed that a monosomic alien chromosome carrying Pg-16 passes through the gametes at a lower-than-normal rate, these ratios indicated that the gene for stem rust resistance was still on the barbata chromosome. Because Pg-16 is likely still on the alien chromosome in these eight families, they were of no further interest. Fourteen BC_1F_2 populations, all of which originated from R_3 line No. 954, segregated for rust reaction with a ratio close to 3R:1S (X^2=13.53 with 14 d.f.; P = 0.50) while another 41 populations derived from the same R_3 line were susceptible. It is likely that plants within this family were carrying a translocation for Pg-16.

Resistant BC_1F_2 plants from one critical population were used as males in test crosses with Rodney O and Dumont. Dumont is a commercial cultivar well adapted to western Canada. Resistant plants were used in three backcrosses to both Rodney O and Dumont. Following the final backcross and after selfing and selecting resistant plants for two more generations, BC_3F_3 seedlings were rust tested. Of the 171 BC_3F_3 lines backcrossed to Rodney O, 45 were homogeneous resistant and 126 segregated. The calculated chi-square statistic (X^2 = 3.79, 1 d.f.; P = 0.05) indicated that this was not significantly different from the expected 1:2 ratio. Of the 126 segregating lines, the segregation ratio within 90 lines was not significantly different (P = 0.05) from the 3R:1S ratio expected in the presence of a translocation and the segregation ratio within 104 lines was not different from the expected 3R:1S ratio at a probability of 0.01. There were 59 BC_3F_3 lines backcrossed to Dumont and of these 13 were homogeneous resistant and 46 segregated. The chi-square test (X^2 = 3,39, 1 d.f.; P = 0.05) indicated that this ratio was not significantly different from the expected 1:2 ratio. Of the 46 segregating lines, the segregation ratio within 39 lines was not significantly different (P = 0.05) from the expected 3R:1S ratio and the segregation ratio within 42 lines was not different from the expected 3R:1S ratio at a probability of 0.01.

20

3. CONCLUSIONS

The BC_1F_2 and BC_3F_3 data indicated that a translocation has been identified in one parental line (R_3 line No. 954). Most of the observed segregation ratios indicated that transmission of the chromosome carrying Pg-16 was near normal. That the transmission was not entirely normal was indicated by the excess of homogeneous susceptible lines in the case of BC_1F_2 populations and by an excess of susceptible plants in some BC_3F_3 lines. Because there was an excess of susceptible plants or lines in some cases, one must assume that transmission of the translocated chromosome through the male and/or female gamete was slightly impeded. In spite of this, both within-line and within-family segregation ratios indicated that gene Pg-16 passed through both male and female gametes of the progenies of this translocation line. Our previous work with Pg-16 revealed that very few progenies of MAS or MAA lines were resistant, indicating that in lines in which the desired gene was still on the monosomic alien chromosome, Pg-16 passed through the male and/or female gamete at a very low frequency. Even though it appeared that the transmission was not completely normal, in contrast with nontranslocation lines transmission was good and this line could be useful in a breeding programme.

REFERENCES
1. Dinoor A and I Wahl: Can. J. Plant Sci. 43, 263-270, 1963.
2. Martens JW, AP Roelfs, RIH McKenzie, PG Rothman, DD Stuthman and PD Brown: Phytopathology 69, 293-294, 1979.
3. Harder DE: Can. J. Pl. Path. (in press).
4. Sears ER: Brookhaven Symp. in Biol. 9, 1-21, 1956.
5. Driscoll CJ and NF Jensen: Genetics 48, 459-468, 1963.
6. Aung T, Hugh Thomas and IT Jones: Euphytica 26, 623-632, 1977.

INTRODUCTION OF DIPLOID WILD AVENA SPECIES INTO A.SATIVA

S.E. FRITZ and M.E. SORRELLS
Cornell University, Ithaca, NY 14853, USA

1. INTRODUCTION
MacKey(1) has suggested that the success of crop plants such as wheat, oats and rice has been due to disomic polyploidisation that has combined the merits of both cross- and self-fertilising systems. The success of these crops may be due to the potential heterotic responses induced by fixed epistatic interactions in disomic polyploids Introgression of wild species into A.sativa via an amphioctoploid could enhance opportunities for evolutionary development and provide a pool of germplasm from which heterotic types could be extracted. Thomas and colleagues(2,3,4) have reported relatively stable octoploids derived from 2x/6x crosses. Diploid Avena accessions have been reported to carry genes for resistance to crown and stem rust, barley yellow dwarf virus, soil mosaic virus(5), powdery mildew and cereal cyst nematode(6). These wild rel tives may provide useful genetic variation for inducing epistatic interactions that are fixed in 2x/6x amphioctoploids.

The objectives of this research were:

1. To produce 2x/6x amphioctoploids;
2. To estimate agronomic potential and cytological stability of newly developed octoploids.
3. To compare octoploids derived from 4x/6x and 2x/6x crosses with parental and locally adapted hexaploid varieties for agronomic traits, and
4. To intermate octoploids to produce a segregating gene pool.

2. MATERIALS AND METHODS
A total of 37 diploid accessions were used as females in interspecific crosses with nine hexaploid cultivars. Embryos were removed 10-14 days post-pollination using sterile technique and placed on Gamborg's B5 medium(7). Somatic chromosome counts were made to verify the hybrid genomic constitution. Colchicine treatment was applied using the inverted-vial technique(8) and the amphiploid seeds were harvested. In 1984 A_1, A_2 and A_3 generations of three amphiploids, F_1 octoploid intercrosses and octoploid lines (supplied by H Thomas and M Leggett) were grown in the glasshouse to study agronomic characteristics and cytological behaviour. An analysis of variance for differences among crosses was calculated on the A_1-A_3 genotypes for all traits. The frequency of chromosomal associations was calculated at metaphase I in pollen mother cells. Cytological studies of somatic chromosome number and agronomic data were recorded on nine octoploid intercrosses. Replicated yield trials were conducted in 1983 and 1984 to compare agronomic characteristics of octoploids derived from 2x/6x and 4x/6x crosses with hexaploid varieties.

3. RESULTS AND DISCUSSION

No significant differences between amphiploid crosses or generations were observed for any of the agronomic traits studied, but a significant cross*generation effect was observed for biological yield. Table 1 presents cross*generation means for biological yield, seed weight and harvest index for these amphiploids. The means of the 8x lines supplied by H Thomas and M Leggett, and the F_1 generation from octoploid intercrosses have also been tabulated for comparison. Biological yield appears to decline in early generations. This trend is evident from the A_1 to A_2 generation in cross 3924/Korwood and the A_2 to A_3 generation in cross 3924/Porter. the A_1 mean (18.2 g) appears to be higher than A_2 or A_3 estimates, 12.4 and 13.4 g, respectively. The means for all traits of the 8x lines and F_1 intercrosses appear significantly higher than the estimates for the A_1-A3 amphiploids. One 8x line, A.longiglumis CW 57 x A.sativa cv. Sun II, yielded 10.6 g of seed on a plant of 32.0 g biological yield. This genotype was significantly better than all the others for all agronomic traits and appeared quite promising in replicated trials.

TABLE 1. Cross*generation means for amphiploids grown in 1984

Cross*generation	BLY (g)	SDWT (g)	HI	N
3935/Korwood-A_1	19.8	3.4	0.17	6
3924/Korwood-A_1	16.6	2.1	0.13	9
3924/Korwood-A_2	10.2	1.8	0.18	13
3924/Porter-A_2	14.6	2.7	0.17	21
3924/Porter-A_3	13.4	2.5	0.19	12
Mean 8x	20.3	5.9	0.28	13
Mean F_1	25.7	7.6	0.29	3
LSD	2.0	0.52	0.026	

BLY = biological yield; SDWT = seed weight;
HI = harvest index;   N = number of observations

Somatic and meiotic chromosome analysis of these octoploids revealed irregularities in chromosomal behaviour. Table 2 presents the mean frequencies of univalents, bivalents, trivalents, quadrivalents and higher order multivalents observed. Since the meiocytes counted in compiling these data did not all have the 56 chromosomes that were expected, the number of chromosomal elements accounted for in this table does not total 56. Cytological analysis of hexaploid cultivars showed no irregularities in chromosome behaviour and near perfect ring bivalent pairing at meiotic metaphase. An analysis of variance of the octoploid cytological data indicated significant differences in the numbers of bivalents and trivalents between amphiploids 3924/Korwood and 3924/Porter. Other configurations were not significantly different. The amphiploid cross 3924/Korwood had significantly higher numbers of bivalents and corresponding lower numbers of univalents and multivalents than did the amphiploid 3924/Porter. High biological yield and seed weight values were not necessarily associated with more regular chromosome pairing behaviour in these amphiploids. Lagging chromosomes were common in both crosses at anaphase. Some asynchrony of chromosome division was observed, as was some precocious separation of chromatids at anaphase I. Cross

3924/Korwood showed a lower proportion of pollen with micronuclei and a lower mean micronuclei number than 3924/Porter. Since generation was confounded in this experiment, it was impossible to determine whether the differences in agronomic traits and meiotic behaviour were due to the cross combination or to the number of generations following amphiploid production. These materials are not true breeding amphiploids, due to somatic and meiotic instability.

TABLE 2. Meiotic chromosome associations in amphioctoploids obtained from 2x/6x hybrids

Cross	I[a]	II	III	IV	H
3924/Porter	1.452	22.234	0.876	0.490	0.272
3924/Korwood	1.118	24.430	0.459	0.363	0.074
3935/Korwood	0.270	27.470	0.200	-	-
A.ventricosa[b] x Manod	0.130	27.200	0.400	0.070	-
A.prostrata[b] x Sun II	3.600	23.130	0.730	0.270	-
A.longiglumis[b] x Sun II	0.500	24.000	0.000	0.330	-
LSD 0.05 = 1.523 D_{1-2} =	0.334	2.106*	0.417*	0.127	0.198

[a]Columns represent the frequency of univalents, bivalents, trivalents, quadrivalents and higher order multivalents
[b]Diploid species

The mean biological yield, seed weight, and harvest index of nine octoploid intercrosses was 38.12 grams, 5.19 grams and 0.1195, respectively. Coefficients of variation ranged from 33 to 96. The considerable variation in plant productivity may reflect differences in the plants' chromosome complement and cytological stability. Means and ranges of somatic chromosome number, standard deviation and the number of cells analysed are presented in Table 3. Somatic instability was observed in all octoploid intercrosses. Cells with 56 chromosomes were observed in most plants, but the overall mean chromosome number for the five crosses was 50.29. The average deviation in chromosome number was 5.07. Seventy-three percent of the observed microspores carried micronuclei. The average number of micronuclei per microspore was 1.4

TABLE 3. Mean somatic chromosome numbers, ranges and standard deviations of 8x-8x intercrosses

Cross	Mean	Range	Standard deviation	Cell number
8402-1	52.34	31-56	5.14	38
8402-2	48.50	41-53	4.35	10
8405-1	52.24	37-56	4.99	17
8406-1	49.53	36-56	4.93	36
8406-2	47.31	31-56	8.97	16
8407-2	51.83	41-56	3.64	23
8409-2	49.27	32-56	6.45	11
8409-5	51.33	49-53	2.08	3
Mean	50.29	37-55	5.07	

Replicated yield trials indicated that 2x/6x derived octoploids performed significantly better for grain yield and test weight than octoploids derived from 4x/6x crosses but yielded, on average, only 70% of hexaploid cultivars. One octoploid - A.longiglumis CW 57 x A.sativa cv. Sun II - yielded significantly higher than Orbit and Astro (two locally adapted cultivars) and outyielded Sun II - its 6x parent - by 350 kg/ha. Test weight of this amphiploid was 2 kg/hl higher than the mean hexaploid cultivar.

4. CONCLUSIONS

Crosses 3924/Porter, 3924/Korwood and 3935/Korwood showed acceptable biological yield, seed weight and harvest indices though yield parameters appeared to decline in early generations. Several segregants yielded more grain than octoploid lines in replicated trials. Meiotic chromosome pairing occurred primarily as bivalents, though irregularities were common. Results from replicated trials indicate that one octoploid performed as good or better than hexaploid checks. This is encouraging since these materials have not undergone intermating and selection. Preliminary agronomic analysis of octoploid intercrosses appears positive, and with further intermating and selection acceptable agronomic types may be identified and exploited.

REFERENCES

1. MacKey J: Hereditas 66, 165-176, 1970.
2. Swami UBS and H Thomas: Can. J. Genet. Cytol. 8, 51-56, 1966.
3. Thomas H and T Rajhathy: Can. J. Genet. Cytol. 9, 154-162, 1967.
4. Thomas H and DA Lawes: Euphytica 17, 404-411, 1968.
5. Zillinsky FJ and RA Derick: Can. J. Plant Sci. 40, 366-370, 1960.
6. Hayes JD: Rep. Welsh Pl. Breed. Stn for 1967, p.68, 1968.
7. Gamborg OL, RA Miller and K Ojima: Exp. Cell Res. 50, 151-158, 1968.
8. Bell GDH: J. agric. Sci. 40, 9-18, 1950.

INTERSPECIFIC HYBRIDS IN <u>AVENA</u>

J. M. LEGGETT
Welsh Plant Breeding Station, Aberystwyth, Dyfed, Wales, UK

1. INTRODUCTION

The genus <u>Avena sativa</u> L. (Poaceae) contains a number of species which form a <u>polyploid series</u>. Some of these species possess agronomic characters which it would be desirable to transfer to the hexaploid cultivated oat.

At the hexaploid level, such transfers can be accomplished using conventional breeding procedures since the representative species in this group are fully interfertile. However, at the diploid and tetraploid level, isolation and sterility barriers exist which frequently prevent the free flow of genes between and or within ploidy levels.

This paper describes the production and initial results of two interspecific hybrids whose agronomically desirable characters could not be introduced into the hexaploid cultivated oat by conventional breeding methods.

2. <u>AVENA MACROSTACHYA</u> HYBRIDS

The autotetraploid perennial oat <u>Avena macrostachya</u> has a number of characters which it would be desirable to transfer into the hexaploid cultivated oat, it is very winter hardy surviving a Quebec winter, and is reported to be tolerant to barley yellow dwarf virus (BYDV), and of course it has a perennial outcrossing habit. It was also of interest cytologically because no hybrid between it and any of the annual oat species had been reported previously.

In an attempt to transfer agronomically desirable characters into the cultivated oat, crosses were made between <u>A.sativa</u> and <u>A.macrostachya</u> using the latter as the male parent. A large number of pollinations were carried out and the seed set was low. Even when seed set did occur, the developing caryopses frequently degenerated before they were large enough to excise for culturing. All the immature seed which were large enough to be handled were removed from the hybrids between 14 and 21 days after pollination the embryos excised and placed in culture tubes on Gamborgs B5 medium without 2,4-D or kinetin. When plantlets were established they were transferred to soil filled pots in the glasshouse.

Three plants were established from the embryo cultures and grown to maturity. Panicles were removed from the hybrids and anthers at the correct stage were stained and examined cytologically. The main features of the chromosome pairing were the high frequency of univalents and the occasional formation of trivalents. A mean of 9.37 bivalents was recorded and of these 64% were ring bivalents (Table 1). The multivalents recorded were of the chain type although two ring quadrivalents were observed in otherwise obscure pollen mother cells. The maximum chromosome pairing recorded was 1 IV + 1 III + 10 II + 8 I and the minimum was 7 II + 21 I.

TABLE 1. Mean chromosome pairing per pollen mother cell at metaphase I
          of meiosis in the hybrids A.macrostachya x A.sativa

Number of cells	Mean values (range)				
	IV	III	II (rings)	II (rods)	I
58	0.04(0-1)	0.24(0-2)	6.02(3-9)	3.35(0-8)	15.43(8-21)

Because A.macrostchya is a quadrivalent forming autotetraploid, hybrids
between it and any of the annual oat species would be expected to form at
least seven bivalents because there would still be a homologous set of
chromosomes from the autotetraploid A.macrostachya parent. However, any
bivalent formation in excess of seven bivalents in hybrids between
A.macrostachya and A.sativa must be due to intergenomic pairing of the
A.sativa chromosomes. The formation of trivalents in this hybrid could
arise either as a result of pairing between the hexaploid homoeologues or
from the pairing of chromosomes from both parents. The formation of
multivalents greater than trivalents must involve chromosomes from both
parents since no multivalents greater than quadrivalents were observed in
the autotetraploid A.macrostachya parent(1).

It is interesting to compare the chromosome pairing of the
A.macrostachya x A.sativa hybrid with that of the synthetically derived
pentaploid hybrid A.hirtula(4x) x A.sativa(2). The synthetically derived
hybrid had a mean of 0.95 III + 6.85 II + 18.45 I indicating that the
chromosomes from the A.hirtula parent had paired but that little
homoeologous pairing between the chromosomes of the hexaploid parent had
taken place (Table 2).

Similarly in the pentaploid hybrid involving the diploidised tetraploid
A.barbata and the hexaploid A.sativa, most of the chromosome pairing can
be accounted for by the pairing of the A.barbata chromosomes or the
partial homology of the A genomes of the two species (Table 2).

TABLE 2. Mean chromosome pairing per pollen mother cell at metaphase I of
         meiosis in A.macrostachya and hybrids between A.sativa and A.hirtula
         and A.barbata

	V	IV	III	II(ring)	II(rod)	I
A.macrostachya	-	3.93	-	-	6.06	0.13
A.hirtula(4x) x A.sativa	-	-	0.95	3.9	2.95	18.45
A.barbata x A.sativa	0.02	0.13	1.14	0.06	6.09	18.61

In the A.macrostachya x A.sativa hybrid a mean of 9.31 bivalents were
formed of which 64% were rings and the multivalent frequency recorded was
lower than either the A.hirtula(4x) x A.sativa or the A.barbata x A.sativa
hybrids. This situation clearly indicates that in the A.macrostachya x
A.sativa hybrid, a number of chromosomes from the A.sativa genomes are
pairing. The maximum chromosome pairing recorded in this hybrid 1 IV +
1 III + 10 II + 8 I demonstrates that 62% of the A.sativa chromosomes are
paired in addition to the expected seven bivalents from the A.macrostachya
parent.

In haploid plants of hexaploid Avena the maximum chromosome pairing
recorded is 2 II + 17 I(3). Clearly then the observed chromosome pairing
in the A.macrostachya x A.sativa hybrid is in excess of expectation from
the known chromosome pairing of the A.macrostachya parent and the haploid

oat. It would thus appear that there is a gene or genes in the A.macrostachya genomes which has an effect on the pairing of homoeologous chromosomes similar to that reported by Rajhathy and Thomas(4) in the diploid oat A.longiglumis (CW 57) which enhances homoeologous chromosome pairing.

From the chromosome pairing observed in this F_1 hybrid it was evident that producing backcross seed would be difficult and this proved to be the case. Less than 0.1% seed set was obtained from hand pollinations.

The chromosome numbers of these first backcross seeds were expected to be 38 or less since the 35 chromosome F_1 hybrid would be expected to contribute 17 or less chromosomes and the recurrent A.sativa parent 21 chromosomes. However, of the BC1 seeds which were chromosome counted chromosome numbers from 40 to 48 were recorded. The logical explanation for this would be that most of the A.sativa chromosomes from the F_1 hybrid plant had been incorporated into the female gamete together with up to 7 chromosomes from the A.macrostachya parent and of course the full complement of 21 from the male parent.

This BC1 progeny was not examined cytologically initially as it was of more importance to ensure that some seed was produced. However, cloned material is flowering and it is hoped to examine these plants cytologically.

Of the BC1 plants the 40 chromosome progeny have not produced any seed to date, but one plant which had 48 chromosomes was almost completely fertile and thus produced quite a lot of seed. The F_2 seed derived from the 48 chromosome BC1 plant is being grown in the glasshouse to multiply the seed for testing and analyses. Preliminary cytological examination of some of these progeny has revealed that of the order of 20 bivalents together with several univalents are formed at metaphase I of meiosis (depending on the chromosome number of the plant) which confirms that the majority of the A.sativa chromosomes did in fact become incorporated into the female gamete which led to the BC1 plants.

The low frequency of quadrivalents formed in the F_1 hybrid between A.macrostachya and A.sativa indicated that there was some homoeology between the two species, but the true homology might have been masked by the presence of strict homologues from the autotetraploid parent A.macrostachya. However, from the preliminary pairing data of the BC1 F_2 it would appear that the A.macrostachya chromosomes remain as univalents even though they have no strict homologues with which to pair. Thus the cytological evidence indicates that there is some residual homology between the chromosomes of A.macrostachya and A.sativa but that the relationship is very weak.

When sufficient F_3 seed is available tests for cold tolerance will be undertaken. If any plants are found to be tolerant to low temperatures then I will endeavour to establish addition lines initially, look for recombinants and possibly produce substitution lines. If the addition/substitution lines are unstable and no natural recombinant can be isolated then I will endeavour to produce an artificially contrived recombinant using the A.longiglumis CW 57 system.

3. SAIA HYBRIDS

The second hybrid I wish to report is that between the diploid species A.strigosa cultivar Saia and the hexaploid cultivated oat. The diploid species is known to be tolerant to BYDV, resistant to some races of rust, tolerant to the wild oat herbicide Hoegrass R, tolerant to cereal cyst nematode and is also thought to have some adult plant resistance to mildew. My initial concern is to isolate plants tolerant to BYDV.

Crosses were made using the diploid A.strigosa as the female parent. A large number of pollinations were made and all developing seeds were removed from the plants and the embryos excised and placed in culture tubes. Of the plants which grew to maturity none produced seed as had been anticipated from the known chromosome pairing of the tetraploid hybrid. In an attempt to restore some degree of fertility, the plants were treated with colchicine using the capping method. After colchicine treatment a number of seeds were produced and chromosome root tip counts were made which indicated that the number of chromosomes was 56.

These 56 chromosome amphidiploids were crossed using A.sativa as the pollen parent and some seed was produced. The resultant BC1 plants were grown to maturity and allowed to self. The progeny were chromosome counted and were found to have chromosome numbers ranging from 43 to 47.

These BC1 F$_2$ plants are undergoing a screening procedure together with the relevant controls in order to assess the degree of tolerance to BYDV. Initial observations indicate a differential response to the presence of BYDV, which is probably due to the different numbers and or combination of chromosomes present. Any plants which show a high degree of tolerance to the virus will be further backcrossed in order to produce addition lines and or substitution lines depending on the stability of the former. If addition/substitution lines are found to be stable then attempts will be made to extract some of the other agronomically desirable characters as well as tolerance to BYDV.

REFERENCES

1. Baum BR and T Rajhathy: Can. J. Bot. 21, 2434-2439, 1976.
2. Thomas H and ML Jones: Chromosoma 15, 132-139, 1964.
3. Nishiyama I and M Tabata: Jap. J. Genet. 38, 311-316, 1964.
4. Rajhathy T and H Thomas: Nature New Biol. 239, 217-219, 1972.

TISSUE CULTURE INDUCED VARIATION IN OATS

H.W. RINES*, S.S. JOHNSON and R.L. PHILLIPS
USDA-ARS* and Department of Agronomy and Plant Genetics, University of Minnesota, St Paul, MN 55108, USA.

1. INTRODUCTION

Genetic and cytogenetic variation appears as a common feature in plants derived from tissue cultures of oats, just as has been reported for cell and tissue culture derived plants of a wide array of species. Hundreds of papers including numerous reviews have been written describing this variation and speculating on its origin(1,2,3,4). The objectives of the studies described here were to further document the occurrence, frequency, and types of variation in oat plants regenerated from tissue culture, to investigate the origin of this variation, and to explore the value of this variation in oat improvement.

The first detailed report of plant regeneration from tissue cultures of oats by Cummings et al. in 1976(5) included accounts of phenotypic and cytogenetic variation among the regenerated plants and their progenies. Striped leaves, fatuoids, shorter plant height, and later maturity were among variants observed. Meiotic cells of seven regenerated plants were analysed; two plants had the normal 21 bivalents at diakinesis while five exhibited unique aberrant meiotic configurations.

2. FREQUENCY OF OCCURRENCE OF VARIATION

As a follow-up to these findings McCoy et al.(6) undertook a systematic analysis of the cytogenetic variation present in tissue culture derived oats and obtained evidence of a high frequency of chromosome breakage among the regenerated plants. Four culture lines were established from immature embryos in each of two cultivars, 'Tippecanoe' and 'Lodi'. At four month intervals plants were regenerated from these cultures and their meiotic cells analysed for chromosome number and structure alterations. The frequency of cytogenetically abnormal regenerants varied with both genotype and the culture age at time of regeneration. Cytological abnormalities were present in 12% of Tippecanoe and in 49% of Lodi plants regenerated from 4-month-old tissue cultures. The frequency of cytological abnormalities increased to 48% in Tippecanoe and 88% in Lodi plants regenerated from 20-month-old tissue cultures. Of a total of 799 regenerated plants analysed, 227 had a cytologically defined alteration. A striking feature of the array of defined cytological abnormalities was that over 85% were due to chromosome breakage and often involved partial chromosome loss. About 20% of the defined alterations were identified as interchanges and 1% as inversions, but the predominant type of variants present were telocentric or near-telocentric chromosomes. This was the case for both Tippecanoe and Lodi.

3. TYPES OF VARIATION AND ITS ORIGIN

The unexpected high frequency of apparent telocentrics among plants regenerated from oat tissue cultures led McCoy et al.(6) to postulate that there may be regions of late replicating heterochromatin flanking the centromeres in oat chromosomes and that this pericentromeric heterochromatin may be involved in chromosome breakage in tissue culture; i.e. in tissue culture cells the heterochromatic DNA which is normally late-replicating may be even further delayed in replication so that at the time of cell division the DNA replication in these regions is still incomplete. This incomplete replication could result in chromosome bridges and breakage events at anaphase. The hypothesis was based on the observations that heterochromatin is usually late replicating(7) and that several plant species have regions of heterochromatin flanking the centromeres of their chromosomes. However, at the time the hypothesis was proposed, it was not known if oats had pericentromeric heterochromatin and, if so, if it were late replicating. Recently completed(8) pachytene analysis has shown that oat chromosomes have extensive regions of densely staining heterochromatin flanking their centromeres (Fig. 1). Heterochromatic regions also were found at the nucleolar organiser regions of the three satellited chromosomes and as telomeric blocks in at least five of the chromosomes. The length of the pericentromeric heterochromatic regions is highly characteristic for many of the oat chromosomes. Measurements of the relative length of these regions together with relative arm lengths and total chromosome lengths have allowed tentative identification of 17 of the 21 chromosomes at the

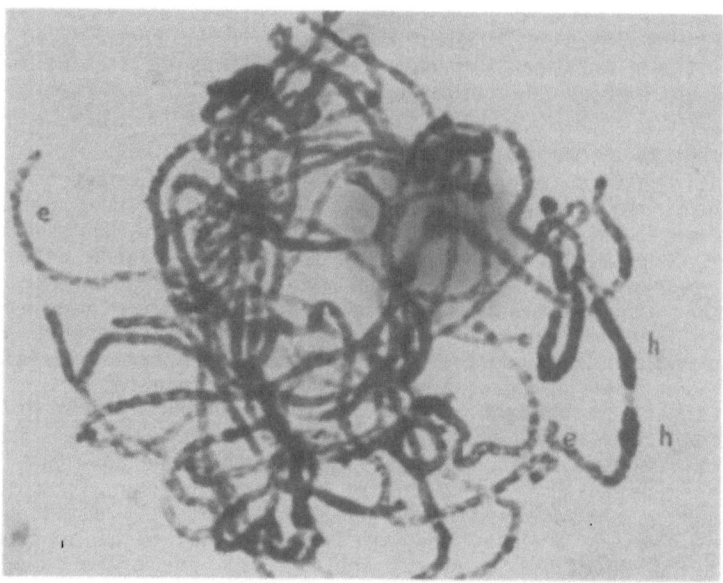

FIGURE 1. A pachytene chromosome preparation with densely staining pericentromeric heterochromatin (h) and lighter staining euchromatin (e). [2060x]

pachytene stage of meiosis(8). Evidence for pericentric heterochromatin as a likely site of chromosome breakage has come from pachytene analysis of progeny of a regenerated plant which had partial loss of a satellited chromosome; the break, presumably induced in tissue culture, had occurred in a region of centromeric heterochromatin(8).

In other studies pertaining to the hypothesis that tissue culture induced breakage is associated with late replication of heterochromatin, labelling patterns in autoradiographs following ^{3}H-thymidine uptake in root-tip cells indicated late replication of DNA in pericentromeric and telomeric regions of oat chromosomes(8). A component of the hypothesis which has not yet been addressed is whether late replicating heterochromatin may occasionally replicate even later in tissue culture cells. However, since the heterochromatic regions are the last to replicate, it is plausible that these regions are more sensitive to perturbations of the nuclear cycle which may occur during culture. If such a cause for the origin of cytogenetic instability in tissue culture could be verified, then factors which may affect cell division rate or DNA replication efficiency may be identified which would improve genetic stability in cultures.

Analysis of cytogenetic variants among tissue culture generated plants in maize(9,10) and in triticale(11,12) also implicate the involvement of heterochromatin in producing chromosome breaks in cultured cells.

Evidence for the activation of transposable elements in tissue culture regenerants of maize(10) indicates that genetic variation may result from action of transposable elements activated by chromosome breakage. Our hypothesis would be that this chromosome breakage, in turn, may trace back to late replicating heterochromatin as the origin of the events producing the variation. Whether oat genomes contain transposable elements that could be activated by chromosome breakage in tissue cultures is yet unknown, but evidence for such elements is being continually reported in more and more species.

4. VALUE FOR OAT IMPROVEMENT

The high frequency of observed chromosome breakage and products resulting from this breakage have been suggested to have potential value for oat genetic analysis and oat improvement(6). Ditelosomic stocks produced from telosomics recovered in tissue culture regenerated plants would be particularly valuable in gene mapping analysis, as are ditelosomic stocks in wheat(13). Our efforts to date to recover ditelosomic progeny from telosomic Tippecanoe, Lodi and Clinton tissue culture regenerants have been unsuccessful. Telosomic progeny are recovered, so the problem may be in a failure of the telocentric chromosome to be transmitted through the male. Genotype effects on transfer of chromosomal abnormalities have often been observed in oats(14), so the use of other parent stocks may permit the recovery of the desired ditelosomics. Monosomic plants are also recovered among tissue culture regenerants and represent an additional type of valuable stocks for genetic analysis and for producing chromosomal substitution lines. Efforts to transfer crown rust resistance from Avena strigosa to A.sativa through tissue culture induced chromosome breakage and exchange have been initiated but are incomplete.

As a follow-up to the studies by McCoy et al.(6) which documented the frequency of defined chromosomal changes present in oat plants regenerated from tissue cultures, we selected plants from that study that were classified as 'cytogenetically normal' and conducted yield tests on lines derived from them. The objective was to determine if during passage through tissue culture the regenerated oat plants acquired genetic changes which, although not cytologically detectable, would alter productivity of the plant. There were two practical reasons for these tests. One was to look for evidence of variation for increased productivity or quality. The other was to determine if productivity could be retained during extended periods of culture as might be needed for in vitro selection for specific traits such as herbicide resistance.

The initial yield tests were done using materials that were readily available from previous studies. They consisted of 53 lines derived from progeny of plants regenerated from 4-month-old cultures of Lodi and 123 lines derived from progeny of plants regenerated from 16- and 20-month-old cultures of Tippecanoe. Controls consisted of bulk seed stocks of the cultivars. Grain yields were measured on hill plots planted with 30 seeds each on a grid pattern with 30.5 cm between hill centres in randomized complete blocks. There were six replications at each of two locations, St. Paul and Rosemount, Minnesota.

The mean yield of the 53 progeny-derived lines from Lodi plants regenerated from 4-month-old cultures was about 6% below that of the non-cultured control. Thirteen lines were significantly lower yielding and none were higher. The mean yield of the 123 progeny-derived lines from Tippecanoe plants regenerated from 16- and 20-month-old cultures was about 25% below that of the non-cultured bulk control. All but four lines were significantly lower yielding and none were higher. The two highest yielding Tippecanoe derived lines were also later maturing.

These preliminary results indicate extended periods in culture produced genetic variation additional to the cytologically detectable changes and also that these changes tend to be detrimental to crop productivity. One problem in this study was that the control was simply a cultivar bulk. The lines analysed all traced back to four embryos out of 100 initially used in attempts to initiate cultures. There may have been selection for culturability such that the four were not representative of the cultivar bulk. Tests are now underway to compare yields of derived lines from 4-month-old cultures of Tippecanoe to yields of derived lines from 12-, 16- and 20-month-old cultures of Tippecanoe to determine more accurately if yield potential does decrease with time in culture in oats due to an accumulation of mutations.

5. IN CONCLUSION

It is evident that genetic and cytogenetic variation is produced in oat tissue cultures. Some types are potentially useful, but much of the variation appears detrimental. A better understanding of the cause of tissue culture induced variation may permit researchers to increase or decrease the frequency and types of variation produced depending on the goals of the research.

REFERENCES
1. Larkin PJ and WR Scowcroft: Theor. Appl. Genet. 60, 197-214, 1981.
2. Larkin PJ and WR Scowcroft: In, T Kosuge, CP Meredith and A Hollaender (eds), Genetic Engineering of Plants. Plenum, New York. pp.289-314, 1983.
3. Orton TJ: Plant Molec. Biol. Rep. 1, 67-76, 1983.
4. Orton TJ: Adv. Plant Pathol. 2, 153-189, 1984.

5. Cummings DP, CE Green and DD Stuthman: Crop Sci. 16, 465-470, 1976.
6. McCoy TJ, RL Phillips and HW Rhines: Can. J. Genet. Cytol. 24, 37-50, 1982.
7. Lima-de-Faria A: In, A Lima-de-Faria (ed.), Handbook of Molecular Cytology. North Holland Publishing, Amsterdam. pp.277-325, 1969.
8. Johnson SS: MS Thesis, University of Minnesota, 1985.
9. Benzion G: PhD Thesis, University of Minnesota, 1984.
10. Benzion G, RL Phillips and HW Rines: In, I Vasil (ed.), Plant Regeneration and Genetic Variability in vitro: Oats and Maize. Academic Press, Orlando, Florida (in press).
11. Armstrong KC, C Nakamura and WA Keller: Z. Pflanzenzüchtg. 91, 233-245, 1983.
12. Lapitan NLV, RG Sears and BS Gill: Theor. Appl. Genet. 68, 547-554, 1984.
13. Sears ER: In, Proc. 2nd International Wheat Genetics Symposium, Stockholm, pp.370-381, 1966.
14. Rajhathy T and H Thomas: Cytogenetics of oats (Avena L.), Misc. Publ. Genet. Soc. Canada No. 2, 1974.

PHYLOGENETIC RELATIONSHIPS BETWEEN AVENA SPECIES REVEALED BY THE RESTRICTION ENDONUCLEASE ANALYSIS OF CHLOROPLAST AND MITOCHONDRIAL DNAS

K. MURAI and K. TSUNEWAKI

Laboratory of Genetics, Faculty of Agriculture, Kyoto University, Sakyo-ku, Kyoto 606, Japan

1. INTRODUCTION

The cytoplasmic variation in the genus Avena has been studied so far by two means, i.e. electrophoretic characterisation of the large subunit of ribulose 1,5-bisphosphate carboxylase (Rubisco)(1), and a comparative study of alloplasmic lines of common oats(2). Since 1976, the restriction endonuclease analysis of organellar DNAs has become a new tool for studying genetic diversity among cytoplasms of different plant species. Due to the maternal transmission of organellar DNAs, this approach is useful to determine the cytoplasm donor to polyploid species(3). We investigated the variability of chloroplast(ct) and mitochondrial(mt) DNAs among Avena species to clarify their maternal lineage.

2. MATERIALS AND METHODS

The species used as sources of ct and mtDNAs are given in Table 1. These stocks were the kind of gift of T. Morikawa, of the University of Osaka Prefecture, I. Nishiyama, Emeritus Professor, Kyoto University,

TABLE 1. Chloroplast (ct) and mitochondrial (mt) genome types
of 19 Avena species

Strain no.	Species	2n	Nuclear genome	Ct genome	Mt genome
1	A.canariensis	14	AcAc	I	1
2	A.damascena	14	AdAd	I	2
3	A.longiglumis	14	A1A1	II	-
4	A.prostrata	14	ApAp	I	3
5	A.hirtula	14	AsAs	I	4
6	A.strigosa*	14	AsAs	I	5
7	A.wiestii	14	AsAs	I	4
8	A.clauda	14	CpCp	III	-
9	A.pilosa	14	CpCp	IV	-
10	A.ventricosa	14	CvCv	V	-
11	A.abyssinica*	28	AABB	I	4
12	A.barbata	28	AABB	I	4
13	A.vaviloviana	28	AABB	I	6
14	A.magna**	28	AACC	I	7
15	A.murphyi	28	AACC	I	7
16	A.byzantina*	42	AACCDD	I	8
17	A.fatua	42	AACCDD	I	8
18	A.sativa*	42	AACCDD	I	8
19	A.sterilis	42	AACCDD	I	8

*Cultivated species; **Synonymous with A.marrocana

Japan, and R. Loiselle, Central Office for Plant Gene Resources of Canada. Chloroplasts were isolated from leaves after the method of Tsunewaki and Ogihara(3) but a discontinuous gradient with 15, 40 and 60% sucrose was used instead of the Percoll gradient. Mitochondria were prepared from seedling leaves by the method of Bonen and Gray(4), to which a step of digesting non-mitochondrial DNAs with DNase I was added. Ct and mtDNAs were purified by centrifugation with a CsCl-ethidium bromide gradient(5). The ctDNA was digested with eight restriction endonucleases, BamHI, EcoRI, HindIII, KpnI, PstI, SalI, SmaI and XhoI, and mtDNA with BamHI. The DNA fragments were separated by electrophoresis, using 0.8% agarose gel in 40 mM Tris, 20 mM NaOAC and 2 mM EDTA. A dendrogram showing the phylogenetic relationships among chloroplast and mitochondrial genomes was constructed using the method of Engels(6).

3. RESULTS
3.1 Chloroplast and mitochondrial genome diversity
   Chloroplast genomes of 19 species were classified into five types, I-V, based on their restriction fragment patterns of the eight endonuclease digests (Table 1). The BamHI restriction fragment patterns of all species are shown in Figure 1. The mitochonrdial genomes of 15 species carrying

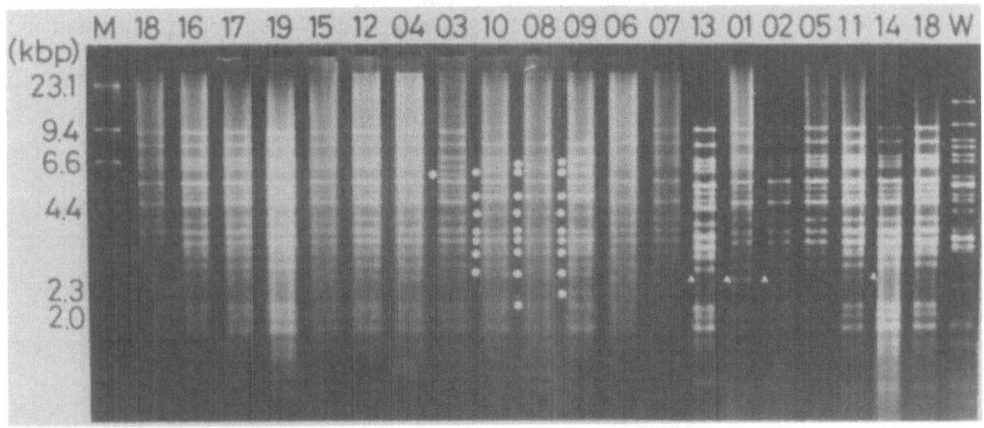

Figure 1. BamHI restriction fragment patterns of ctDNAs from 19 Avena species. The lane number corresponds to the code number of species given in Table 1. M and W indicate HindIII-digested λ DNA and BamHI-digested common wheat ctDNA, respectively. ●: Fragments differed from the corresponding fragments of A.sativa (No. 18). ▲: Intraspecific variation (data not shown).

the Type I chloroplast genome were classified into eight types, 1-8, from their BamHI patterns (Table 1 and Fig. 2). Based on the number of different fragments between each pair of chloroplast or mitochondrial genomes, the genetic distance, $\hat{p}$, was calculated according to Engels' method (Tables 2 and 3). Using these $\hat{p}$ values, a phylogenetic tree for the chloroplast and mitochondrial genomes was drawn (Fig. 3).

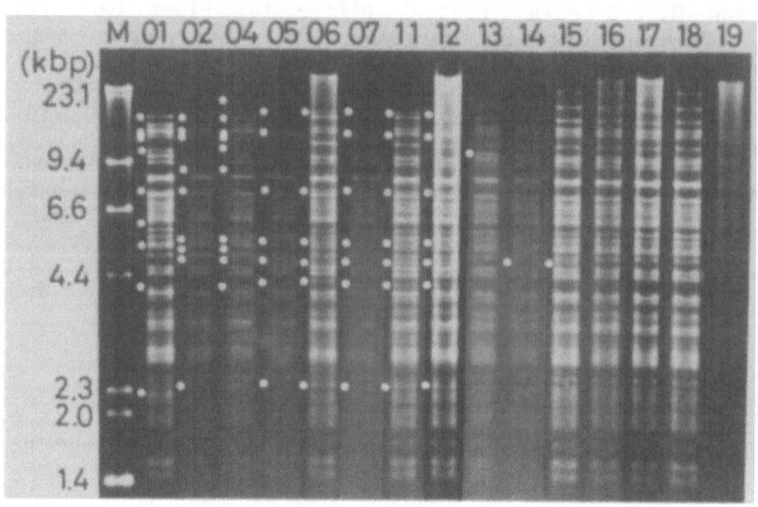

FIGURE 2.   BamHI restriction fragment patterns of mtDNAs from 15 Avena species having the Type I chloroplast genome.   The lane number corresponds to the code number of species given in Table 1.   M indicates HindIII-digested λ DNA and HaeIII-digested øx174DNA. ●: Fragments differed from the corresponding fragments of A.sativa (no. 18).

TABLE 2.   Genetic distance, $\hat{p}$ ($\times 10^3$), among five chloroplast genomes

Chloroplast genome	Chloroplast genome				
	I	II	III	IV	V
I	0				
II	13	0			
III	124	124	0		
IV	102	111	44	0	
V	93	102	44	44	0
Average	83.0	87.5	84.0	75.3	70.8

The differentiation of chloroplast genomes is limited to the diploid species.   Two chloroplast genome types, I and II, found in A genome diploids, are closely related with each other, and greatly differ from the other three types, III-V, found in C genome diploids.   The divergence of chloroplast genomes has occurred in concert with that of nuclear genomes in diploids.   The genetic distance between chloroplast genomes of C genome diploids ($\hat{p}$ = 0.0044) is larger than that between A genome diploids ($\hat{p}$ = 0.0015), indicating that C genome diploids evolved earlier than A genome diploids.   A.longiglumis diverged from other A genome diploids. All three C genome diploids appear to have differentiated at almost the same time.

TABLE 3. Genetic distance, $\hat{p}$ $(\times 10^3)$, among eight mitochondrial genomes

Mitochondrial genome	Mitochondrial genome							
	1	2	3	4	5	6	7	8
1	0							
2	75	0						
3	74	30	0					
4	93	78	107	0				
5	93	78	107	32	0			
6	93	78	76	96	64	0		
7	93	78	76	64	64	16	0	
8	78	94	93	81	81	16	16	0
Average	85.6	73.0	80.4	78.7	74.1	62.7	58.1	65.6

The fact that mitochondrial genomes of 15 species carrying the same chloroplast genome, Type I, are classified into eight types, indicates that the mitochondrial genome is more changeable than the chloroplast genome.

On the origin of polyploid species: The data on mitochondrial genome diversity demonstrate that the mitochondrial genome of A.barbata is identical to those of A.hirtula and A.wiestii (Type 4). This fact supports the previous suggestion that the barbata group originated from the hirtula-wiestii complex(7). Our data also agree with the previous view that a progenitor of the magna-murphyi complex contributed the A and C genomes to hexaploid species(7). The previous morphological and isozyme

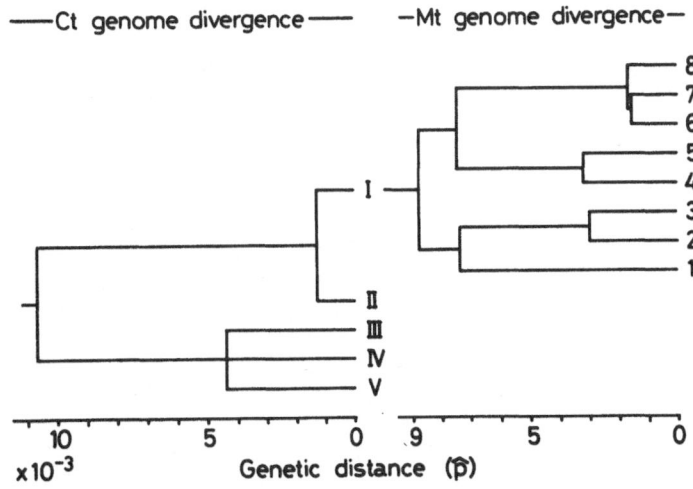

FIGURE 3. A dendrogram showing the phylogenetic relationships among chloroplast and mitochondrial genomes in Avena. I-V and 1-8: Chloroplast and mitochondrial genome types, respectively (see Table 1).

38

studies(8,9) suggested that the magna-murphyi complex received the A genome from A.canariensis. The present results, however, revealed a great difference between their mitochondrial genomes. In fact, the meiotic studies of their hybrids proved little homology between the chromosome complements of A.canariensis and A.magna(10). These facts suggest that the maternal parent of the magna-murphyi complex is not A.canariensis but the hirtula-wiestii complex. Furthermore, we propose that the cytoplasm of hexaploids, including A.sativa, was derived from the hirtula-wiestii complex. This proposal does not contradict the previous data on the Rubisco large subunit(1) and the effects of six alien cytoplasms on two characters of A.sativa(2).

4. ACKNOWLEDGEMENTS

This work was supported in part by a Grant-in-Aid (No. 60400005) from the Ministry of Education, Science and Culture, Japan.

REFERENCES
 1. Steer MW and H Thomas: Can. J. Genet. Cytol. 18, 769-771, 1976.
 2. Leggett JM: Can. J. Genet. Cytol. 26, 698-700, 1984.
 3. Tsunewaki K and Y Ogihara: Genetics 104, 155-171, 1983.
 4. Bonen L and MW Gray: Nucl. Acid Res. 8, 319-335, 1980.
 5. Kolodner R and KK Tewari: Biochem. Biophys. Acta 402, 372-390, 1975.
 6. Engels WR: Proc. Natl. Acad. Sci. USA 78, 6329-6333, 1981.
 7. Rajhathy T and H Thomas: Misc. Pub. Genet. Soc. Canada 2, 1-90, 1974.
 8. Baum BR, T Rajhathy and DR Sampson: Can. J. Bot. 51, 759-762, 1973.
 9. Craig IL, BE Murray and T Rajhathy: Can. J. Genet. Cytol. 16, 677-689, 1974.
10. Leggett JM: Can. J. Genet. Cytol. 22, 287-294, 1980.

This paper is Contribution No. 475 from the Laboratory of Genetics, Faculty of Agriculture, Kyoto University, Japan.

MOLECULAR BIOLOGY OF OAT STORAGE PROTEIN GENES: PROBES FOR EVALUATING GERMPLASM VARIATION

I. ALTOSAAR and S. F. FABIJANSKI
Biochemistry Department, University of Ottawa, Ottawa, Ontario, Canada

1. INTRODUCTION

The advent of recombinant DNA methodologies is now allowing for the more rapid improvement of crop parameters. Seeds are now being 'engineered' with new foreign genes being added to them or deleterious genes being inactivated by targeted insertion of DNA elements. The agri-food industry shall only remain competitive by continuing to produce cultivars that are custom-tailored for agronomic conditions and food processing requirements. To do this the plant breeding community and cereal geneticists must rapidly assimilate and master the newly available techniques of plant molecular biology such as Ti-plasmids, tissue culture, protoplast fusion, viral vectors, transposons and gene signals.

There are already many examples of successful genetic engineering of plants. The major seed protein in French beans called phaseolin, is a globulin of the vicilin type. The phaseolin gene has been isolated, cloned, and inserted into the tumour-inducing (Ti) plasmid of Agrobacterium tumefaciens. This newly-constructed plasmid vector was used to genetically transform tobacco plants. The foreign gene, now under the control of the Ti plasmid's transferred DNA (T-DNA) promoter sequence, was inserted into the tobacco chromosomes, maintained, and remained functional through meiosis(1). Detailed transcriptional and translational analysis showed that all the genetic signals specifying tissue-specific expression and molecular processing have been evolutionarily conserved across these two diverse plant families.

The past three years has seen a revolution in our understanding and use of the Ti-plasmid for moving specific genes in and out of plants. The T-DNA vector system has worked well for inserting new genes into dicotyledonous plants but, since Agrobacterium did not cause tumours in monocots (e.g. cereals), it always was believed that Ti could not be used for transferring new traits to this major group of industrially-important crop plants. Recent work has shown however that T-DNA does in fact become incorporated into monocots(2). Protoplasts of Triticum monococcum have been transformed with foreign DNA and selectable markers(3).

For wheat and other cereals, regeneration of fertile transformed plants has not been possible from single cell or protoplast tissue cultures. But the availability of new plant growth 'elicitors' may help to overcome this problem as well. For rice, wheat, millet and oats, selection of embryogenic callus now yields long-term, high frequency plant regeneration(4). Consequently many labs have transformation experiments in progress where various T-DNA vector constructs of selectable marker genes are being incubated and infected into wheat cereal cultures and leaf discs directly. Ultimately, genetic engineering of cereals will involve things like a) changing the number of copies of a certain gene in a cultivar, b) varying the expression of a gene by altering the DNA sequences which regulate its transcription into messenger RNA or its translation into protein or enzyme, and c) inserting genes from other species. Genes which are already available for insertion into crops, or soon will be available for transfer include storage protein genes, and insect, herbicide and viral resistance genes.

2. OAT GENES AND THEIR GENETIC CONTROL MECHANISMS

To increase our knowledge of oat genes and their genetic control mechanisms, we have studied several 'on-off' systems. One model system is the lipase gene which actively makes enzyme in oats in large quantities while this gene-product is virtually absent in wheat, barley and rye(5). The lipase gene is turned on at a specific time during the six-week time period of seed development when all grain quality characteristics are established(6). A second lipase gene may also be active separately during grain germination. Looking at the tissue specificity of this marker enzyme (and potential marker gene), it was found primarily localised in the aleurone cells(7). Together, these data suggested that the lipase gene in oats may provide useful information on specific regulatory mechanisms and thus its DNA sequences might be obtained via cloning technology. To this end, we have purified oat lipase and hope to obtain some amino acid sequence data from this molecule. Appropriate DNA probes will then be synthesised and hybridised to our oat and wheat cDNA and genomic libraries (see below) to isolate the corresponding lipase gene(s).

Grains are processed into a variety of foods depending on the biochemical properties of the flour. Genes representative of the important wheat proteins have been isolated using recombinant DNA techniques in readiness to insert directly into wheat chromosomes, via transformation of elite wheat cultures(8). To better understand and manipulate cereal grain quality, the molecular evolution of seed storage proteins in wheat, barley and rye is being studied through the DNA sequence and structure of cloned genes(9). The tribe Aveneae is closely related to these cereals and the characterisation of these storage protein groups in oats may shed more genetic light on these important grain proteins. We found common oat cultivars of different protein content to exhibit molecular weight and charge heterogeneity of their prolamins similar to wheat but the level of expression is greatly reduced(10). Instead of forming 60-70% of grain protein as in other cereals, oat prolamin synthesis is stopped at 10-15%. The genetic expression of the glutelin fraction in oat is virtually shut down(11). Even the residual protein, left in the extracted flour after removal of all the Osborne fractions, we have now found to consist of insoluble globulin by Western immuno-blotting techniques(12). Therefore, rather than consisting of mainly alcohol-soluble proteins as in the Triticeae, 75-85% of oat grain protein is composed of salt-soluble globulin instead(13). During grain development, the α (40 KD) and β-globulin polypeptides (20 KD) appear to be under strict genetic control as larger subunits are synthesised 2-3 weeks after flowering, contrasting markedly with soy glycinins or pea legumins.

To better understand the genetic differences between wheat and oat, we started studying the genes being transcribed during grain development. Authentic messenger RNA could only be obtained from endosperm after dehulling green oats in liquid nitrogen(14). Oat mRNA was translated in vitro in a wheat germ extract and found to synthesise a globulin subunit of 60,000 molecular weight, rather than the two 40 KD and 20 KD polypeptides present in mature grain(15). We confirmed that the α-β polypeptides are disulfide linked in unreducing conditions suggesting that the 60 KD subunit is homologous to pea legumin. We verified this by direct in vivo labelling, finding endoplasmic reticulum (ER) to be the site of synthesis for the 60 KD globulin subunit(16). The labelled globulin was then shown to be transported to protein bodies intact, but there it was proteolytically cleaved to form the dissociable α-β dimer composed of the 40 KD and 20 KD polypeptides respectively(17). This striking evolutionary conservatism between dicot legumes and monocot cereals was directly corroborated by DNA-DNA hybridisation using a pea legumin cDNA clone(18).

We therefore further fractionated oat protein to see if its globulin fraction also contained vicilins characteristic of legumes (e.g. phaseolin). Sucrose density centrifugation did in fact yield a 7S oat component as well(19). Our affinity chromatography data indicates that oat glycosylates the vicilins in a manner similar to legumes as well(20).

If legumin and vicilin genes have indeed been maintained in the monocots, their protein products may also be present in cereals other than oat. ^{125}I-labelled antibodies, raised against both oat 12S globulin holoprotein and the α-subunit, gave us positive signals in Western immuno-blots against globulin extracts from wheat, barley, rye, corn and rice(21). The legumin levels were of course not high except in rice. We therefore analysed each rice fraction by Western immuno-blotting and found that all rice glutelin polypeptides cross-react, thus confirming that the legumin gene is actively expressed in rice(22).

Similarly, we raised antibodies against oat 3S and 7S vicilins. These were also radio-iodinated and used to challenge extracts from wheat, rye, barley, rice and corn. Again, our studies showed that polypeptides antigenically related to pea and oat vicilins do occur in the globulin fractions of all cereals(23). The molecular weight variability we observed in the genetically homologous proteins among these different cereals can be expected in view of the variant subunit forms which have been observed intraspecifically with pea and oat globulin. In oat, the legumin:vicilin ratio is about 7:1 whereas in the members of the Triticeae it is about 1:8, almost the reverse (21,22). This striking example of differential gene expression among cereals parallels the prolamin:globulin difference discussed above. With recombinant DNA gene libraries from oat and wheat now in hand one can analyse the precise gene signals responsible for controlling these individual grain proteins. We studied wheat mRNA and translated it in vitro showing that the 60 KD legumin subunits also arise from a precursor message(24). The bulk of the mRNA, however, codes for prolamins. Fractionating oat mRNA carefully on sucrose gradients, we were surprised to find that the bulk of it corresponds to the major size class of mRNA as in wheat(25). When translated in vitro, the major peaks of oat mRNA (12S and 15S) yield prolamin-like polypeptides. Prolamins, however, only constitute 10-15% of oat protein as detailed above. Only a minor peak of oat mRNA 18S in size, was found to code for the globulins, the abundant protein constituting 75-85% of oat protein(25). This suggested that 18S mRNA is preferentially translated on oat ribosomes and that some form of translational control is responsible. We have started to investigate how prolamin mRNA translation is blocked in oat endosperm by simple reconstruction experiments. These show that the polysomal protein factors do influence the translational specificity of the 18S oat globulin poly A[+] mRNA relative to the more abundant prolamin mRNAs present(26). Most recent experiments with ^{32}P-in vivo labelling indicate that there are also differences in the phosphorylation states of certain ribosomal proteins between endosperm and leaves (S. Fabijanski, unpublished work). We hope eventually to have specific gene constructs of hybrid (prolamin/globulin) mRNAs to test in both oat and wheat endosperm systems. This may show more clearly which cereal DNA sequences influence protein synthesis in grain(27). This becomes valuable information, indeed essential data, when foreign genes are transformed into cereals and need to be expressed at precise levels in order for specific grain quality characteristics to be achieved.

To develop this prolamin:globulin model system in wheat:oat endosperm, oat endosperm RNA was reverse transcribed to produce complementary DNA (cDNA). This cDNA was cloned in both plasmid pBR322 and lambda phage gt10 to form several cDNA libraries. Screening with 12S (prolamin) or 18S (globulin) end-labelled mRNA on plaque filters, we have isolated both 12S-specific and 18S-specific clones. Our prolamin cDNA clone, p3B3, has been used to probe for the expression of this gene during the 6-week period of grain development by Northern blotting. Similarly, we have used it to probe the DNA from nineteen species of Avena to look at how prolamin gene copy number and chromosomal rearrangement vary with ploidy level. The p3B3 cDNA insert is about 1,100 nucleic acid base pairs (bp) long and hybrid selects a 16 KD prolamin. We are now sequencing this cDNA by Sanger's dideoxy method. The prolamin protein fraction shows many differences among species in the mature protein, possibly a result of gene copy number which varies: about 30 copies per haploid genome in A.sativa as determined by reconstruction on Sac 1 digests; 20-30 copies in A.barbata, 30-40 copies in A.clauda and only 15-20 copies in A.magna.

To further evaluate germplasm variation, repeated DNA sequence clones have also been generated to probe restriction fragment length polymorphisms (RFLPs) in the oat polyploid series (Figs. 1,2). We have investigated the relationships between the species in the genus Avena by Southern blot analysis of species DNA probed with cloned repeated sequences. Our results were generated using primarily two probes, pTa-71, a rRNA clone isolated from wheat(8), and RS-1, a repeated sequence clone which we isolated from a Ch 4A genomic library of Avena DNA. The data in Figure 1 demonstrates that it may be possible to distinguish the different genomic karyotypes on this sort of analysis. In addition to the dominant common band seen in all the species, it seems that some species have homogenised a group of different variants or homologous sequences through the course of evolution. The amount of variation seen in this RFLP pattern may indicate the rapidity in which variant genes have been homogenised. Using the oat repeated sequence clone RS-1 on the other hand, quite different hybridisation patterns were observed in the Avena species (Fig. 2). It became apparent that the different genomic karyotypes displayed distinctive bands. A southern blot hybridisation signal designated Band 1 appeared in all species containing an A genome and was absent in others, such as A.clauda, a C diploid. Similarly, RS-1 probe produced a C-karyotype specific band at about 4.3 kilobase position, absent in A.barbata and A.hirtula as shown. These figures demonstrate that genomes that have been classically grouped into different karyotypes can also be distinguished using molecular probes. It is very likely that the differences in the restriction digests and subsequent blots can also be associated with chromosome morphology.

3. CONCLUSION

To summarise, oat lipase, prolamins, legumins and vicilins have been

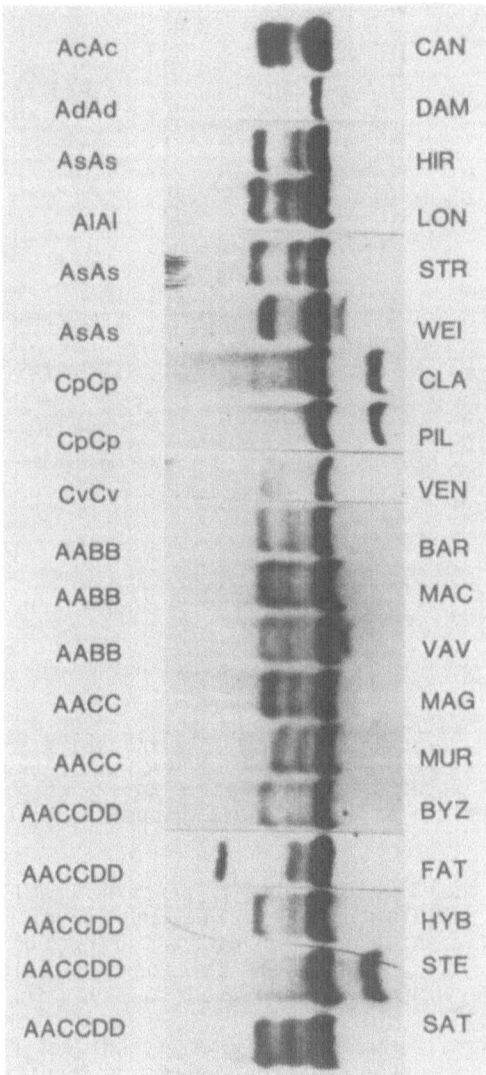

FIGURE 1. Germplasm evaluation of 19 oat species using Southern blot analysis: 15 µgms per lane of species DNA restricted with Sac 1, separated on agarose gel, transferred to nitrocellulose filter and probed with radioactively labelled wheat rDNA repeated sequence clone pTa71. Autoradiograph of the hybridisation filter is shown. Species are indicated with their corresponding karyotypes as follows: CAN, canariensis; DAM, damascena; HIR, hirtula; LON, longiglumis; STR, strigosa; WEI, weistii; CLA, clauda; PIL, pilosa; VEN, ventricosa; BAR, barbata; MAC, macrostachya; VAV, vaviloviana; MAG, magna; MUR, murphyi; BYZ, byzantina; FAT, fatua; HYB, hybrida; STE, sterilis; SAT, sativa.

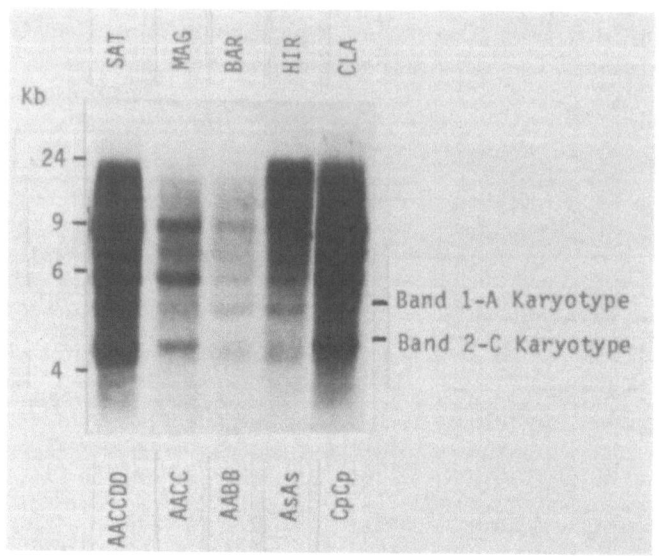

FIGURE 2. Southern blot hybridisation signals using oat repeated sequence clone RS-1 against 15 µgms per lane of genomic DNA fragmented with Bgl II. Karyotype specific bands can be seen in the 5 kilobase (Kb) region of A-containing species and at 4.3 Kb for C-genome containing species.

characterised at the level of their protein chemistry, immunological cross-reactivity, cellular localisation and developmental regulation. Several marked differences between oat and rice, on the one hand, and wheat, barley and rye, on the other, have been detailed at the molecular level. These differences have been used to study the genetic control of certain grain quality characteristics. To better understand these regulatory mechanisms, recombinant DNA techniques have been used to isolate some of these genes so that their molecular structure can be determined. The transcriptional and translational controls that may be determining final levels of gliadins, glutelins, legumins and vicilins in various genotypes will prove invaluable in transformation experiments aimed at engineering specific grain properties. The use of molecular probes to study the genomic relationships of Avena can provide some clues as to the origin of the commercially grown Avena species and can therefore provide breeders with directions for the efficient transfer of desirable traits of wild oats into commercial varieties.

REFERENCES

1. Sengupta-Gopalan C, NA Reichert, RF Barker, TC Hall and JD Kemp: P.N.A.S. 82, 3320-3324, 1985.

2. Hooykaas-Van Slogteren GMS, PJJ Hooykaas and RA Schilperoort: Nature 311, 763-764, 1984.
3. Lorz H, B Baker and J Schell: Mol. Gen. Genet. 199, 178-182, 1985.
4. Rines HW and HH Luke: Theor. Appl. Genet. 71, 16-21, 1985.
5. Matlashewski GJ, AA Urquhart, MR Sahasrabudhe and I. Altosaar: Cereal Chem. 59, 418-422, 1982.
6. Urquhart AA, CA Brummell, I Altosaar, GJ Matlashewski and MR Sahasrabudhe: Cereal Chem. 61, 105-108, 1984.
7. Urquhart AA, I Altosaar, GJ Matlashewski and MR Sahasrabudhe: Cereal Chem. 60, 181-183, 1983.
8. Flavell RB, PI Payne, RD Thompson and CN Law: Biotech & Gen. Eng. Rev. 2, 157-173, 1984.
9. Kreis M, BG Forde, S Rahman, BJ Miflin and PR Shewry: J. Mol. Biol. 183, 499-502, 1985.
10. Robert L, C Nozzolillo and I Altosaar: Cereal Chem. 60, 438-442, 1984.
11. Robert L, A Cudjoe, C Nozzolillo and I Altosaar: J.C.I.F.S.T. 16, 196-200, 1983.
12. Robert L, C Nozzolillo and I Altosaar: Cereal Cehm. 62, 276-279, 1985.
13. Robert K, GJ Matlashewski, K Adeli, C Nozzolillo and I Altosaar: Cereal Chem. 60, 231-234, 1983.
14. Garson K, GJ Matlashewski, K Adeli, LS Robert and I Altosaar: Analyt. Biochem. 134, 512-516, 1983.
15. Matlashewski GJ, K Adeli, I Altosaar, PR Shewry and BJ Miflin: FEBS Lett. 145, 208-212, 1982.
16. Adeli K and I Altosaar: Plant Physiol. 73, 949-955, 1983.
17. Adeli K, P Allan-Wojtas and I Altosaar: Plant Physiol. 76, 16-20, 1984.
18. Matlashewski GJ: Ph.D. Thesis, Univ. Ottawa Biochem. Dept., 1983.
19. Burgess SR, PR Shewry, GJ Matlashweski, I Altosaar and BJ Miflin. J. Exp. Bot. 34, 1320-1332, 1983.
20. Adeli K and I Altosaar: Plant Physiol. 75, 225-227, 1984.
21. Robert L, C Nozzolillo and I Altosaar: Biochem. J. 226, 847-852, 1985.
22. Robert L and I Altosaar: Biochim. Biophys. Acta 829, 19-26, 1985.
23. Robert L, K Adeli and I Altossar: Plant Physiol. 78, 812-816, 1985.
24. Fabijanski S, I Altosaar, M Lauriere, J-C Pernollet and J Mosse: FEBS Lett. 182, 465-469, 1985.
25. Fabijanski S, GJ Matlashewski and I Altosaar: Plant Molec. Biology 4, 205-210, 1985.
26. Fabijanski S and I Altosaar: Plant Molec. Biology 4, 211-218, 1985.
27. Adeli K and I Altosaar. FEBS Lett. 178, 193-198, 1984.

CALLUS INDUCTION AND PLANT REGENERATION FROM OAT CULTIVARS

I.P. KING, H. THOMAS and P.J. DALE
Welsh Plant Breeding Station, Aberystwyth, Dyfed, Wales, UK

1. INTRODUCTION

Variation is observed among plants regenerated from callus cultures. This 'somaclonal' variation has been recorded in a wide range of plant species. The heritable variation produced in this way can be for agriculturally important characters and therefore may be of value for crop improvement. Changes in chromosomes may also give an opportunity for genetic analysis and gene transfer.

The aim of this investigation was to use a culture system already devised for oats (1) to determine a) the optimum size of immature embryos for callus induction and plant regeneration under our conditions b) the effect of plant genotype on callus induction and plant regeneration using 10 oat cultivars and c) the extent of variation in plants regenerated from a disomic addition line.

2. MATERIALS AND METHODS

The cultivars were grown from early December in a glasshouse heated to 20°C during the 16 h light period (with supplementary lighting) and 10°C during the 8 h dark period. Embryos were excised from surface sterilised caryopses and measured along their root-shoot axis. One hundred immature embryos per cultivar (in 5 replicates) were cultured in 25-compartment dishes (10 x 10 cm) containing Murashige and Skoog's basal medium (MS) with 2 mg l^{-1} 2,4-D and 30 g l^{-1} sucrose. The medium was adjusted to pH 6, solidified with 5 g l^{-1} Sigma agar and autoclaved at 121°C for 15 min. Calluses were transferred to new medium after 4 weeks and callus growth was determined by estimating the callus diameter 3 weeks later. Plant regeneration was assessed at 12 weeks after culturing. When plantlets were large enough they were transferred to culture tubes containing MS medium with 0.5 mg l^{-1} 2,4-D and approximately one month later to MS medium without 2,4-D.

The disomic addition line, cultured in a similar way, was Avena sativa with a pair of chromosomes (2n = 42 + 2 chromosomes) from A.prostrata carrying a homozygous dominant gene for mildew resistance.

3. RESULTS AND DISCUSSION

Over the ten cultivars, the embryo sizes cultured ranged from 1-5 mm long (Table 1a). There were differences in callus growth from the various embryo sizes (P = < 5%), with greatest callus growth from embryos 3.5 - 4.5 mm long. The proportion of calluses regenerating plants was also influenced by embryo size. There was a positive correlation (r = 0.83; P = < 5%) between embryo size and the proportion of calluses regenerating plants. The highest level of plant regeneration was from embryos 4 - 4.5 mm long.

There were no significant differences between cultivars in the mean embryo sizes cultured or in the amount of callus produced (Table 1b). The rate of regeneration from the different cultivars, however, ranged from 3-42% suggesting that, in agreement with many other reports, there are genotypic differences in the ability to regenerate plants from callus cultures.

TABLE 1. Callus production and plant regeneration from (a) a range of embryo sizes (* = no. embryos cultured) and (b) ten oat cvs.

(a) Embryo size (mm)	Callus size (mm)	Regeneration (%)
1.1 - 1.5	10.5 (7)*	0 (4)*
1.6 - 2.0	11.2 (42)	12 (24)
2.1 - 2.5	11.3 (139)	22 (101)
2.6 - 3.0	12.0 (180)	25 (129)
3.1 - 3.5	12.3 (168)	26 (118)
3.6 - 4.0	14.1 (158)	25 (109)
4.1 - 4.5	13.9 (85)	41 (59)
4.6 - 5.0	13.1 (19)	32 (11)
(b) Cultivar		
Trafalgar	12.9	30
Rollo	12.3	28
07408 1n III/2	12.3	3
Rhiannon	13.5	20
Dula	12.1	24
Avalanche	12.8	42
Caron	14.4	20
Pennal	12.0	14
Cabanna	13.4	28
Margam	12.8	34

The progeny of 21 plants regenerated from callus of the disomic addition line were analysed cytologically and for mildew resistance. Of the 21 lines examined at meiosis, 7 had a single multivalent (A.prostrata chromosomes pair only rarely with A.sativa), 3 had double multivalents, 2 had heteromorphic bivalents, 3 had bridges at anaphase, 18 had univalents and 18 had fragments at meta-anaphase. Tetrads were also examined for micronuclei. The proportion of tetrads with micronuclei in control plants was approximately 1% but in the 21 lines, 10 lines had 2-10%, 9 lines had 11-20%, 1 line had 21-30% and 1 line had 31-40%.

Testing the 21 lines for mildew resistance revealed that 8 lines had some susceptible plants. All the resistant plants examined had 44 chromosomes and the susceptible plants 42 chromosomes. No resistant plants had 42 chromosomes; there was no evidence therefore for the translocation of the mildew resistance gene from A.prostrata to A.sativa chromosomes. There was evidence, however, for very frequent changes in chromosome structure in cultured cells and for somatic instability and preferential loss of A.prostrata chromosomes. Somatic elimination of A.prostrata chromosomes in the intact plant occurs rarely, if at all. Their non-transmission through meiosis and seed set is also at a low frequency (approximately 2%) (2).

REFERENCES

1. Rines HW and RJ McCoy: Crop Sci. 21, 837-842, 1981.
2. Bahari JB: Chromosome addition lines in Avena. MSc Thesis, University of Wales, 1979.

DIFFERENTIAION OF SPIKELET TYPE IN <u>AVENA</u> IN RELATION TO THE PLOIDY

I. NISHIYAMA and T. MORIKAWA
University of Osaka Prefecture, Sakai, Osaka, Japan

The spikelet type of <u>Avena</u> is usually characterised by different combinations of two pairs of floret characters, hulled versus hull-less and separation versus non-separation. In the present study, four spikelet types were found in 20 <u>Avena</u> species and some induced polyploids, as follows:

1. Wild type (abb. W type): All hulled florets separate from their pedicels or one another when mature, as in <u>A.fatua</u>. Included 7 species.
2. Subwild type (S type): Spikelets separate from their pedicels but all hulled florets remain attached, as in <u>A.sterilis</u>. Included 5 species.
3. Cultivated type (C type): Spikelets and all hulled florets remian attached to their pedicels or each other, as in <u>A.sativa</u>. Included 3 species.
4. Naked type (N type): Spikelet and all hull-less florets remain attached. The spikelet is always multiflorous and the rachilla segment is elongated, as in <u>A.nuda</u>(6x) and <u>A.nudibrevis</u>(2x).

In about 70 reciprocal F_1 hybrids, intra- and interploid hybrids, the spikelet types showed a long sequence of recessive < dominance or incomplete dominance in relation to the ploidy as follows:

$$N^1 \; < \; C^{1,2} \; < \; W^{1,2} \; < \; S^{1,2} \; < \; W^3 \; < \; S^3 \; < \; C^3 \; < \; N^{3(1)}$$

Superscripts of 1, 2 and 3 indicate the spikelet types
of 2x, 4x and 6x species, respectively,

It is remarkable that the spikelet types of 6x species show a different order of dominance sequence to that of 2x and 4x, and that they are dominant to all spikelet types of 2x and 4x species. This might be due to the different mutational mechanisms by which they occurred. Whereas the spikelet types of 4x species might just be due to duplication of the factors of 2x species.

On the other hand, based on data of nullisomic analysis of 6x species, Nishiyama(1) assumed that chromosome MK-9 of the 3rd genome (or D) and its homoeologues of the 2nd(C) and the 1st genome(A) carry major genes for the spikelet type characters (Table 1).

The naked type spikelets of 2x and 6x species appear to be controlled by the same chromosome of the 1st genome(A) but show a marked difference in dominance sequence. This might be due to a difference in origin.

TABLE 1. Spikelet types affected by genes located in chromosome MK-9 or its homoeologues

MK-9 or its homoeologue	Spikelet type			
	A.fatua	A.sterilis	A.sativa	A.nuda
MK-9 of the 3rd genome(D)	W^{3*}	S^3	$\underline{C}^3$	C^3
Homoeologues of the 2nd genome(C)	$\underline{W}^2$	$\underline{W}^2$	$\underline{W}^2$	W^2
Homoeologues of the 1st genome(A)	W^{1**}	W^{1**}	W^{1**}	$\underline{N}^{3(1)}$

*Underlined symbols indicate phenotypes of 6x species
**Probably W^1 though possibly C^1 or N^1

The spikelet type characters are closely associated with pubescence at the base and on the rachilla segment of the grain, as well as probably awnedness. Either a few or no cross-overs were observed between these characters. The evidence suggests that genes for all of the associated characters are located very closely together or even form a compound gene.

MK-9 carries a gene for pairing of meiotic chromosomes but its homoeologues do not. Accordingly nullisomics lacking a pair of MK-9 chromosomes were asynaptic(1). The genetic situation may be favourable for the diploidisation of chromosome pairing though 6x species consist of partially homologous genomes.

Thus, 6x species may show permanent heterozygosity for spikelet type.

REFERENCE

1. Nishiyama I: Bull. Research Inst. Food Sci., Kyoto Univ. 4, 67-85, 1951.

50

THE IDENTIFICATION OF <u>AVENA</u> CHROMOSOMES BY MEANS OF C-BANDING

J. POSTOYKO AND J. HUTCHINSON
Department of Biological Sciences, Manchester Polytechnic, Manchester, UK

1. INTRODUCTION
   Giemsa C-banding has been applied to many crop species to visualise the
constitutive heterochromatic component of the chromosomes. The banding
patterns obtained are often sufficiently distinctive to allow individual
chromosomes to be identified both within and between species.
   This paper sets out to demonstrate the potential of C-banding as an
initial screening technique to identify the constituent chromosomes of an
<u>Avena</u> interspecific hybrid, thus possibly reducing the number of
backcrosses which need to be made to produce chromosome addition lines.

2. MATERIALS AND METHODS
   The following plant stocks were used: <u>Avena sativa</u> cv. 'Sun II',
<u>A.macrostachya</u> Bal., ex. Cos., et Dur., and an <u>A.sativa</u> x <u>A.macrostachya</u>
hybrid(1). All the material came from the Welsh Plant Breeding Station.
   Mitotic metaphase spreads were prepared and C-banded using the protocol
described by Teoh and Hutchinson(2), except that the staining was in 4-8%
v/v. Giemsa in Gurr's buffer, pH 6.8.

3. RESULTS AND DISCUSSION
   Figure 1 shows that the cultivated oat, <u>A.sativa</u>, has relatively little
heterochromatin. The most prominent C-bands are located at the secondary
constrictions of the nucleolar organising chromosomes. Other
heterochromatic bands are either smaller in size, or stained less

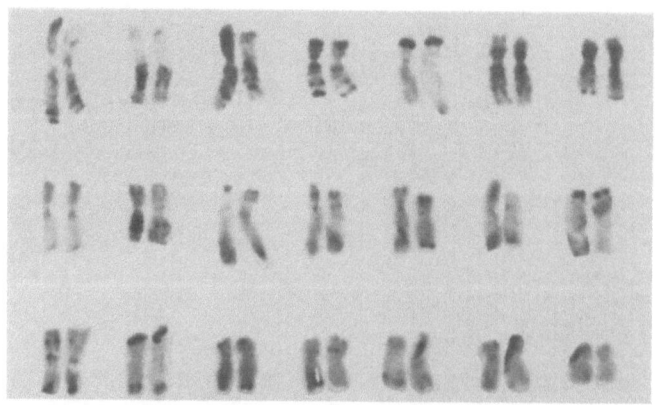

FIGURE 1. C-banded chromosomes of <u>A.sativa.</u>

intensely. However, nine further pairs of chromosomes may be identified by their banding patterns in most preparations. The remaining, smaller chromosomes of the complement have very fine C-bands, and may be distinguished only with difficulty.

In contrast to A.sativa, the wild perennial species A.macrostachya is highly heterochromatic. Large, predominently centromeric or intercalary bands are found in most chromosomes (Fig. 2). A.macrostachya is described in the literature as an autotetraploid(3) yet, interestingly, although the chromosomes may be grouped into sets of four on the basis of size and arm-ratio, there are differences in the banding patterns between chromosomes within some of the sets. This may suggest an allotetraploid origin for A.macrostachya, or may simply reflect variation in heterochromatin which may exist between homologous chromosomes.

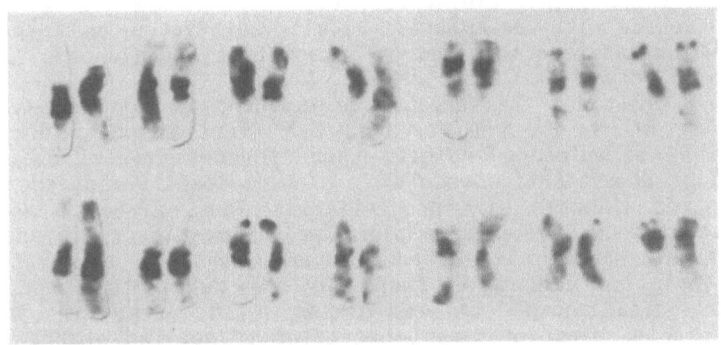

FIGURE 2. C-banded chromosomes of A.macrostachya.

However, despite the heterogeneity within the A.macrostachya genome, examination of an interspecific hybrid shows that it is possible to distinguish the chromosomes of A.macrostachya from those of A.sativa. It therefore seems likely that the technique of C-banding will prove to be a useful tool for the study of Avena chromosomes.

REFERENCES
1. Leggett JM: Can. J. Genet. Cytol. 27, 29-32, 1985.
2. Teoh SB and J Hutchinson: Theor. Appl. Genet. 65, 31-40, 1983.
3. Baum BR and T Rajhathy: Can. J. Bot. 54, 2434-2439, 1976.

SESSION I. GENETIC RESOURCES AND THEIR USE IN THE BREEDING OF OATS
CHAIRMAN'S COMMENTS AND SUMMING UP

G. LADIZINSKY
Faculty of Agriculture, The Hebrew University, Rehovot, Israel.

Advance in plant breeding depends upon the availability of genetic variability. Land races and primitive varieties were used in breeding programmes in the past, but more recently may have been neglected when better adapted and high yielding varieties have become available. Another potential source of variability within cultivated germplasm may be material which has been obtained by irradiation, mutagenesis and recently also by somacloning.

Although a great deal of variability is found in most cultivated crops, the majority of it has been created by selection under domestication. Various lines of evidence indicate that 'Founder Effect' played a major role in plant domestication resulting in much lower variation in the crop plant compared with its wild progenitors. Thus, although the immediate gene pool of the cultivated crop plant is the most accessible and easy to use, the plant breeder is forced to deal with wild germplasm when the required variation cannot be found in the cultivated gene pool. A conspicuous disadvantage of exploiting wild germplasm in breeding programmes is the need for many backcrosses to get rid of the undesirable traits which have been introduced by the wild species. Another feature of utilizing wild species is, quite often, the need to overcome crossability barriers and low fertility of the F_1 hybrids and segregating generations.

It is obvious then that utilization of wild species for breeding purposes is a long process and usually requires prebreeding projects to transfer traits of economic value of the wild species into a cultivated background.

A prerequisite of exploiting wild germplasm is an adequate knowledge of the genetic affinities between the crop plant and its wild relatives. Knowledge of this kind is now available in oats and the prospects and the difficulties of exploiting various wild species for breeding purposes have been identified. Furthermore, the geographic distribution and the ecological preferences of the various wild species are well known and additional collections of wild material can be made if necessary.

Wild species have been the main source of disease resistance in oats in the last few years, more than that, quality improvement and increase of total yield has been made by utilizing wild germplasm. It is thus a reasonable assumption that this trend will continue in the future. There is no way of guaranteeing that a particular wild species will be valuable as a source of useful traits but it can be speculated, on theoretical grounds, that unique genetic variability exists in the secondary gene pool of A.sativa, namely, in the tetraploid and diploid species. However, exploitation of that variability will require specific methods to overcome crossability and sterility barriers.

Rapid development of biotechnological methods suggests that genetic engineering in plants might be feasible in the future. A lot has to be learned about the mode of action of various genes in oats before such methods can be used but theoretically it may be possible to insert a small segment of alien DNA, which carries a gene of economic value, into the A.sativa genome and by that means overcome reproductive and sterility barriers. This would avoid the need for so many backcrosses which are typical of gene transfer via interspecific crosses.

SESSION II

Pests and diseases of oats

Chairman: J. Sebesta

DISEASES OF OATS

M.D. SIMONS
Agricultural Research Service, US Department of Agriculture,
Iowa State University, Ames, Iowa 50011, USA

1. INTRODUCTION

This paper will briefly describe each of the major oat diseases and summarise current research thrusts. 'Current' here means work that has been published within the last several years. Highly specialised work on oat pathogens, judged to be of less general interest, will not be included.

2. CROWN RUST

Over time and worldwide, crown rust is probably the most important disease of oats. It is caused by the heteroecious, long cycle rust fungus Puccinia coronata Cda. The complete life cycle includes stages on shrubs of the genus Rhamnus, which are infected in the spring and which serve to initiate the disease on the young oats. For this reason eradication of Rhamnus is commonly recommended as a control measure. Its effectiveness, however, is limited because urediniospores from oats growing in mild climates can initiate infection in most major oat-growing areas of the world. The disease can be effectively controlled with fungicides, but this is usually not economically feasible. The only effective and practical means of control is by the use of resistant varieties.

Identification of effective sources of resistance requires knowledge of the virulence of the fungus population in the area for which resistant varieties are to be produced(1). Virulence surveys are now being carried out at a number of locations around the world. Such surveys from Poland(2) suggest a recent increase in the rate of change in pathogenicity of the crown rust population. Certain resistance genes would probably be more effective in some central European countries than in others(3). Combinations of host resistance genes that would give protection from all variants of the fungus studied were described, along with a strategy for incorporating such multigenic resistance into oat varieties.

It is generally believed that changes in pathogenicity of P.coronata usually occur in a single step from avirulence to virulence. Recent work in the USA(4), however, has shown that virulence toward plants with a specific resistance locus can increase through a number of small steps. This phenomenon will complicate breeding procedures because of the difficulty of identifying and handling variants of the rust having intermediate virulence.

The first work on the specialisation of the fungus at the level of the host genus was done in 1894(5,6). Using a limited number of cultures, it was observed that urediniospores from a given grass genus would not infect species of other genera. Consequently, the fungus was divided into 'forma speciales' on the basis of pathogenicity. Recent work in Israel(7), however, has raised serious questions about the validity of this concept. Cultures of P.coronata isolated from six grass genera and from cultivated oats were used to inoculate seedlings of 22 grass species in eight genera and one oat cultivar. Many cultures infected species of two or more genera. The fact that a culture from one host infected a second host did not necessarily mean that a culture from the second host would infect the first one. For example, the culture isolated from Festuca was infectious to hosts in six other genera, while Festuca itself was susceptible to only one additional culture. At the other extreme, the culture isolated from Arrhenatherum was avirulent on all other host genera, while Arrhenatherum itself was susceptible to at least some extent to all cultures studied. Thus, differentiation of P.coronata into forma speciales seemed meaningless. Abandonment of recognition of formal divisions within P.coronata was suggested.

Crown rust has been traditionally controlled by genes for specific resistance (roughly synonymous with vertical, oligogenic, seedling, etc.), which is highly effective as long as the fungus does not change in pathogenicity. The 'breakdown' of such resistance, particularly in the USA and Canada in the 1940s and 1950s, resulted in greater emphasis being placed on general resistance (roughly synonymous with horizontal, polygenic, slow rusting, field resistance etc.). A recent study(8) showed the average apparent infection rates for P.coronata on Fulghum (susceptible) and Burt (intermediate in reaction) to be 0.4 and 0.35 units per day, respectively. These rates were significantly faster than that for Red Rustproof (field resistant), which had a rate of only 0.2 units per day. These studies, however, were laborious and new techniques are being developed to detect slow rusting. Slow rusting varieties are often characterised by low receptivity, and methods to accurately measure low receptivity have been recently developed(9,10). These methods should simplify handling this trait in breeding programmes.

Representative of current work on the genetics of resistance to P.coronata is a report(11) showing that a certain resistant cultivar had two complementary genes for resistance. These genes showed full dominance to eight races and incomplete dominance to two races. The use of genetic analysis to differentiate between two oat lines that appeared to have identical resistance(12) was recently demonstrated. The appearance of a new isolate revealed that one of these lines also contained a group of minor genes that controlled infection severity and frequency of infection. Other recent work(13) has also demonstrated combinations of major and minor genes.

Field resistance is generally regarded as polygenic, but genetic analysis of an adult-resistant strain of Avena sterilis from Israel showed the resistance to be conditioned by a single major gene(14). In another recent study(15), the field resistance of certain strains of A.sterilis appeared to be conditioned by several additive genes. Another study(16) involving 14 strains of A.sterilis, all of which lacked seedling resistance to P.coronata, showed that one or two generations of backcrossing could be used effectively in transferring presumably polygenically inherited field resistance to cultivated oats. The use of backcrossing simplifies the problem of obtaining agronomically satisfactory segregates when transferring resistance from A.sterilis.

Tolerance, defined here as the ability of one variety to suffer less loss from a given amount of disease than another variety, was recognised in a general way before the turn of the century. Subsequent work has been hampered by the difficulty of accurately measuring the amount of disease involved. Recent work(10) using spore counts from large plots in conjunction with yield and seed weight data from hill plots indicated, for example, that Otter was indeed tolerant relative to Cherokee because the two had similar yield and seed weight responses to infection, but more spores were produced on Otter. In another study(17), "areas under the disease progress curve" were used as the measure of disease severity. Under severe epidemic conditions yields of some moderately resistant varieties equalled or exceeded the yields of certain highly resistant varieties even though more rust developed on the moderately resistant varieties.

Multilines are known to reduce losses from P.coronata, and cross protection may partially account for their superior performance. Recently published experiments(18) showed that oats inoculated with a virulent race of P.coronata 12 h or more after inoculation with P.graminis from wheat or an avirulent race of P.coronata resulted in less elongation of intercellular hyphae and in reduced development of haustoria. Since all the plants in a functioning multiline are ordinarily resistant to some of the incoming inoculum, cross protection may account for some of the resistance that has been observed. Accumulation of phytoalexins in oat leaves occurs after exposure to noncompatible races of the fungus(19), suggesting that these compounds may function in the resistance of multilines. Approaching a different aspect of the problem, the effect of genotype unit area (the ground area occupied by a genetically homogeneous unit) on the efficacy of multilines was tested(20). Contrary to what might be expected, the results indicated that the size of the unit, at least in sizes varying from 0.003 to 0.84 square metres, had no significant effect.

3. STEM RUST

Stem rust is a potentially destructive disease of oats that occurs to at least some extent almost everywhere that oats are grown. In certain areas, it is more important than crown rust. The causal fungus, Puccinia graminis Pers., has a life cycle very similar to that of the crown rust fungus except that certain species of Berberis and closely related genera serve as alternate hosts. Also like crown rust, it can be a problem in the absence of the alternate host. It can be readily controlled with fungicides, but this is generally not economically feasible, leaving resistance as the only practical means of control.

While the extremely widespread and damaging epidemics reported earlier in the century have not occurred in recent times, stem rust still occurs with sufficient frequency and severity to make it a potential hazard to oat production. In 1982, for example, moderate to heavy infection (40-80%) occurred in the Canadian provinces of Ontario, Manitoba and Saskatchewan, and caused heavy losses in late sown fields(21,22).

The relatively large number of papers published in recent years dealing with various aspects of pathogenic specialisation show the importance that oat pathologists attach to this work. A study of changes in avirulence-virulence combinations of the pathogen population during a 57 year period in Canada(23) indicated that the fungus was highly dynamic even when the host population was relatively static. The early years of this period were characterised by the presence of races virulent only on universally susceptible hosts or varieties with a single resistance gene. A gradual increase in virulence and a differentiation of populations of eastern and western Canada began in 1943. By 1950 races having no known specific genes for virulence or only one gene were being displaced by races with combinations of virulence genes in eastern Canada. This did not occur in the west until 1961. Since 1961 races with two or three virulence genes have been replaced with races combining a still greater number of virulence genes. Studies conducted over several years in New South Wales, Australia(24,25) showed considerable annual variation in the abundance, racial diversity, and virulence of the P.graminis population.

Fundamental relationships between different races of P.graminis have been studied recently using the technique of electrophoresis(26). More than 280 polypeptides were detected in each race, and at least 22 polypeptides varied among the races. Some races had identical patterns and some differed by 4 to 16 polypeptides. Certain polypeptides were shown to be associated with certain patterns of pathogenicity.

Nomenclature of the pathogenic variants of P.graminis on oats has long been a problem. American and Canadian oat workers(27) interested in stem rust recently devised a formula system that requires no complicated keys. The formulas are self-explanatory; differential lines can be added or dropped at will, and new avirulence/virulence combinations are easily described by writing new formulas. Anyone anywhere in the world can readily add data from local or new differential varieties.

As in the case of crown rust, use of 'forma speciales' may be inappropriate. In Japan(28,29) the host range of P.graminis from oats included 9 genera of the Festuceae, 13 of the Avenae, 6 of the Triticeae, and 1 of the Stipeae. This suggests that the use of the Latin trinomial term for the stem rust fungus on oats does little more than specify that the isolate is virulent on oats. On the other hand, electrophoretic analyses(26), showed that there were greater differences in polypeptides among the different forms of P.graminis than there were between different races isolated from oats.

Resistance to P.graminis has been more difficult to find than resistance to other major cereal rusts. In a recent study(30) in which over 1400 strains of oats from Iran, Iraq and Turkey were screened, resistance to P.graminis occurred infrequently and then was found only in A.sterilis and A.barbata. In Japan, Tajima(31) also found resistance to P.graminis to be rare. The general paucity of specific resistance to P.graminis has led to greater emphasis on searching for and studying field resistance. In a recent attempt to locate such resistance(32), 240 visibly susceptible lines of oats were exposed to severe stem rust infection in the field. Statistically significant variation occurred, with 46 lines exceeding the overall mean for resistance measured as reduction in seed weight due to infection. In Israel, it was demonstrated that mechanisms controlling this type of resistance in A.sterilis involved restricted colonisation of host tissue by mycelium and lower spore production per pustule(33). These factors functioned against all the oat stem rust races tested, suggesting that this moderate form of resistance was general rather than race specific.

In a basic study of the epidemiology of oat stem rust(34) a relatively severe, but geographically limited epidemic in the north central USA was compared with other epidemics in earlier years. There was little relationship between yield loss and temperature or rainfall during the epidemic period, but date of onset of rust was related to yield loss.

In one example of recent genetic studies(35), plants derived from a cross between Kyto and A.sterilis were highly resistant in the adult stage. Resistance was much better than that of either parent and clearly demonstrated transgressive segregation.

4. SMUT

Loose and covered smut of oats, caused by Ustilago avenae (Pers.) Rostr. and U.kolleri Wille., respectively, probably occur to some extent everywhere that oats are grown. Spores are borne on the seed and systemically infect the germinating seedling, eventually replacing floral parts of the panicle with masses of black spores. The symptoms are highly visible and damage is obvious. Oat smut can be readily and economically controlled by chemical seed treatment, but development of resistant varieties is the most favoured approach to control.

Several decades ago oat smut was much more severe than it is now; however, reports of economically significant occurrence are still not uncommon. Within recent years, smut in the Kirov region of the Soviet Union reached 94% severity(36). Local varieties were susceptible, but good sources of resistance were available for use in breeding programmes. Pathotypes of U.avenae with virulence on several previously resistant oat varieties have appeared in recent years in the north central USA(37). Widely grown varieties ranged from susceptible to resistant.

Taxonomic relationships of the causal fungi of the oat smuts and closely related fungi causing smuts of barley have long been controversial. Hybridisation of barley smuts with oat smuts, using a species of Agropyron as a common host, showed that the echinulation of the spore walls of U.nigra from barley was conditioned by the same two dominant complementary genes that had been previously identified in U.avenae from oats(38).

5. POWDERY MILDEW

Powdery mildew of oats is potentially important in the more cool, humid oat growing regions of the world. The causal fungus (Erysiphe graminis DC.) is, like the rusts, an obligate parasite that shows a high degree of pathogenic specialisation. The disease can be controlled with fungicides or resistant varieties.

Priestley and Bayles(39) summarised the incidence and severity of powdery mildew of oats in Wales and England over the past two decades. In most years at least 10% mildew occurred on susceptible varieties. The mean mildew severity on the most susceptible variety over the 20 year period ranged from 26 to 65% at the different locations. High intensity of the disease was associated with high temperatures and heavy rainfall during November, December and January.

A comparison of the infection process on resistant and susceptible varieties showed that conidial germination was not related to adult plant resistance(40). Frequencies of penetration pegs, haustoria and papillae did vary among varieties and among leaves of different ages, and were related to some extent to adult resistance.

The development of resistant varieties has been hampered by a scarcity of good sources of resistance. Much of the best resistance that is known occurs in diploid and tetraploid species of oats and is, therefore, difficult to transfer to cultivated varieties. A method of interfering with regular meiotic behaviour that facilitated transfer of a dominant gene for resistance from the tetraploid Avena barbata to cultivated oats was recently described(41). In Pakistan(42), monosomic analysis indicated that the gene controlling resistance to one race of the fungus was not located on any of the monosomes tested.

62

6. SEPTORIA BLIGHT

Septoria avenae Frank (Leptosphaeria avenaria Weber) infects the leaves, culms (black stem), and seed of oats. Although well known early in the century, it was not recognised as a widespread and potentially destructive disease of oats until the 1950s. It can be controlled with fungicides, but resistance is the favoured means of control.

Systemic fungicides gave good control of the disease in eastern Canada(43) under an intensive management system that included high nitrogen fertilisation. Improvement in seed weight, protein percentage and grain yield were noted. In long term experiments on the interaction of Septoria blight and manganese nutrition, leaf blighting developed on both resistant and susceptible varieties under growth chamber conditions(44). Symptoms, however, differed markedly with different levels of manganese.

Locating and utilising effective sources of resistance has been difficult. Little resistance was found in oat seedlings from various sources including wild species from the Middle East(45). A few strains did develop fewer leaf symptoms and black stem lesions than others. Preliminary genetic studies of progeny from a cross between resistant and susceptible varieties showed a wide and continuous range of reaction suggestive of polygenic inheritance.

7. HELMINTHOSPORIUM BLIGHT

Helminthosporium avenae Eidam (Pyrenophora avenae Ito and Kurib.) causes a seedling blight and leaf blotch of oats that is widespread and has been reported to cause serious damage in western Europe, the southern USA, India and Japan. It is sometimes seedborne and sometimes carried in crop residue. Seed treatment is thus relatively effective under some conditions and ineffective under others. In Italy(46), maneb and other fungicides gave good protection against seedborne inoculum. In Ireland(47) Erwinia herbicola applied to the seed improved emergence and controlled coleoptile streak and leaf stripe symptoms in pot tests.

In the past, locating a high degree of resistance to the fungus has been difficult and recent work from Bulgaria(48) substantiates this. Less than 5% of a large number of accessions were resistant under field conditions. A potentially useful system of screening for resistance was suggested by a study from Italy(49), which showed that root growth was markedly reduced by culture filtrates of the fungus even at very low concentrations. Resistant varieties were less sensitive to the toxic effects of these filtrates than were susceptible varieties.

Recent work(50) in India showed that spore germination was complete in four hours, that appressoria formed in eight hours and were often over the junction of epidermal cell walls, and that infection was by direct penetration.

8. FUSARIUM DISEASES

The disease complex of oats attributable to species of Fusarium is widespread, particularly in cooler, northern climates. Development of this disease may involve invasion and killing of the young plant after seed germination and either before or after emergence. If the young plants are not killed, infection may later reach the crown resulting in stunted plants. Head blight, although not commonly severe, has been reported to cause digestive disturbances of livestock.

Recent research in Canada(51) showed that infection of oat samples harvested in Quebec averaged 15%. Significant differences in infection percentage were found among cultivars, corroborating earlier suggestions that a reasonable degree of control may be obtained by developing resistant cultivars.

9. RED LEATHERLEAF

The red leatherleaf disease of oats, caused by Spermospora avenae (Sprague and AG Johnson) Sprague, is a disease that is not widespread nor well known, but which has recently appeared in the literature. It was reported in the 1930s and 1940s to occur in the northwestern USA and in Turkey. Diseased plants are stunted, rigid and the surface of the leaves has a somewhat leathery appearance, which suggested the common name. It can become quite prevalent under conditions of cool, damp weather.

Red leatherleaf was recently reported from Australia(52) where it occurred in Victoria. It was not found in other states. All varieties tested were susceptible.

10. BACTERIAL BLIGHT

Halo blight, caused by Pseudomonas coronafaciens (Elliott) Stevens is the most common bacterial disease of oats, probably occurring to at least some extent in most years in the major oat growing regions of the world. The bacterium overwinters on the seed or on plant refuse in the field. The disease is generally favoured by cool, wet weather and is checked by the advent of warm, dry weather. It is capable of causing economic losses when conditions are favourable for its development.

Different strains of the causal bacterium differ markedly in the lesion development they induce(53). In the United Kingdom(54), varieties regarded as resistant were indistinguishable from susceptible varieties when subjected to heavy artificial inoculation. Since the levels of inoculum used in these experiments were far beyond what might be expected in the field, the use of resistant varieties still appears to be a satisfactory means of control where control is necessary.

11. YELLOW DWARF

Yellow dwarf, caused by the Barley Yellow Dwarf Virus (BYDV) first appeared in the 1950s as a widespread and destructive disease. It now rivals crown rust as the number one oat disease worldwide and, in a given year, may be much more important than crown rust over wide areas. The virus, of which there are many strains, is transmitted by various species of aphids. Infected plants are stunted, and depending on variety and environmental conditions, the leaves may range in colour from yellow to a striking bright red. Panicles show greater or lesser degrees of blasting. It is theoretically possible to control the disease by controlling the aphid vectors, but the use of resistance is ordinarily recommended.

Recent work in Wales(55) involving live trapping of aphid vectors, showed that 10 species transmitted the virus to test plants. Some of these aphid species were much more important than others. The use of such trapping techniques was suggested as a tool in forecasting epidemics of the disease. In Canada(56), young oat plants infested with different numbers of aphids showed a strong dosage response. With increased numbers of aphids per plant, more plants became infected, the incubation period decreased, symptoms were more severe, and yield decreased progressively.

64

This work suggests that control measures designed to reduce aphid populations might be quite effective on a practical basis, even though they failed to completely control the disease.

The use of aerial photography as a means of detecting and estimating yellow dwarf and other cereal diseases has been explored(57). Test plots were infested with viruliferous aphids; controls were treated with insecticides. Sequential infra-red photographs starting early in the season showed that plots with yellow dwarf could be identified from the air.

Yellow dwarf often occurs in conjunction with fungal diseases, and the relationship of yellow dwarf to infection by Puccinia coronata and Erysiphe graminis has been recently reported(58,59). Numbers of uredia of P.coronata were negatively correlated with yellow dwarf severity. E.graminis was initially suppressed on plants that had been previously infected with the virus, but was subsequently enhanced, suggesting that the two diseases interact in some complex fashion.

Work carried out in New Zealand(60) on the relationship of nutrient level to yellow dwarf suggested that the ratios of different nutrients affected the severity of the disease. Virus infection delayed emergence of the heads and reduced fresh and dry weights markedly in nutrient solutions high in Ca $(NO_3)_2$ and low in KH_2PO_4.

Work related to the location and utilisation of potentially valuable sources of resistance to the virus has recently been reported in the USA(61) and Sweden(62) where the relationships of different insect vectors and strains of the virus have been studied in relation to specific oat varieties. Although these studies indicate no high degree of resistance, there are significant amounts of tolerance available to oat breeders. Similar results have been reported from Canada(63).

A knowledge of the inheritance of resistance to yellow dwarf is essential to efficiently breeding resistant varieties. Good basic information on this subject has come from a recent study in Canada(64). Seven resistant lines of Avena sterilis and one resistant cultivated oat were crossed with a susceptible cultivated oat. Additive gene action explained most of the genetic variability observed. Heritability estimates were generally high, suggesting that progress could be made in breeding programmes. Data from four of the eight crosses fit relatively simple genetic models. This modifies the prevailing earlier view that resistance to the virus is generally polygenic, and suggests that somewhat different procedures might be employed in developing resistant varieties.

An example of a potential application of modern biotechnology to the solution of a practical problem was furnished by a study very recently published in the USA(65). The virus content of extracts from oats grown in the glasshouse, growth chamber or field was measured by enzyme-linked immunosorbent assay. In some cases, symptomatic resistance to the virus as previously ascertained by plant breeders was associated with reduced virus content in infected plants. Although additional research is needed, this technique could be a potentially valuable adjunct in breeding for yellow dwarf resistance.

REFERENCES

1. Sebesta J and DE Harder: Plant Disease, 67, 56-59, 1983.
2. Mazaraki M: Biul. Inst. Hodow. Aklim. Roslin. No. 150, pp.27-32 (In Polish. Abstracted in Rev. Pl. Path. 64, p.166, 1985), 1983.

3. Sebesta J, DE Harder and B Zwatz: Cereal Rusts Bull. 12, 7-15, 1984.
4. Simons MD and ME McDaniel: Phytopathology 73, 1203-1205, 1983.
5. Eriksson J: Ber. Deutsch. Bot. Gesellsch 12, 292-231, 1894.
6. Eriksson J and E Henning: Zeitschr. Pflanzenkrank. 4, 257-262, 1894.
7. Eshed N and A Dinoor: Phytopathology 70, 1042-1046, 1980.
8. Berger RD and HH Luke: Phytopathology 69, 1199-1201, 1979.
9. Luke HH, PL Pfahler and RD Barnett: Plant Disease 65, 125-127, 1981.
10. Politowsi K and JA Browning: Phytopathology 68, 1177-1185, 1978.
11. Sebesta J: Euphytica 28, 807-808, 1979.
12. Sebesta J: Euphytica 32, 857-861, 1983.
13. Simons MD: Phytopathology 69, 450-452, 1979.
14. Harder DE, RIH McKenzie and JW Martens: Phytopathology 74, 352-353, 1984.
15. Harder DE and RIH McKenzie: Can. J. Plant Path. 6, 135-138, 1984.
16. Simons MD: Phytopathology 75, 314-317, 1985.
17. Singleton LL, MB Moore, RD Wilcoxon and MK Kernkamp: Phytopathology 72, 538-540, 1982.
18. Tani T, Y Yamashita and H Yamamoto: Phytopathology 70, 39-42, 1980.
19. Ueno T, T Nakshima and H Fukami: Mem. Coll. Agr., Kyoto Univ. 122, 93-129, 1983.
20. Mundt CC and JA Browning: Phytopathology 75, 607-610, 1985.
21. Martens JW: Can. J. Plant Path. 2, 253-255, 1980.
22. Martens JW and DE Harder: Can. J. Plant Path. 5, 189-193, 1983.
23. Martens JW and RIH McKenzie: Can. J. Bot 57, 952-957, 1979.
24. Burdon JJ, JD Oates and DR Marshall: J. Appl. Ecol. 20, 571-584, 1983.
25. Oates JD, JJ Burdon and JB Brouwer: J. Appl. Ecol. 20, 585-596, 1983.
26. Kim WK, JW Martens and NK Howes: Can. J. Plant Path. 6, 111-118, 1984.
27. Martens JW, AP Roelfs, RIH McKenzie, PG Rothman, DD Stuthman and PD Brown: Phytopathology 69, 293-294, 1979.
28. Tajimi A: Bull Nat. Grassland Res. Inst. (1981) No. 20, 76-92. (In Japanese. Abstracted in Rev. Plant Path. 61, p.404, 1982), 1981.
29. Tajimi A: J. Jap. Soc. Grassland Soci. 28, 433-438. (In Japanese. Abstracted in Rev. Plant Path. 63, p.54, 1984), 1983.
30. Martens JW, RIH McKenzie and DE Harder: Can. J. Gen. Cytol. 22, 641-649, 1980.
31. Tajimi A: J. Jap. Soc. Grassland Sci. 26, 372-376. (In Japanese. Abstracted in Rev. Plant Path. 60, p. 594, 1981), 1981.
32. Simons MD and PG Rothman: Phytopathology 74, p.878, (abstr.), 1984.
33. Sztejnberg A and I Wahl: Phytopathology 66, 74-80, 1976.
34. Roelfs AP and DL Long: Phytopatholgy 70, 436-440, 1980.
35. Martens JW, PG Rothman, RIH McKenzie and PD Brown: Can. J. Gen. Cytol. 23, 591-595, 1981.
36. Suslova YN and KI Veselkova: Zashchita Rastenii 12 (In Russian. Abstracted in Rev. Plant Path. 60, p.326, 1981), 1980.
37. Arny DC, DT Caine and FB Diez: Phytopathology 71, p.858, (abstr.), 1981.
38. Huang HQ and J Nielsen: Can. J. Bot. 62, 603-608, 1984.
39. Priestley RH and RA Bayles: J. Nat. Inst. Agric. Bot. 15, 55-66, 1979.
40. Douglas SM, RT Sherwood and FI Lukezic: Physiol. Plant Path. 25, 219-228, 1984.
41. Thomas H, W Powell and T Aung: Euphytica 29, 635-640, 1980.
42. Hafiz HMI and AS Larik: Pak. J. Bot. 15, 38-42, 1983.
43. Johnston HW, JA MacLeod and JDE Sterling: Can. J. Plant Path. 3, 215-218, 1981.

44. Clark RV: Can. J. Plant Path. 5, 89-96, 1983.
45. Clark RV: Can. J. Plant Path. 2, 213-216, 1980.
46. Pezzali M and A Porta-Puglia: Informatore Fitopathol. 33, 47-49, 1983.
47 Downes MJ: Abstracted in Rev. Plant Path. 9, p.209, 1980.
48. Kunovski ZH and T Breshkov: Rasteniev dni Nauki 18, 118-123. (In Polish. Abstracted in Rev. Plant Path. 61, p.287, 1982), 1981.
49. Graniti A and AP Puglia: Phytopathol. Medit. 23, 39-42, 1984.
50. Arora RK, CL Mandahar and RK Pahwa: Indian J. Mycol. Plant Path. 10, 8-11, 1980.
51. Couture L: Can. J. Plant Sci. 62, 29-34, 1982.
52. Pascoe I and T Woodcock: Austral. Plant Path. 10, 60-61, 1981.
53. Schaad NW, DR Sumner and GO Ware: Plant Disease 64, 481-483, 1980.
54. Smith JJ and JW Mansfield: Physiol. Plant Path. 18, 345-356, 1981.
55. A'Brook J and AM Dewar: Ann. Appl. Biol. 96, 51-58, 1980.
56. Burnett PA and CC Gill: Phytopathology 66, 646-651, 1976.
57. Clark RV, DA Galway and YC Paliwal: Phytopathology 71, p.867 (abstr.), 1981.
58. Potter LR: Ann. Appl. Biol. 94, 11-17, 1980.
59. Potter LR: Ann. Appl. Biol. 100, 321-329, 1982.
60. Thomson AD and JD Ferguson. New Zealand J. Agric. Res. 23, 367-372, 1980.
61. Jedlinski H: Phytopathology 69, p.540 (abstr.), 1979.
62. Ryden K: Vaxtskyddsnotiser 48, 14-19, (In Swedish with English summary), 1984.
63. Comeau A: Euphytica 33, 49-55, 1984.
64. Landry B, A Comeau, F Minvielle and CA St-Pierre: Crop Sci. 24, 337-340, 1984.
65. Skaria M, RM Lister, JE Foster and G Shaner: Phytopathology 75, 212-216, 1985.

This is a cooperative contribution from ARS, US Department of Agriculture and Iowa Agriculture and Home Economics Experiment Station Project 2447, and is Journal Paper J-119500

PATHOGENICITY OF CROWN RUST, STEM RUST AND POWDERY MILDEW ON OATS IN EUROPE AND SOURCES OF RESISTANCE

J. SEBESTA*, D.E. HARDER*, I.T. JONES**, M. KUMMER*, B.C. CLIFFORD** and B. ZWATZ*

1. INTRODUCTION

Oat crown rust, stem rust and powdery mildew are endemic in Europe but their severity varies depending on region and year. Crown rust and stem rust are more common in southern countries and powdery mildew in more humid ones, where it is considered to be the most important disease of oats. In the period 1979-1983, the highest occurrences were in the UK (Wales, Rothwell), German Democratic Republic (GDR), Poland and Yugoslavia. In Czechoslovakia the highest incidence was in 1981.

2. MATERIALS AND METHODS

Crown rust samples were collected from the European Oat Disease Nursery(1), farm crops and from wild oats (Avena fatua L.). Powdery mildew data came from the European Oat Disease Nursery trials. The virulence analyses of crown rust on standard differentials(2) and/or Pc-lines(3,4) with A.sterilis L. resistance genes, were carried out in the seedling stage(5). On the Pc- and Pg-lines sets, the rust isolates were differentiated into avirulence/virulence combinations using the formula method of Green(6). The percentage of the total number of isolates from each country, which showed virulence on each of the Pc- and Pg-lines, was taken as a criterion for breeding importance(4).

Investigations of inheritance of Pc- and Pg-resistance in F_2 and F_3 generations were conducted at both seedling and adult plant stages by the methods described by Sebesta(7). Powdery mildew reaction tests were carried out mainly at the seedling stage. Tests on leaf segments of both seedlings and adult plants on benzimidazole were used to confirm their applicability to detect some resistance types.

3. RESULTS AND DISCUSSION

3.1 Virulence of rusts and powdery mildew on oats in Europe

The number of virulences detected in crown rust in particular appeared to be related to disease severity. In both stem and crown rust, it is supposed that these virulences are dispersed as inoculum moves between different parts of the continent(8,9).

*J. SEBESTA   RIPP-Plant Protection Institute, Ruzyne 507, 16106 Prague 6, Czechoslovakia

*D.E. HARDER   Research Station, Winnipeg, Manitoba, R3T 2M9, Canada

*M. KUMMER   Institut für Züchtungsforschung der Landwirtschaftswissenschaften der DDR,4300 Quedlinburg, GDR

*B. ZWATZ,   Bundesanstalt für Pflanzenschutz, 1021 Wien II, Austria

**Welsh Plant Breeding Station, Aberystwyth, Dyfed, Wales, UK

In crown rust, in eleven European countries, the highest frequency of virulence was on resistance genes Pc 35, 40, 45 and 47, averaging 38.8, 47.6, 41.9 and 43.9%, respectively (Table 1).   Average virulence frequencies for all resistances differ widely between countries, e.g. Switzerland 10.7, Federal Republic of Germany (FRG) 9.5%, Denmark 10.5%, Italy 38.9% and Portugal 35.8%. No virulence corresponding to genes Pc 58 and Pc 59 has been detected.

TABLE 1. Average frequency of virulences of P.coronata avenae on Pc-resistance genes as found on the European continent in the years 1977-1980

Pc-resistance gene	Percentage of virulent isolates of P.coronata avenae
Pc 35	38.8
Pc 38	25.5
Pc 39	4.9
Pc 40	47.6
Pc 45	41.9
Pc 46	29.5
Pc 47	43.9
Pc 48	6.4
Pc 50	6.7
Pc 54	33.6
Pc 55	4.1
Pc 56	17.7
Pc 58	0.0
Pc 59	0.0
Pc 60	16.6
Pc 61	16.9
Pc 62	12.8
Pc 63	23.7

In stem rust, in Austria, Czechoslovakia and Poland, virulence to genes Pg 1, 2, 3, 4, 8, 9 and 12 occurs and in addition, in Poland the race initially designated P1 (= E 10) was isolated in 1978, possessing virulence to Pg 13.   Virulences to Pg-1, 2, 3, 8, 15 and less to Pg 9 occurred in the FRG and Switzerland in 1980 and no virulence has been found on Pg 4 and Pg 13 there.   In Austria, Czechoslovakia, FRG, Switzerland and Poland in the years 1978-1984 ten avirulence/virulence combinations were isolated (Table 2), although the limited number of isolates did not allow conclusions about their distribution or virulence frequencies.

In powdery mildew, at Aberystwyth, five virulence combinations have been differentiated on four specific resistances(10) as shown in Table 3.

In addition, virulence has been detected to Cc 6490, a translocation line carrying mildew resistance (designated OMR 4) from Avena barbata(11), effective against all previously known virulences. Virulence OMV 4 (found at relatively very low frequency) has been identified in combination with OMV 1 and 2 and also OMV 1, 2 and 3, the latter attacking all specific resistances available in commercial oats.

TABLE 2. Avirulence/virulence combinations of isolates of Puccinia graminis f.sp. avenae from Austria (A), Czechoslovakia (CS), Federal Republic of Germany (FRG), Poland (P) and Switzerland (CH) in 1978-1984

European numbers (correspondence to NA nos.)	Avirulence/virulence combination of Pg genes	No. of isolates				
		A	CS	FRG	CH	P
E 1	1,2,3,4,8,9,13,15,16,a/	1	1	-	-	-
E 2	1,2,4,8,13,15,16,a/3,9	1	-	2	-	-
E 3 (= NA 5)	1,2,4,8,9,13,16,a/3,15	-	6	-	-	-
E 4 (= NA 50)	3,4,9,13,15,16,a/1,2,8	2	-	-	-	1
E 5 (= NA 23)	4,9,13,15,16,a/1,2,3,8	1	15	4	1	-
E 6 (= NA 21)	3,9,13,15,16,a/1,2,4,8	2	6	-	-	-
E 7 (= NA 18)	2,4,9,13,16,a/1,3,8,15	-	-	-	-	-
E 8 (= NA 27)	9,13,15,16,a/1,2,3,4,8	-	5	-	-	-
E 9	4,9,13,16,a/1,2,3,8,15	-	1	7	1	-
E 10 (= P 1)	2,3,4,15,a/1,8,9,13,16	-	-	-	-	1
Total no. of isolates		7	35	13	2	2
Mean virulence capability		2.9	3.7	4.2	4.5	4.0

Mean virulence capability = (a x b) x c, where a = no. of isolates of virulence combination, b = no. of ineffective host (Pg) genes in that combination, and c = total no. of isolates

TABLE 3. Oat mildew resistance (OMR) groups, their respective differential cultivars and corresponding oat mildew virulence (OMV) groups

OMR group	Differential cultivar	Other cultivars	Oat mildew virulence group (OMV)				
			0 [1]*	1 [2]	1,2 [3]	1,3 [4]	1,2,3 [5]
0	Milford	Selma, Leanda	+	+	+	+	+
1	Manod (01747/10/7)**	Peniarth	-	+	+	+	+
2	Cc 4146	Trafalgar, Cabana	-	-	+	-	+
3	9065 Cn 6/3/74 (Cc 4346)	Mostyn, Avalanche	-	-	-	+	+

+ = Compatible reaction (type 3 and 4); - = Incompatible reaction (type 0-2).
* = Physiologic race designation having the particular virulences is given in parenthesis; ** = Original source of resistance in parenthesis.

3.2 Sources of resistance to rusts and powdery mildew recommended for current breeding programmes

3.2.1 Oat crown rust
Of the standard differentials, the most effective were Trispernia, C.I. 7008 (A.byzantina), Bondvic, C.I. 7009 (A.sativa), Landhafer,

C.I. 7005 (A.byzantina) and Santa Fé, C.I. 7006 (A.byzantina), being attacked only by three or four virulent races. The resistance of Victoria, C.I. 7002 (A.byzantina) was overcome by 8 races. Combinations of the resistance of Landhafer and Victoria, incorporated into cvs Dodge and Garland and presently being transferred into several Czechoslovak advanced lines of oats, are of continuing great importance.

Of the Pc resistance genes from A.sterilis L., Pc 39, 48, 50, 55, 58 and 59 appear to be the most effective.

Cultivars Delphin and Pirol, bred in FRG recently, appear to possess two complementary dominant genes conferring resistance to sixteen isolates of ten crown rust races. In Delphin, the complementary genes showed full dominance to eight races and incomplete dominance to two races(12).

In reselected lines Pc 50-2 and Pc 50-4, which are highly resistant to many European isolates of crown rust and ineffective only against several Lithuanian isolates, two different resistance genes were identified. It is postulated that in Pc 50-2 a major resistance gene operates together with minor genes responsible for severity and, maybe, for frequency of different infection types(13).

3.2.2 Oat stem rust

From the virulence of stem rust races isolated from Austrian, Czechoslovak, Swiss, German and Polish populations it is obvious that the most effective gene is Pg-a. In addition, resistance genes Pg 4, 9, 13 and 16 are important in cultivars with multigenic resistance. The combination of genes Pg 4 + Pg 9 in Rodney ABDH (Pg 1 + Pg 2 + Pg 4 + Pg 9) remains effective.

3.2.3 Oat powdery mildew

In general, the effectiveness of specific resistance of oats to mildew is low. All the relevant sources of specific resistance are overcome in the UK(10,11) and, perhaps, in Poland. Mostyn's resistance was presumably overcome in Spain. In Czechoslovakia and, probably also in GDR, resistances of Mostyn and A.barbata (Cc 6490) remain effective. Manod's resistance has been overcome since the beginning of the study (1979), and the resistance of Cc 4146 since 1982 in Czechoslovakia.

On the other hand, adult plant resistance (APR) to mildew in oats seems to be stable and effective(14,15,16,17) especially in cvs Maldwyn, Maelor and Roxton.

Moderate APR is also present in the A.sterilis derivative Pc 54 and this was effective against the spectrum of mildew races tested in Wales in 1982.

3.3 Potential sources of rusts and powdery mildew resistance
3.3.1 Oat crown rust

There is a large number of diploid, tetraploid, and in particular, hexaploid oats in which crown rust resistance has recently been found(18).

A.sterilis CAV 2648 appears to possess 3 genes, and lines differing in infection types [0-1 and 1, 2] have been derived. Other moderately resistant lines, differing in teliospore production in the seedling stage were isolated.

3.3.2 Oat stem rust

The stem rust resistance in cvs Minrus and Jostrain seems to be wider if compared to their derivatives, Rodney D (Pg 1) and Rodney E (Pg3)(19).

A new and highly effective resistance was found in A.sterilis L. WYR 343 from the Oat Collection of the All-Union Institute for Crop Production in Leningrad, USSR.

3.3.3 Oat powdery mildew

A new specific resistance was found in A.sterilis CAV 2648 (Algeria), effective against OMV-1 and OMV-2 but not OMV-3.

At the Research Institute for Breeding Methods in Quedlinburg (GDR), in cooperation with the Research Institute for Crop Production in Prague, resistance has been transferred from A.pilosa, CAV 0128, No. 264. A synthesised stable octoploid pilosa/sativa amphidiploid was used for backcrossing with A.sativa and the hybrids evaluated for resistance. Up to now it has been impossible to stabilize the mildew resistance at the hexaploid level as several pilosa chromosomes are involved in the genetic control of resistance.

4. SUMMARY

In crown rust, the highest frequency of virulence was on resistance genes Pc 35, 40, 45 and 47. Virulence frequencies differ widely between countries. Of the standard differentials, the most effective were Trispernia, Bondvic, Landhafer and Santa Fé. The resistance of cvs Dodge and Garland, presently being transferred into Czechoslovak advanced lines of oats, are of continuing great importance. Furthermore, genes Pc-39, 48, 50, 55, 58 and 59 were indicated to be the most effective in Europe. In stem rust in Austria, Czechoslovakia, Germany, Switzerland and Poland in 1978-1984 ten avirulence/virulence combinations were isolated. Pg-a was the most effective and Pg 4, 9, 13 and 16 are also of value. In powdery mildew, seven groups of virulences have been differentiated and the effectiveness of specific resistance is low but adult plant resistances are available.

REFERENCES

1. Sebesta J and B Zwatz: Pflanzenschutz-Berichte (Wien) 46, 1-41, 1980.
2. Simons MD and LJ Michel: Plant Dis. Rept, Beltsville 48, 763-766, 1964.
3. Fleischmann G and RJ Baker: Can. J. Bot. 49, 1433-1437, 1971.
4. Sebesta J and DE Harder: Plant Disease 67, 56-59, 1983.
5. Sebesta J: Infektionsmethoden zur Selektion auf Rostresistenz des Getreides im Gewächshaus und im Freiland. Bericht Arbeitstagung 1972 der Arbeitsgemeinschaft der Saatzuchleiter in Gumpenstein (Austria) November 1972, 185-197, 1972.
6. Green GJ: Can. Plant Dis. Surv. 45, 23-29, 1965.
7. Sebesta J: Genetics and Plant Breeding (Prague) 13, 71-79, 1977.
8. Sebesta J: Pl. Prot. 9, 89-94, 1973.
9. Sebesta J: Pl. Prot. 9, 1-6, 1973.
10. Jones IT and ERL Jones: Rep. UK Cereal Pathogen Virulence Survey for 1978, pp.59-63, 1979.
11. Jones IT and ERL Jones: Rep. UK Cereal Pathogen Virulence Survey for 1980, pp.78-84, 1981.
12. Sebesta J: Euphytica 28, 807-809, 1979.
13. Sebesta J: Euphytica 32, 857-861, 1983.
14. Jones IT and JD Hayes: Ann. appl. Biol. 68, 31-39, 1971.
15. Jones IT: Ann. appl. Biol. 86, 267-277, 1977.
16. Jones IT: Ann. appl. Biol. 90, 233-239, 1978.
17. Carver TLW and AJH Carr : Ann. appl. Biol. 86, 29-36, 1977.
18. Sebesta J, B Zwatz and M Kummer: Tag.-Ber., Akad. Landwirtsch.-Wiss. DDR, Berlin (1983), 523-540, 1982.
19. Sebesta J.: Probleme Monogen-resistenter Getreidesorten gegenüber Rostkrankheiten. Bericht Arbeitstagung 1980 gehalten in Gumpenstein (Austria) November 1980, 149-161, 1980.

ADEQUATE RUST RESISTANCE IN OATS

P.G. ROTHMAN
USDA-ARS University of Minnesota, St. Paul, MN 55108, USA

1. INTRODUCTION

For the past 60 years, the use of single genes for specific resistance has formed the basis for most conventional breeding programmes for oat stem rust improvement. While somewhat successful in keeping ahead of the rust pathogen the new resistant cultivars eventually had to be replaced after brief stints of production. Oat breeders, too, were plagued by a paucity of specific resistances to Puccinia graminis f.sp. avenae. In 1965 the super-virulent race 94 (NA-30) surfaced which was capable of attacking all the stem rust resistance genes then known in the cultivated oats in North America. Since that period new sources for stem rust resistance have been reported, many of which were transferred from the wild species of Avena to the cultivated species. Their exotic source of origin, however, does not necessarily assure a more prolonged existence in the cultivated oat if the appropriate race should evolve. One choice for the breeders, however, is to combine these new sources of resistance into various combinations.

2. SOURCES OF RUST RESISTANCE

A literature review will show that the diploid A.strigosa has received the majority of attention as a source of alien germplasm for numerous traits in oat improvement. Some crown rust resistance in present day North American cultivars can be traced to this species. Several approaches to this germplasm utilization will be discussed at this conference, but it was by chance that within an interspecific oat cross between an autotetraploid of A.strigosa (4x) and A.sativa (6x) a rust resistant octoploid line was developed(1). The line appeared as one of a twin seedling; the smaller twin with 35 chromosomes and sterile and the larger twin with 56 chromosomes and fully fertile. Progenies of the octoploid line have been identified with both stem rust and/or crown rust resistance attributed to the A.strigosa parent. No fully fertile line, however, has been found among the progeny which exhibits mammoth florets, seeds, leaf and stem characteristics. Obee(2) is a germplasm release from this origin which crosses readily with the hexaploid oats. The rust resistance is readily transferred although sterility within the early filial progenies is very apparent. A second germplasm line, Delredsa(2), is also a selection derived from an interspecific cross of a diploid parent A.strigosa and a hexaploid A.byzantina (6x) breeding selection. The stem rust resistance is stable and similar to that of the diploid parent. The selection is cross compatible with the hexaploid oat.

Extraction of rust resistance from other oat species includes, A.abyssinica, A.longiglumis, A.magna, A.fatua and A.sterilis.

A.abyssinica: CI 4751 is a typical representative of the tetraploid oat species in that the apex of the lemma is divided into four bristles. The line was obtained by the USDA World Oat Collection in 1947 and was recorded as coming from the Smolensk district of Russia. It has seedling resistance to the St. Paul buckthorn composite of crown rust races and to all of the prevalent races of stem rust with the exception of race NA 17. To my knowledge, this species has not been tapped as a source of stem rust in any oat breeding programme. An interspecific cross between CI 4751 and a derived tetraploid obtained by colchicine treatment of the diploid A.strigosa glabrescens:CI 3815 resulted in progenies which segregated for rust reactions to selected key races of stem rust since the latter line lacks resistance for stem rust. Among many crosses attempted with the resistant selections a successful cross was obtained with Minhi No. 8 which was the smaller of a twin seedling found in the cultivar Minhafer. The twin plant was short and partially fertile suggesting it might have a polyploid number other than hexaploid. Low fertility and poor seed set was encountered through the first four filial generations of the cross but the progenies appear similar to the sativa parent. Resistance to all tester races of stem rust was found among its progenies but crosses with hexaploid selections have not yet been successful.

The Ethiopian crown rust resistant A.abyssinica: CI 7233 has also been used in interspecific crosses with the Iowa/Canada rust resistant derived tetraploid Abd. 54SP101: CI 7232 (which itself is an interspecific cross between A.abyssinica: CD 4549 and A.strigosa: CD 3820). Abd. 54SP101 was widely used as a source of crown rust resistance in many breeding programmes. Its source of stem rust resistance was never determined or utilized.

Within the progeny of a cross between CI 7233 and CI 7232, backcrossed once to CI 7232, lines have been identified which are resistant to all of the prevalent NA stem rust races as well as to the races of crown rust within the St. Paul buckthorn composite, some of which parasitise CI 7232.

The germplasm line Amagalon(2) is by far the most interesting of all my interspecific oat crosses. It resulted from a cross between the crown rust resistant tetraploid A.magna:CI 8330 and the stem rust resistant diploid A.longiglumis: CW 47. The sterile triploid hybrid set a single seed after application of a 0.25% aqueous solution of colchicine was applied to the plant crown. The colchicine derived plant was partially fertile and proved to be a stable hexaploid which successfully crosses with the cultivated oats. Within the progeny can be found resistance to the prevalent races of stem and crown rust. Progeny lines vary widely for the morphological characters but possess the common trait of lateness.

While a myriad of crown rust resistant genes have been identified and now occur in released cultivars from the oat species A.sterilis, its utilization as a source of stem rust resistance has been very limited. The two germplasm releases Alpha: CI 9211(3) and Omega: CI 9139(4) originated as a cross between the oat cultivar 'Kyto': CI 8250 and a slow rusting line of A.sterilis: CI 8377. While not expressing a specific type of stem rust resistance, the sterilis line was characterised by a lagging development of pustules culminating in a susceptible-type of stem rust reaction. Kyto has seedling resistance to the prevalent North American stem rust races but when crossed, the resulting two germplasm lines were found to have resistance which is superior to either parent. The genetic mechanism of this resistance is not fully understood but more recent work suggests that the seedling gene Pg 12 from Kyto is interacting with minor

genes of the sterilis line. Virulence to this type of resistance has now
been found, but by mass screening of both germplasm lines - which were
composites of bulked progeny rows - component lines with added resistance
have been identified which are resistant to the races which parasitise the
original released lines, suggesting an additive accumulation of possible
minor genes.

Voluminous amounts of literature today are devoted to the wild oat
species A.fatua. Most, of course, deals with its obliteration rather than
its utilisation. The USDA has collected over 2200 samples mostly from the
northcentral part of the US in an effort to preserve and evaluate its
potential as a germplam source for the improvement of the cultivated oat.
Little work outside of that of Suneson's in California deals with
utilizing the wild oat. A.fatua is a very compatible rust host with
rampant rusting of the entire plant once infected by the pathogen. The
plants' ability, however, to develop sound kernels under these adverse
conditions suggests a desirable trait which might be manipulated in our
struggle with the rust pathogens. Numerous A.fatua/A.sativa crosses
studied have not fixed any approaches to management of this attribute of
A.fatua, but the occurrence of specific rust resistance in some progenies
needs to be confronted. Its origin is unclear but suggests that A.fatua
may carry useful genetic material that can be expressed under certain
circumstances. The resistance found within the A.fatua hybrid progenies
is effective against our prevalent crown and stem rust races.

Genes for oat stem rust resistance, in the main, are called seedling
resistance genes for they condition resistance throughout the life of the
plant. (However, rust reactions can be modified by high temperatures). In
1962, the Canadians(5) reported an 'adult plant' type of resistance in
which the plants in the seedling stage are susceptible to all races of oat
stem rust except for the race 8 group which are avirulent on the Pg 1 gene
possessed by this line. As the plants develop however, they become
resistant to all the prevalent races of oat stem rust. The origin of this
type of resistance is blurred in that CI 3034 was received from Rhodesia
in 1926 as a rust resistant selection out of a field of heavily rusted
'Burt' oats. The variety Burt had its origins in Alabama, USA, in 1878
when a farmer named Burt harvested a single plant out of a field of 'Red
Rustproof' oats. The single plant was increased under the name Burt and
by 1919 was sown on about 500,000 acres in the USA. The variety was noted
for its great variability in plant and grain characteristics. CI 3034 has
a yellow-green plant colour which may be genetically linked with the rust
resistance and results in a poor productive plant. Among the various
crosses made with CI 3034 at St. Paul, MN, there is always an independent
segregation of these two characters resulting in green, rust resistant
progenies. MN 806464, a cross between Marvellous and CI 3034, had the
fourth and first lowest coefficients of infection in the 1982 and 1983
International Oat Rust Nursery, respectively. This broad spectrum of
resistance at the many locations of the IORN warrants the use of this
unique type of resistance. The high level of adult plant resistance is
maintained in the progenies of the cross, and many of these progenies also
have seedling resistance to all races coupled with the adult type.

3. UTILISATION OF RESISTANCE GENES

While it is not possible to say with certainty that the various reported
sources of rust resistance are indeed diverse, merely because of their
origin in various species of Avena, their prudent use seems warranted

since the maintenance of a diverse collection of resistant germplasm seems a primary defence against the genetic vulnerability to the rust pathogens. Combinations of resistant sources thwart rapid changes in the rust population which usually follows the incorporation of single genes for resistance. With this in mind, these sources of exotic resistances have been crossed in all possible combinations, followed by the intercrossing of the combinations, in an attempt to obtain two or more sources of resistance in a common germplasm line (Table 1).

TABLE 1. Combinations of diverse sources of oat germplasm lines with effective stem rust resistance to the widespread North American race 27 are listed below. Adequate protection to race NA-28 is obtained in many of the combinations so adequate resistance is available for all known North American oat stem rust races

Germplasm rust lines	Minn. source	North American stem rust race			St. Paul Buckthorn composite crown rust
		27	28	30	
Alpha	CI 9221	R	S	R	S
Alpha selection	845476	R	R	R	R
Alpha/Amagalon	833736	R	R	R	R
Amagalon selection	842138	R	R	R	R
Amagalon selection	833749	R	S	R	R
Amagalon/Black mesdag	846011	R	R	R	R
Amagalon/Marvellous	836346	R	R	R	R
Amagalon/Black mesdag/Aojss	848362	R	R	R	R
Amagalon/Obee/A.sterilis/3/Amagalon/ Black mesdag/Froker/Omega	847445	R	R	R	S
Amagalon/Marvellous/Amagalon/Black Mesdag	848183	R	R	R	R
Amagalon/A.nuda/Marvellous	846703	R	R	R	R
Aojss	845502	R	S	R	R
A.fatua/Marvellous	11811-1	R	R	R	R
A.fatua/A.sterilis	841471	R	R	R	R
A.abyssinica/CI 14751/ASG-660/MINHI No.8	843827	R	R	R	S
A.abyssinica CI 7233/2*Abd.SP101	839926	R	R	R	R
Delredsa	848383	R	S	R	S
Delredsa/Aojss	848521	R	R	R	S
Delredsa/Amagalon/Black mesdag	846952	R	R	R	R
Delredsa/Aojss/3/A.abyssinica/2*Abd.SP101/ DAL/ALPHA	844651	R	R	R	R
Marvellous/CI 13034	848071	R	R	R	R
Obee selection	846302	R	S	R	R
Obee selection	11828-1	R	R	R	S
Obee/Obee/A.fatua	846291	R	R	R	R
Obee/Obee/A.fatua/3/Amagalon	837654	R	R	R	R
Obee/Sajapago/Amagalon/Black mesdag	846991	R	R	R	R
Obee/3/Gopher/MN 2629/Fulghum/Fla.500	847989	R	R	R	S
Omega	CI 9139	R	S	R	S
Omega selection	846921	R	R	R	S
Sajapago(4x)	6172-B	R	R	R	R

Granted that precise testing of lines for multiple resistance might be open to doubts, the resistant germplasm lines identified under this regime of crossing and testing have, at the very least, a single effective source of stem rust resistance to our most widely prevalent stem rust race. Because only 16 Pg genes have been identified for stem rust resistance, this is a measure taken in advance to avert the loss of new resistant germplasm through the single gene/sudden loss-of-effectiveness syndrome of the past. How well these sources of resistance stop of delay the rust pathogen will, of course, be measured in retrospect.

REFERENCES
1. Rothman PG and K Sadanaga: Crop Sci. 8, 635-636, 1968.
2. Rothman PG: Crop Sci. 24, 1217-1218, 1984.
3. Rothman PG: Crop Sci. 16, p.315, 1976.
4. Martens JW, PG Rothman, RIH McKenzie and PD Brown: Can. J. Genet. Cytol. 23, 591-595, 1981.
5. McKenzie RIH and JW Martens: Crop Sci. 8, 625-627, 1968.

GENETIC ANALYSIS OF CROWN RUST RESISTANCE AND
PHYTOALEXIN PRODUCTION IN OATS

T. MORIKAWA* and S. MAYAMA**
*College of Agriculture, University of Osaka Prefecture
**Faculty of Agriculture, Kagawa University, Japan

1. INTRODUCTION

Antifungal compounds accumulate in resistant oat cultivars in response
to crown rust infection.   They are regarded as phytoalexins and are
referred to as avenalumins I, II and III(1).   Mayama et al.(2) examined
the relationships between resistance gene expression and avenalumin
accumulation employing 21 Pc lines. It is well known that the Pc-2 (Hv-1)
gene controls susceptibility to victorin, the host-specific toxin produced
by Helminthosporium victoriae.   The toxin is also known to be an elicitor
of avenalumin(3).

In order to study the genetical basis of avenalumin accumulation,
intercrosses were made between four oat cultivars.   The genetic behaviour
of avenalumin accumulation was then examined by inoculation with crown
rust races 203, 226 and victorin.

2.   MATERIALS AND METHODS

2.1 Plant materials and fungi

The four cultivated oats (Avena sativa L.) Victoria, Shokan 1, CW 491-4
and Kanota were used in this study.   Reactions to two races of Puccinia
coronata f.sp. avenae and the host specific toxin, victorin, are given in
Table 1.   Victoria and Shokan 1 are resistant to race 226 and susceptible
to race 203.   Victoria which has the Hv-1 gene, is susceptible to
victorin.   Shokan 1, a Japanese cultivar alleged to be resistant to crown
rust, is highly resistant to race 226 but its resistance gene or genes
were not identified by testing on this set of Pc lines. CW 491-4, which
has Pc-38, is resistant to race 203 and susceptible to race 226.   Kanota
is susceptible to both races.   The degrees of resistance of these four
cultivars towards races 203 and 226 were determined by length of
intercellular hyphae and final infection type.

2.2 Inoculation methods

Inoculation of 7 day old seedlings was achieved by dusting on
uredospores with a brush and incubating in a moist chamber for 16 h(2).

2.3 Estimation of avenalumin content

Quantitative estimation of avenalumin I accumulation in treated leaves
was conducted as described by Mayama(4).   A primary leaf was extracted
with hot methanol and analysed by high-pressure liquid chromatography
(HPLC).

2.4 Treatment with victorin. Purified toxin was obtained from culture
filtrates of H.victoriae(3).   Primary leaves from plants growing on
vermiculate were cut into 0.3 cm segments.   Each segment was then floated
on various concentrations of victorin in wells of microplates.   The

TABLE 1. Specific reactions of oat cultivars to crown rust fungi and victorin (H.victoriae)

Pathogen	Cultivar			
	Victoria (Pc-2)	Shokan 1 (unknown)	CW 491-4 (Pc-38)	Kanota (none)
P.coronata				
Race 226	R	R	S	S
Race 203	S	S	R	S
H.victoriae				
victorin	S	R	R	R

R = resistant; S = susceptible

genotype for the Hv-1 locus was then determined by the severity of necrosis or chlorosis in response to the toxin.

3. RESULTS AND DISCUSSION

The quantitative assay for avenalumin I was carried out on four parental oat cultivars and three F_1 hybrids to clarify the relationship between resistance to crown rust and avenalumin accumulation (Table 2). Shokan 1, which was highly resistant to race 226, showed the highest avenalumin

TABLE 2. Induction of avenalumin I accumulation by inoculation of crown rust races 203 and 226 into cultivars and their F_1 hybrids

Cultivar and F_1	Race 226					Race 203	
	36[a]	48	72	96	120	48	96
CW 491-4 x Shokan 1	11.6[b]	1124.2	-	-	-	-	66.1
Kanota x Victoria	-	-	20.7	41.3	41.5	-	-
Shokan 1 x Victoria	-	546.7	469.4	-	-	-	-
Shokan 1	-	2099.1	2271.4	-	-	-	tr
CW 491-4	-	tr	tr	tr	tr	45.7	243.2
Victoria	-	-	65.6	147.5	140.2	-	-
Kanota	-	tr	tr	tr	tr	-	-

[a] = time after inoculation(h); [b] = avenalumin I (ug/g fresh wt); tr = trace

accumulation amongst the parental and hybrid plants at 48-72 h after inoculation. Victoria, which was moderately resistant to race 226, showed about $1/30$ of the avenalumin accumulation in Shokan 1. Maximum avenalumin accumulation in Victoria was observed at 96 h after inoculation. CW 491-4 is susceptible to race 226 and little avenalumin accumulated, but a relatively larger amount of avenalumin was detected after inoculation with race 203, to which the cultivar is moderately resistant. In Kanota, which is susceptible to both races, no accumulation of avenalumin was detected.

The results suggest that avenalumin accumulation is closely related to the degree of resistance expression and is controlled by the genes for resistance to crown rust. The F_1 hybrids derived from the crosses CW 491-4 x Shokan 1 and Kanota x Victoria showed resistance to race 226 and avenalumin accumulations were about half and one-third of those of their resistant parents, respectively. The same trend was observed in the F_1 hybrid CW 491-4 x Shokan 1 when infected by race 203. The F_1 hybrid Shokan 1 x Victoria has two resistance genes against race 226, whereas avenalumin accumulation was only half of that of CW 491-4 x Shokan 1 infected by race 226. This suggests that the resistance gene in Shokan 1 might be affected by other genes for resistance to crown rust. The allelism between the two resistance genes is not clarified and thus it is difficult to explain the cause of the decrease of accumulation in the F_1 hybrids.

The sensitivity of Kanota and Victoria, and their F_1 hybrid, to victorin is shown in Table 3. In the F_1s, the expression of sensitivity to victorin began at 10 ng/ml, became maximum at 250-500 ng/ml and was 5-10 times less than that of Victoria. Kanota had no expression of sensitivity to victorin at any concentration. It was possible to distinguish the genotype at the Hv-1 locus by reaction to the various concentrations of victorin.

TABLE 3. Sensitivity to victorin of Kanota and Victoria and their F_1 hybrid

Parent and F_1	Concentration of victorin (ng/ml)							
	0	1	5	10	50	100	250	500
P1 Kanota (hv-1hv-1)	-	-	-	-	-	-	-	-
P2 Victoria (Hv-1Hv-1)	-	+	+	+	++	++	+++	+++
F_1 (Hv-1hv-1)	-	-	-	+	+	+	++	++

- = no reaction; + = slight chlorosis; ++ = moderate chlorosis;
+++ = severe chlorosis

Segregation of resistance to crown rust race 226 and victorin in the F_2 and BC_1 populations from Kanota x Victoria and CW 491-4 x Shokan 1 is given in Table 4. The segregation ratio for sensitivity to victorin in 120 F_2 plants from Kanota x Victoria fitted the expected ratio of 1 sensitive [28] : 2 intermediate [64] : 1 resistant [28]. The same F_2 plants were then checked for resistance to crown rust race 226. All plants which were either sensitive or intermediate to victorin were resistant to crown rust in· infection type. The remaining plants which were resistant to victorin were mostly susceptible to race 226 but 3 plants were resistant. These 3 plants, which it is suggested have no Hv-1 gene, showed moderate resistance. However, the segregation ratio for resistance to race 226 fitted the expected ratio for single factor inheritance. In the BC_1, the segregation ratio for both characters fitted the expected ratios of 1 (resistant to toxin and susceptible to crown rust) : 1 (intermediate to both toxin and crown rust). In the cross between CW 491-4 and Shokan 1, the F_2 progeny segregated in a ratio of 1 resistant [55] : 2 intermediate [84] : 1 susceptible [55], indicating that resistance in Shokan 1 is controlled by a single dominant gene.

TABLE 4. Segregation of resistance to crown rust infection
and victorin in the F_2 and BC1

Cross	Pathogen /toxin	Generation	No. of plants			Total	χ^2	P
			R	I	S			
Kanota x Victoria	Victorin	F_2	28	64	28	120	0.533[1:2:1]	0.50-0.80
	Race 226	F_2	28	64(3)	25	120	1.783[1:2:1]	0.30-0.50
Kanota x Victoria	Victorin	BC 1	25	34		59	1.373[1:1]	0.20-0.30
	Race 226	BC 1		34	25	59	1.373[1:1]	0.20-0.30
CW 491-4 x Shokan 1	Race 226	F_2	55	84	55	189	3.485[1:2:1]	0.10-0.20

R = resistant; I = intermediate; S = susceptible.

It has been reported that the resistance expression of the Pc genes is associated with the restriction of growth of the intercellular hyphae, and occurs at different times after inoculation. The degree of crown rust resistance is thought to be correlated with the rapidity and quantity of avenalumin accumulation(2). In this study, avenalumin accumulation varied with the kind of resistance genes and their combinations, and was consistent with the genetic behaviour of the resistance genes in the F_1, F_2 and BC1. The results indicate that the induction of avenalumin accumulation is governed by a single incomplete dominant resistance gene.

Attempts to combine the rust resistance of Pc-2 with resistance to H.victoriae have not been successful(5). Luke et al.(6) reported that the Hv-1 locus is composed of two or more alleles exhibiting semidominance and is directly related to resistance to crown rust. Simons(7) in his review states that there are other alleles at the Pc-2 locus amongst oat cultivars from South-America which are all susceptible to H.victoriae, but vary in their reaction to different races of P.coronata. In the F_2 and BC_1 of Kanota x Victoria, Hv-1 was either pleiotropic or very tightly linked to Pc-2. Victoria has at least two other genes for resistance to crown rust(8) and thus it is most likely that the resistance of the 3 selections to both victorin and crown rust race 226 in the F_2 is due to the other loci controlling rust resistance.

REFERENCES

1. Mayama S, T Tani, Y Matsuura, T Ueno and H Fukami: Physiol. Plant Path. 19, 217-226, 1981.
2. Mayama S, Y Matsuura, H Iida and T Tani: Physiol. Plant Path. 20, 189-199, 1982.
3. Mayama S and NT Keen: Phytopathology 74, p.850, 1984.
4. Mayama S: Mem. Fac. Agri. Kagawa Univ. 42, pp.64, 1983.
5. Day PR: Genetics of host-parasite interaction. WH Freeman Co., San Francisco, pp.238, 1974.
6. Luke HH, HC Murphy and FC Petr: Phytopathology 56, 210-212, 1966.
7. Simons MD: Mono. No. 5, Amer. Phytopath. Soc., Hefferman Press Inc., Worcester, Mass. pp.47, 1970.
8. Simons MD, FJ Zillinsky and NF Jensen: USDA ARS Publ. 34-85, pp.22, 1966.

VARIATION AND DISTRIBUTION OF THE OAT CROWN RUST-FUNGUS, PUCCINIA CORONATA AVENAE, IN BRITAIN IN RELATION TO BREEDING RESISTANT OAT CULTIVARS

B. C. CLIFFORD
Welsh Plant Breeding Station, Aberystwyth, Dyfed, Wales, UK

1. INTRODUCTION

Puccinia coronata avenae, the cause of oat crown rust, is endemic in Britain and occurs most commonly in the warmer southern and south western parts of England and Wales and in particular in Devon, Cornwall, Kent and S. Wales. The fungus appears to overwinter in the uredial stage on autumn-sown crops, on volunteer oat plants and on wild Avena spp. The alternate host, Rhamnus cathartica, although grown as an ornamental shrub, is unimportant in the epidemiology and biology of the fungus, and survival of P.coronata depends on survival of the oat host.

The need to carry out disease surveys to evaluate crop losses and to monitor virulence relates to the perception of breeders of the need to develop resistant cultivars. With the declining oat acreage in the last twenty years and the need to improve yield and quality, this perceived need has been understandably low. Consequently, there has been a relatively low level of research on oat crown rust and much is not known about the epidemiology and economic importance of the disease. Yield losses can be in the order of 25% as indicated by data from a trial in Pembrokeshire, South Wales, where the resistant spring cv. Manod was compared with the susceptible cv. Condor(1). Rust infection resulted in a 28% reduction in grain yield in cv. Condor largely from a 21% reduction in 1000 grain weight and a 35% increase in husk. These effects on yield and quality are further illustrated by the results of a trial conducted at the Welsh Plant Breeding Station in 1974 where yield losses in the order of 25% in susceptible cultivars such as Maris Tabard and Mostyn were largely due to reductions in 1000 grain weight(2).

2. CULTIVAR RESISTANCE AND BREEDING OBJECTIVES

Although crown rust resistance has a relatively low breeding priority at present, there was an active programme at the Welsh Plant Breeding Station during the 1930s to 1950s. This was largely based on the specific resistances of cvs Bond, Victoria, Landhafer and Ascencao. The spring cv. Manod, which carries the Victoria resistance, came from this programme and was popular in the 1960s. Incidentally, the Victoria blight pathogen, Cochliobolus victoriae, is not known to occur in Great Britain. In 1974, the Swedish cv. Nelson was released in the UK and its successor cv. Trafalgar in 1978; both of these carry resistance from cv. Landhafer. The recently released cvs Pinto and Avalanche carry unidentified resistances but all of the above are susceptible to the common race 265. All other spring and winter cultivars released in the last decade are susceptible.

From the above considerations, it can be argued that the best mainstream breeding objective is to avoid undue susceptibility. This is actively pursued in WPBS breeding programmes by evaluating advanced breeding populations and yield trial material in disease nurseries uniformly inoculated with specific pathogen isolates. This ensures the release of cultivars with acceptable levels of resistance and also, as such material is fed back into the breeding programmes, the general improvement of the breeding gene pool. It also appears desirable to occasionally develop cultivars carrying specific major gene resistance by rapid backcrossing. These are necessary to respond to any increased threat by crown rust that may be realised in the event of a rapid increase in the oat acreage. In the longer term, attention should be paid to the identification of non-specific resistance factors which are expressed as 'slow rusting' in the field.

3. SURVEY OF VIRULENCE IN PUCCINIA CORONATA AVENAE

A necessary adjunct to resistance breeding is the montoring of corresponding virulence in the pathogen population. This has been carried out at the WPBS since 1967 under the auspices of the UK Cereal Pathogen Virulence Surveys with the objectives of identifying virulence of immediate commercial importance and potential concern to breeders. The standard international differential series of resistant cultivars has formed the basis of the surveys as these resistances are agriculturally relevant. Standard procedures have been used which involve the classification of seedling reactions into resistant, intermediate and susceptible for each isolate tested. Numbers of samples received are generally low and so interpretation of virulence frequency analysis should be with caution. Results for years where sample numbers were reasonable are given in Table 1.

TABLE 1. Virulence frequencies to standard differential cultivars

Code	Differential cultivar	Virulence frequency (%)				
		1969	1971	1972	1981	1981 (Eire)
1	Anthony	12	9	4	38	71
2	Victoria	4	4	0	0	0
3	Appler	64	61	63	88	93
4	Bond	72	30	48	100	100
5	Landhafer	4	4	0	88	93
6	Santa Fé	4	4	0	12	21
7	Ukraine	40	30	11	0	71
8	Trispernia	8	9	0	6	14
9	Bondvic	4	4	0	31	21
10	Saia	96	96	100	100	100
No. of samples		25	23	27	16	14

In 1981, samples were received from the crown rust population in Eire which allows comparison with that in Great Britain. The lowest virulence frequencies are for cvs Victoria, Santa Fé, Trispernia and Bondvic and the highest for cvs Appler, Bond, Landhafer and the diploid Saia. Of these latter four, only the Landhafer resistance appears to be present in commercial cultivars (see above). Virulence to cv. Ukraine is interesting in fluctuating from year to year and apparently at random.

REFERENCES

1. Griffiths, D.J.: Rep. Welsh Pl. Breed. Stn for 1961, pp.61-62, 1962.
2. Clifford, B.C.: Rep. Welsh Pl. Breed. Stn for 1974, p.33, 1975.

TRANSGRESSIVE SEGREGATION FOR INCREASED LEVELS OF ADULT PLANT RESISTANCE TO MILDEW IN OATS

I.T. JONES and H.W. RODERICK
Welsh Plant Breeding Station, Aberystwyth, Dyfed, Wales, UK

1. INTRODUCTION

Powdery mildew (Erysiphe graminis avenae) is the most serious foliar disease of oats in Britain. It can cause up to 40% loss in grain yield(1,2), and probably accounts for an average loss of 5-10% annually.

Because major gene controlled, hypersensitive type resistance provides effective control for only a short period of time, attention has been given to adult plant resistance as a more durable source. Certain older cultivars have maintained, up to the present time, a moderate level of partial resistance. In this paper the results of experiments carried out on progeny lines from crosses involving one such variety, namely Maldwyn, are reported.

The resistance of Maldwyn has already been extensively investigated. Jones and Hayes(3) described in detail the pattern of resistance as fairly susceptible at the seedling stage but exhibiting a non-hypersensitive type resistance on the upper leaves of the adult plant. Histologically, Carver and Carr(4,5) have shown that the Maldwyn resistance is due to a reduction in successful primary infections or penetrations, reduced haustorial efficiency and number of haustoria per colony. Recently Douglas, Sherwood and Lukezic(6) have confirmed the complex nature of this resistance. Jones, O'Reilly and Davies(7) and Jones(8) studied the inheritance of this resistance in a cross with the susceptible cultivar Selma and found it to be quantitative and continuous with an estimated maximum of nine effective factors segregating, Maldwyn contributing at least seven of these factors. This resistance has remained effective since its release in 1947 to all known races and as far as is known is race non-specific or horizontal.

Jones(9) described how he had developed lines from a cross of Maldwyn with Mostyn which exhibited a higher level of resistance to mildew than Maldwyn. In this paper we describe the results of our observations on one of these lines, OM 1621, although the experiments included other lines, data from which will be presented elsewhere.

In addition to an assessment of infection levels carried out under field conditions, the detached leaf technique was applied to glasshouse grown plants at different stages of growth to identify which components of resistance were accounting for this transgression. Furthermore, there was an attempt to assess the durability of this enhanced resistance by testing with 'trained' isolates.

2. MATERIALS AND METHODS

In the three experiments reported here the genotypes are confined to a) the Mostyn parent with resistance factor OMR 3(10) or Eg-3(11); b) the Maldwyn parent with no specific resistance factors but with a moderate level of adult plant resistance; c) OM 1621, an F_{10} line derived from the cross Mostyn^2 x Maldwyn and possessing the OMR 3 major gene donated by Mostyn.

2.1 Experiment 1
A field experiment comprising 5 x 0.75 m drills of each genotype was grown according to a randomized complete block design with two blocks. A natural mildew infection occurred and obviously had the corresponding virulence to the OMR 3 resistance in view of the high level of infection on Mostyn. The percentage leaf area infected was estimated on each drill at 10 day intervals from 31 May onwards.

2.2 Experiment 2
In this first laboratory experiment 3 cm long detached segments of leaves 4 (lower), 5 and 6 (proximal and distal parts) of each genotype replicated four times were placed on benzimadazole/agar in compartmented boxes as described by Jones and Hayes(3), and inoculated on 28 November 1984 with mildew race 6 (OMV 1,2,3,4) with virulence to the OMR 3 host gene. Inoculation was by shaking infected plants over the open boxes. The boxes were then transferred to a growth chamber and incubation occurred at 10 ± 2°C, 8 h light/16 h dark photoperiod.

2.3 Experiment 3
In this second laboratory experiment carried out when plants were at GS 47-50, three separate sets of segments of the four uppermost leaves (Flag-3, F-2, F-1, Flag) replicated four times were inoculated on 18 June 1985 with three different mildew isolates all with virulence to the major resistance gene OMR 3. The first was a neutral isolate but virulent on Mostyn, Maldwyn and OM 1621. The second had been multiplied on the cv. Maldwyn for over 50 transfers and was also virulent on Mostyn and the third had been multiplied on the cv. Mostyn for 10 transfers. After inoculation the segments were transferred to the same growth chamber as for experiment 2, and were incubated under the same conditions.

The resistance components studied in the laboratory experiments were a) latent period i.e. number of days from inoculation to the first symptom expression; b) percentage leaf segment area infected and c) the number of spores produced per mm^2 of leaf surface, or a sporulation index based on a 1-10 scale.

In experiment 2, percentage leaf segment area infected was assessed for each individual segment 7 days after its first mildew symptom had been recorded. Five spore counts per segment were then taken on the same day on two of the four replicates, using a Coulter Counter (Model ZF6) and a leaf area meter. In experiment 3, spore counts were not taken but rather a sporulation index (1 = no sporulation, 10 = profuse sporulation) given to each segment at the same time as percentage area infected was recorded i.e. 14 days after inoculation.

3. RESULTS
Table 1 gives the results of the field experiment. The percentage leaf area infected with mildew on Maldwyn was significantly lower than Mostyn at all stages of growth.

Line OM 1621 showed a marked increase in level of resistance at the later stages of growth, being significantly more resistant on the last two scoring dates.

TABLE 1. Transgressive segregation for increased resistance to mildew in a cross
involving the durable resistant parent Maldwyn (% leaf area covered with
mildew, tested in 1985. Mean of 10 x 0.75 m drills)

Genotype	Scoring date Growth stage	31 May 15,22	10 June 31	20 June 43	1 July 57	10 July 65
Mostyn		3.0^{a+}	27.5^a	29.0^a	31.0^a	35.5^a
Maldwyn		1.3^b	10.5^b	14.9^b	20.3^b	25.0^b
Line OM 1621 (F_{10}, Mostyn^2 x Maldwyn)		1.6^b	10.5^b	8.6^b	10.1^c	13.5^c

[+]Means within scoring dates followed by the same letter are not
significantly different at P = 0.05 according to Duncan's Multiple Range Test

The results of the detached leaf assessment for leaves, 4, 5 and 6
(Table 2) showed that Maldwyn had a significantly longer latent period,
less area infected and fewer mean spores/mm^2 than Mostyn. However, at
this stage of growth, OM 1621 did not differ significantly from Maldwyn
for any of the three components measured.

TABLE 2. Components of partial resistance to mildew in parents and a
selected line as expressed in leaves 4(lower), 5 and 6(proximal
and distal parts). (Means of 4 leaves x 4 replicates = 16 items)

Genotype	Latent period (days)	Area infected (%)	Spores/mm^2
Mostyn	6.2^{b+}	18.5^a	215^a
Maldwyn	7.6^a	7.2^b	40^b
OM 1621	7.2^a	10.2^b	80^b

[+]Means within components followed by the same letter are not significantly
different at P = 0.05 according to Duncan's Multiple Range Test

At near heading stage (GS 47-50) in experiment 3 and combining the data
for the upper four leaves and the three isolates, Maldwyn showed
significantly higher levels of resistance than Mostyn for each component
(Table 3) as with the earlier growth stage (Table 2).

TABLE 3. Components of partial resistance to mildew at GS 47-50 in parents
and a selected line as expressed in the four uppermost leaves.
(Means of 48 observations = 3 isolates x 4 leaves x 4 replicates)

Genotype	Latent period (days)	Area infected (%)	Sporulation index
Mostyn	5.6^{c+}	8.8^a	4.6^a
Maldwyn	8.9^b	2.0^b	2.1^b
OM 1621	9.6^a	1.4^b	1.8^b

[+]Means within components followed by the same letter are not
significantly different at P = 0.05 according to Duncan's Multiple Range Test

Furthermore, OM 1621 at this later stage showed a slightly higher level of resistance for all three components than the most resistant parent Maldwyn, but it was only for latent period that this difference proved significant.

The latent period is, therefore, looked at in more detail in Table 4. Line OM 1621 is compared with that of its two parents when inoculated with three isolates separately. Again Maldwyn had a significantly longer latent period than Mostyn when inoculated with the three isolates. Also OM 1621 showed a slightly longer latent period than Maldwyn, the difference being significant in the 'Maldwyn trained' isolate. After inoculation with the 'Mostyn trained' isolate, OM 1621 showed a significant decrease in latent period compared with that in the 'Maldwyn trained' isolate. There is also a considerable although not a statistically significant decrease in the length of the Maldwyn's latent period after inoculation with the 'Maldwyn trained' isolate compared with that in the 'neutral' isolate.

TABLE 4. Latent period (days) at growth stage 47-50 of OM 1621 compared with its parents Mostyn and Maldwyn after inoculation with three mildew isolates. (Means of 4 leaves x 4 replicates = 16 items)

	Isolates		
Genotype	'Neutral'	'Maldwyn trained'	'Mostyn trained'
Mostyn	5.7^f	5.5^f	5.5^f
Maldwyn	9.5^{abc}	8.5^{cde}	8.7^{bcde}
OM 1621	9.6^{ab}	10.2^a	9.1^{bcd}

+Means followed by the same letter are not significantly different at P = 0.05 according to Duncan's Multiple Range Test

4. DISCUSSION

Field assessment confirmed that a higher level of adult plant resistance is possessed by Maldwyn relative to Mostyn (Table 1), the latter cultivar being fairly susceptible to races with corresponding virulence as evident in this trial. This data also confirmed a previous report(9) of transgressive segregation for resistance in the line OM 1621 compared with its most resistant parent Maldwyn. The enhanced level becomes significant under small plot trial conditions at about growth stage 57, i.e. when the panicles are almost completely emerged, but under large plot or crop conditions this advantage would be further increased and appear earlier.

The detached leaf experiment 2 established that at a fairly early stage (GS 16-17) Maldwyn had significantly longer latent period, less area infected and less spores produced than Mostyn. The enhanced level of OM 1621 compared with Maldwyn, however, was not evident at this stage. At near heading stage (Table 3) again Maldwyn showed a significantly higher level of resistance than Mostyn for the three components measured. In addition OM 1621 now showed a significant increase in latent period compared with Maldwyn, but the other two components, although giving slightly more resistant values, were not significantly different. Latent period thus seems to be an important component contributing to the enhanced resistance of the transgressive line OM 1621.

When these genotypes were challenged by three different isolates, two of which had been 'trained' on each of the parents, Mostyn and Maldwyn showed slightly, but not significantly, reduced latent periods after inoculation with their 'own host trained' isolates, suggesting the possibility of some erosion. However, these differences are relatively very small compared with changes from major gene hypersensitive type resistance to susceptibility. This gives added confidence to the expectation of durability that is conferred by small mainly additive type factors, especially under field conditions where at least the initial inoculum may not be so specialized as in these tests.

The most extended latent period was recorded in the transgressive line OM 1621 (10.2 days) after inoculation with the 'Maldwyn trained' isolate, significantly longer than for Maldwyn itself. This enhancement was probably due to the accumulation of factors from both Maldwyn and Mostyn parents, illustrating the possibility of increasing levels of partial resistance through critical selection in crosses of parents which themselves are moderately susceptible.

The latent period of OM 1621 was, nevertheless, significantly reduced when inoculated with the 'Mostyn trained' isolate (9.1 v 10.2 days) indicating erosion of this component almost to the level of Maldwyn itself (8.7 days). This suggests that the resistance conferred by at least some of the factors derived from Mostyn to the line OM 1621 is more easily eroded than that from Maldwyn. Thus, all additive or 'minor' resistance factors do not appear to confer equal stability to the overall resistance, and tests should be performed to evaluate the likely durability of any new sources of partial resistance before inclusion in breeding programmes if maximum durability in the progeny is to be safeguarded. It is however, encouraging to find that even after some apparent erosion of the latent period, it is still considerably longer than in Mostyn and would probably not account for a large reduction in the level of partial resistance under crop conditions.

REFERENCES

1. Lawes DA and JD Hayes: Pl. Path. 14, 125-128, 1965.
2. Jones IT: Ann. Appl. Biol. 86, 267-277, 1977.
3. Jones IT and JD Hayes: Ann. Appl. Biol. 68, 31-39, 1971.
4. Carver TLW and AJH Carr: Ann. Appl. Biol. 86, 29-36, 1977.
5. Carver TLW and AJH Carr: Ann. Appl. Biol. 88, 171-178, 1978.
6. Douglas SM, RT Sherwood and FL Lukezic: Physiol. Pl. Path. 25, 219-228, 1984.
7. Jones IT, A O'Reilly and IJER Davies: Rep. Welsh Pl. Breed. Stn for 1982, 103-105, 1983.
8. Jones IT: Ann. Appl. Biol. (in press).
9. Jones IT: Euphytica 32, 499-503, 1983.
10. Jones IT and ERL Jones: Rep. UK Cereal Pathogen Virulence Survey for 1978, 59-63, 1979.
11. Simons MD, JW Martens, RIH McKenzie, I Nishiyama, K Sadanaga, J Sebesta and H Thomas: USDA Agric. Handbook 509, 1-40, 1978.

THE NATURE OF HORIZONTAL RESISTANCE OF OATS TO CROWN RUST

H. H. LUKE
University of Florida, Gainesville, FL 32611, USA

Development of _Puccinia coronata_ in an oat cultivar with horizontal resistance (HR) was assessed using fluorescence microscopy. The first visible response was fluorescence in mesophyll cell walls. This reaction occurred about 18 h after inoculation, and seemed to depend upon development of substomatal vesicles. The growth of the parasite at individual infection sites was inversely related to the degree of host response, which was expressed as the number of fluorescing cells. A rapid and severe host reaction resulted in the arrest of growth of the parasite before haustorial mother cells formed, but slow host response retarded hyphal growth. The arrest of growth was associated with a reduction in the number of uredinia, while the retardation of hyphal growth reduced size of the uredinia and increased the latent period. Thus, fluorescence of the mesophyll cell walls was correlated with these three components of HR. In some sites, the host did not react to invasion by the parasite. When this happened, the hyphae grew at a normal rate. Seven days after inoculation, the average area of the uredinia of a susceptible type was about three times larger than that of the cultivar with HR. My results suggest that the substomatal vesicle produces a compound that is recognised by mesophyll cells, and that the time of host recognisation and the speed of host response condition three major components of HR to crown rust.

RESISTANCE AND TOLERANCE IN OATS, <u>AVENA</u> SPP. TO THE CEREAL CYST NEMATODE
<u>HETERODERA</u> <u>AVENAE</u>

A.R. BARR* and A.J. DUBE**
*Plant Breeding and **Biological Sciences Sections,
 Department of Agriculture, Adelaide, South Australia.

1. INTRODUCTION

Two aspects of plant reaction to cereal cyst nematode (CCN) invasion are
of importace to plant breeders. The first is 'resistance', which is
defined as the inability of the host to support nematode multiplication.
Resistance is required to reduce CCN populations for subsequent crops.
The second is tolerance, which is defined as the ability of the host to
withstand nematode attack with little yield loss. Tolerance is required
to ensure that the resistant crop produces an economic return. It is
important to note that resistant varieties are invaded by CCN and may
sustain serious yield losses(1). Resistance to CCN in oats has been known
since 1938(2) and researched widely since 1960. However, only one
study(3) has reported a high level of tolerance in oats.

Mathison(3) and Fisher <u>et al</u>.(4) assessed tolerance by examining the
yield of genotypes in the presence and absence of CCN damage, the infested
site being physically separated from the uninfested site. This method was
successful for Mathison(3) and Fisher <u>et al</u>.(4) as CCN damage was so
severe that it became the major determinant of yield at the infested site
and the genotypes under test were adapted to the uninfested test
environments. CCN tolerance could then be partitioned from
genotype-environment interaction. However, this partitioning will be
difficult if, either heavily infested CCN test sites are not available, or
the genotypes under test are poorly adapted, diverse and unfamiliar as,
for example, parental selection trials may be.

The development of a commercial applicator for broad area control of
CCN(5) makes it possible to limit the effects of CCN in infested sites.
Split plot experiments can be sown with and without the effects of CCN by
treating plots with nematicide. The comparison of yield differences
between the nematicide treated and untreated plots allows an estimate of
tolerance.

This study examines the resistance and tolerance of a number of oat
genotypes relative to wheat and barley in field and laboratory
experiments.

2. MATERIALS AND METHODS

Two wheat and two barley cultivars with the different combinations of
resistance and tolerance were chosen to use as 'benchmarks' for the <u>Avena</u>
<u>strigosa</u> and 14 <u>A.sativa</u> genotypes under test. Egret wheat is a
susceptible, intolerant type(4) and Festiguay is a resistant(6),
moderately intolerant type (Dube, pers.comm.). Clipper barley is a
susceptible-tolerant type(7) while Galleon is a resistant-tolerant
type(8). The resistance and tolerance of some of the oat varieties is

documented(3,6,7) but many were unknown. 'JB76-22' is a breeders line developed by JAM Brown and JB Brouwer, Department of Agriculture, Victoria from the cross Swan/Cc 4658//Swan. 'OXB 178' is a breeders' line developed by MR Krause at Roseworthy Agricultural College, South Australia from a complex cross which includes Avon as a major parent. The remaining two breeders' lines were developed by AR Barr, Department of Agriculture, South Australia. TAM-0-301 and TAM-0312 are cultivars released by Texas Agricultural Experiment Station which were found to be highly resistant to CCN in laboratory tests (Dube and Barr, unpublished data). The 1983 trial included most of the varieties released in Australia since 1975.

Trials were sown in known CCN infested soil at five locations in South Australia (Table 1). Nemadi(R) (2060 g/kg ethylene dibromide) was applied through a Jectarow applicator at seeding. Trials were of a split plot design with five replicates (1980, 1981, 1982) and four replicates (1983). Plot size was 4 row x 5 m (1980) and 6 row x 4.5 m (1981, 1982, 1983). Herbicides were applied as required.

TABLE 1. Trial details for CCN tolerance experiments 1980-1983

Site	Year	Seeding date	Sironem bioassay(9)	Nemadi(R) rate
Whitwarta	1980	18 July	2.25	6.0 L ha^{-1}
Roseworthy	1980	9 July	-	6.0 L ha^{-1}
Pinery	1981	9 June	1.4	6.0 L ha^{-1}
Pinery	1982	4 June	3.5	6.0 L ha^{-1}
Pinery	1983	19 July	3.0	5.17 L ha^{-1}

Field assessment of resistance at Pinery, 70 km north of Adelaide, was made by digging up five plants per plot on 13 November 1981 and washing the roots and soil through a 16 mesh sieve onto a 60 mesh sieve and counting the females on both the roots and on the sieve.

Laboratory testing for resistance was performed using the method described by Fisher(10). Seven to ten seeds of each line were assessed. The number of females per plant after 8-10 weeks growth is used as a measure of plant resistances.

3. RESULTS AND DISCUSSION

The results of the split plot tolerance tests show that Egret wheat responded significantly to nematicide in all four experiments, with treated plots averaging 23% greater yield than controls (Table 2). In contrast, no significant response in Festiguay was recorded in any trial. The barley varieties showed modest responses ranging from 2% (not significant) to 25% (significant at 0.1%).

A very wide range of tolerance levels was observed amongst the oat varieties. New Zealand Cape was very tolerant to CCN damage with an average response of -0.5% which means that New Zealand Cape is at least as tolerant, and probably more so than the tolerant benchmarks in wheat and barley. Two selections from the cross West/New Zealand Cape proved to be tolerant. This is encouraging for oat researchers wishing to combat CCN by breeding, as oats are traditionally considered the most intolerant of cereals(11).

At the other end of the spectrum, most of the cultivars which are widely grown in southern Australia are very intolerant. Examples include West,

Swan, Moore and Avon. The variety Cooba is harder to categorise as it responds differently to most other oats (Table 2). TAM-0-301 and TAM-0-312 were resistant to CCN in laboratory tests. While the field trials confirmed their resistance, both varieties proved to be intolerant removing hopes that a further source of tolerance could be found.

TABLE 2. Results of tolerance and resistance tests 1980-1982

Species	Genotype	Tolerance tests				Resistance tests	
		Yield response to nematicide[a]					
		Pinery 1982	Pinery 1981	Roseworthy 1980	Whitwarta 1980	Laboratory[b] test	Field[c] test
Oats	New Zealand Cape	3 NS	-8 NS	2 NS	1 NS	0.3	1.9
	Avon	28 **	27 *	7 NS	15 *	0.3	1.3
	JB76-22	29 **	1 NS	0 NS	4 NS	0.2	7.5
	Swan	33 ***	29 NS	7 NS	4 NS	0.6	6.6
	Cooba	18 NS	4 NS	14 *	18 NS	0.5	2.1
	Moore	41 ***	9 NS	-4 NS	2 NS	6.1	62.5
	Stout	62 ***	9 NS	-	-	3.4	27.8
	West	75 ***	37 **	-6 NS	12 *	16.2	23.2
	TAM-0-301	19 *	30 *	-	-	-	1.5
	TAM-0-312	39 ***	10 NS	-	-	-	0.9
	OXB 178	27 **	29 **	4 NS	11 NS	0.1	5.9
	(West x NZ Cape)/9	10 NS	0 NS	-	-	0.2	1.8
	(West x NZ Cape)/23	8 NS	8 NS	-	-	0.2	3.4
Sand oat	Saia	-	0 NS	3 NS	35 NS	0.4	17.0
Barley	Clipper	25 ***	19 *	2 NS	4 NS	17.2	12.4
	Galleon	11 *	16 *			1.0	6.3
Wheat	Egret	25 *	23 ***	22 *	23 **	16.3	32.8
	Festiguay	10 NS	8 NS	-9 NS	-2 NS	2.5	1.9

[a] Response expressed as % yield difference between nematicide treated and controls. Significance level described by the following NS, *, **, *** (not significant, Sig. at 5%, 1%, 0.1% respectively.

[b] Tests of significance on log transformed data; due to unequal samples, LSD 5% ranges from 3.1 to 3.9 for treatments with 10 replicates and 7 replicates respectively.

[c] Averaged over plus and minus nematicide. Tests of significance based on log transformed data; LSD 5% = 3.9.

The range of resistance measured in oats paralleled that in the wheat and barley, with the most resistant oats (New Zealand Cape, Avon, JB76-22, Swan, Cooba, OXB178, TAM-0-301, TAM-0-312) similar or better than the resistant barley, Galleon and Festiguay wheat. Resistance to CCN is not a rare trait amongst Australian oat varieties, but unfortunately it is rarely coupled with tolerance. Only New Zealand Cape combines high levels of both tolerance and resistance. It is therefore the most suitable parent for breeding improved varieties.

One other possible source of tolerance is Avena sterilis, CC 4658. It has been shown to be resistant to all known biotypes of CCN(1) and is widely used in breeding programmes in Europe. The breeders line JB76-22, which has Cc 4658 as a parent, proved to be intolerant and as Cc 4658 itself is a wild type, it is not practical to test it in the field. However, second backcross derivatives of Cc 4658 using West as the recurrent parent, entered widespread testing in 1984 in South Australia. In these trials, Cc 4658 derivatives produced similar yields to New Zealand Cape derivatives and much higher yields than West at CCN infested sites (Table 3) indicating a high level of tolerance is conferred by some Cc 4658 genes. Tolerance was rare in segregants from second backcross progeny even though resistance was common. Thus Cc 4658 offers a second source of combined resistance/tolerance to oat breeders.

TABLE 3. Yields of CCN tolerant breeders lines in advanced trials, South Australia, 1984

	Yield (kg/ha^{-1})		
Genotype	Mean over 3 CCN sites[a]	Mean over 8 low CCN sites	Tolerance index[b]
West	866	1999	43
New Zealand Cape	1558	1573	99
BC$_2$F$_7$ Cc 4658 derived Selection A	1592	2124	75
Selection B	1233	2235	55
Selection C	1689	2112	80
Selection D	1548	2237	69
Mean of 11 BC$_1$F$_7$ West NZ Cape derivatives	1497	2168	69

[a]All three sites were SIRONEM bioassay = 5.
[b]Yield at CCN sites expressed as a percentage of yield at low CCN sites.

Unfortunately, all of the recently released varieties which are adapted to southern Australia are highly intolerant (Table 4). The differential in yield potential between CCN resistant/tolerant lines such as New Zealand Cape and susceptible/intolerant varieties such as Echidna, is so large (Table 4) that New Zealand Cape will only outyield Echidna when CCN is very severe. Similar problems exist with Cc 4658 derivatives, so considerable resources must be invested before CCN resistant/tolerant varieties with high yield potential will be produced.

TABLE 4. Response of recently released oat varieties to nematicide applications, Pinery, 1983

Variety	Yield plus nematicide (kg ha^{-1})	Yield minus nematicide (kg ha^{-1})	Yield response to nematicide (% diff. from control;level of sig.)
West	3137	2517	25 ***
NZ Cape	2187	2035	7 NS
Sual	2918	2147	36 ***
Cooba	2627	2383	10 NS
Mortlock	2944	2237	32 ***
Hill	2911	2390	22 ***
Bulban	3551	2604	36 ***
Barmah	2879	2064	39 ***
Dolphin	3311	2522	31 ***
Echidna	4069	3053	33 ***

Varieties which fall into all four combinations of resistance (low vs. high) and tolerance (low vs. high), occur in the cereals examined (Table 5). No oats were found which fit the low resistance high tolerance category although preliminary evidence indicates that progeny from the first backcross between West and New Zealand Cape may fit this category (Barr and Dube, unpublished data).

TABLE 5. Resistance and tolerance of cereal varieties tested

	Resistant	Susceptible
Tolerant	Oats: NZ Cape, (West*NZ Cape)/9, (West*NZ Cape)/23 Barley: Galleon Wheat: Festiguay	Barley: Clipper
Intolerant	Oats: Avon, Swan, Tam-0-301, Tam-0-312, Cooba, OXB 178, JB76-22	Oats: West, Moore, Stout Wheat: Egret

The split plot method for estimating tolerance used for this study gave similar rankings for the oat and wheat varieties tested by other methods(3,4). It is probably a better method when:

1. CCN damage is not the major determinant of yield;
2. an estimate of actual loss due to CCN is required;
3. CCN damage is variable across the trial site;
4. in early generation of parental selection trials, few locations are used in testing

Its main disadvantage is that twice as many plots must be sown at the CCN site.

Where genotypes are being evaluated at a wide range of locations the method of Mathison(3) and Fishr et al.(4) is a better option.

REFERENCES
1. Cook R and TD Williams: Ann. appl. Biol. 71, 267-271, 1972.
2. Millikan CR: Journal of Department of Agriculture Victoria 36, p.452, 1938.
3. Mathison MJ: Aust. J. Exp. Ag. Anim. Husb. 6, 179-182, 1966.
4. Fisher JM, AJ Rathjen and AJ Dube: Aust. J. Agric. Res. 32, 545-551, 1981.
5. Gurner PS, AJ Dube and JM Fisher: Nematologica 26, 448-454, 1980.
6. McLeod RW: Proc. Linnean Soc. N.S.W. 100, 195-201, 1976.
7. O'Brien PC and JM Fisher: Aust. J. Exp. Agric. Husb. 14, 399-404, 1974.
8. Sparrow DHB and AJ Dube: Proc. 4th Int. Barley Gen. Symp., pp.410-417, 1981.
9. Simon A: Plant Dis. 64, 917-919, 1980.
10. Fisher JM: Proceedings of the European Plant Protection Organisation Colloqium on Cereal Cyt Nematodes, 12, 445-449, 1982.
11. Dixon GM: Plant Pathol. 18, 109-112, 1970.

TRANSMISSION OF BARLEY YELLOW DWARF VIRUS ISOLATES BY THE
APHID RHOPALOSIPHUM INSERTUM (WALKER)

H. JEDLINSKI* and D. VOEGTLIN**
*United States Department of Agriculture, University of Illinois
**Section of Faunistic Surveys and Insect Identification,
 Illinois Natural History Survey, Illinois, USA

1. INTRODUCTION

Barley yellow dwarf virus (BYDV) is persistently transmitted by at least
18 species of aphids throughout the world(1,2,3). There are many strains
of the virus that vary in level and spectrum of virulence, host range and
vector specificity(1,3). Because of the close taxonomic similarities of
Rhopalosiphum insertum (Walker) (RI) and R.padi (L.) (RP), the two vector
species have been confused or lumped together as a complex(2,3,4,5). In
view of their importance in the epiphytology of BYDV, we have investigated
their ability to transmit different vector-specific and non-specific BYDV
isolates.

2. MATERIALS AND METHODS

The aphid species used were RI, RP and Sitobion (=Macrosiphum) avenae
(Fab.) (SA). Fundatrigenae of RI were collected from apical leaves of
hawthorn (Crataegus sp.) in the spring before their migration to their
summer hosts. Single alates caged on 'Michigan Amber' wheat (Triticum
aestivum L. em Thell.) gave rise to colonies of virginoparae. They
initially exhibited underground habitat feeding on roots and subsequently
either a very slow (¼ of the RP rate) or a rapid multiplication (equal to
that of RP) on stems and leaves. Both types of colonies were maintained
on wheat and used in the transmission tests. The other two species were
cultured on 'Hudson' barley (Hordeum vulgare L.) as described
previously(2). The test plants were 'Coast Black' oats (Avena byzantina
Koch.). The transmission tests consisted of 2-day acquisition feeding on
detached leaves in a dark incubator at 15°C and 4-day inoculation test
feeding on seedlings in the one-leaf stage in a growth chamber supplied
with fluorescent and incandescent illumination for a 16 h daylength at
25°C. The BYDV isolates were vector nonspecific (PAV-like) of local
origin(2) and RP-specific (RPV) and SA-specific (MAV) obtained from
W.F. Rochow. All other materials and methods were essentially the same as
those reported previously(2).

3. RESULTS AND DISCUSSION

The ability of RI of slow and rapid multiplication types on wheat to
transmit different BYDV isolates was tested in two separate experiments
and compared with that of other known vector species. The results
presented in Table 1 indicate that RI of two types transmitted the
RP-specific RPV isolate and a vector nonspecific PAV isolate but failed to
transmit the SA-specific isolate MAV to 'Coast Black' oats. The
transmission rates were highest with the RPV isolate, equal to those of RP
with PAV and RPV isolates, and lowest (about 50%) with the PAV isolate.
Under the same conditions in parallel tests, vectors of specific isolates

transmitted their respective isolates efficiently. None of the 48 plants serving as aphid controls and the 19 plants as plant controls became infected in these experiments.

TABLE 1. Comparative transmissibility of three barley yellow dwarf virus (BYDV) isolates by Rhopalosiphum insertum(RI), R.padi(RP) and Sitobion avenae(SA) to 'Coast Black' oats

BYDV isolate	No. of plants infected/infested			
	RI_1	RI_2	RP	SA
PAV (vector nonspecific)	7/12	3/12	12/12	9/12
RPV (RP specific)	12/12	12/12	12/12	0/12
MAV (SA specific)	0/12	0/12	0/12	12/12
Aphid control	0/12	0/12	0/12	0/12
Plant control	0/19			

RI_1 - slow reproduction on Michigan Amber wheat
RI_2 - normal reproduction on Michigan Amber wheat
Alate originated from hawthorn (Crataegus sp.)

4. DISCUSSION

The genus Rhopalosiphum includes aphid species important as pests and vectors of luteoviruses, of which BYDV is a member(2,3). RI was shown to be an efficient vector of BYDV. It efficiently transmitted RPV and PAV isolates, but failed to transmit the MAV isolate. The transmission pattern is similar to that of the rice root aphid R.rufiabdominalis Sasake, reported previously(2,6). These results strengthen the contention expressed by Paliwal(6) that RPV-like isolates should no longer be considered truly vector-specific since they are readily transmitted by three different species. From the vector standpoint it focuses attention on Doncaster's(7) contention that these species may be related.

In 1855, Fitch(8) named Aphis prunifoliae (=padi) from specimens taken on Prunus. This name became widely used for aphids from both Prunoidea and Pomoidea and thoroughly confused scientists over the next 60 years. Baker and Turner(9) stated that the aphid 'winters upon the apple and migrates to grains and grasses'. Theobald's reference to prunifoliae(10) confused padi and insertum and Borner(11) suggested that what was known by this name was probably two species. Rogerson(12) in a very detailed paper sorted the species into padi (migrating from Prunus padus L., P.persica L. and P.virginiana L. to oats, wheat, barley and Poa annua L.) and crataegellum (=insertum) (migrating from Pyrus malus L., other species of Pyrus, Crataegus, Cotoneaster, Mespilus and Cydonia to various Gramineae including Poa annua). Palmer(13) used the name fitchii(Sanderson) for the apple grass aphid and Hille Ris Lambers(14) synonymised fitchii with insertum. This was not accepted or used by scientists of North America(4,15) until recently.

In retrospect it is easy to understand the confusion generated by these two species. They are extremely similar morphologically both alive and on slide preparations, and they can be cultivated on the same secondary hosts. The work of Sun and Robinson(16) demonstrated that padi has eight chromosomes and insertum has ten and provided an unambiguous way to separate them. The results of the field work of Robinson and Chen(15), using this distinction to name the aphids they collected, are most interesting. Throughout July all the aphids they cllected in the field were padi even though there had been a substantial spring build up of insertum on Crataegus in the area. In late August, when they located ten colonies on the roots of barley, five were padi and five were insertum. The tendency of insertum to feed on the lower stem or below ground level on its host has been noted several times in the literature(9,12,14,15) but most of this work has been done on cereals. We hypothesise that insertum may not preferentially select cereals in a natural setting while padi may select cereals. This could account for the lack of records of insertum from summer hosts(14,15,17). Different fecundity rates observed in our transmission study may reflect different degrees of adaptation to cereals among different clones of insertum. This hypothesis would also help to explain the results obtained in the transmission experiments discussed above. The RPV isolate is transmitted very efficiently by insertum but is rarely found in cereals(18). This suggests that there is very little RPV movement from grasses to cereals by this species and observations indicated that transmission occurs primarily in late summer after insertum has built up on grasses and produced alatae. It may be concluded that the biology and population dynamics of RI require more concentrated study in relation to BYDV epiphytology before the whole picture is fully understood. Below are some of the synonyms and combinations of names used for RI and RP:

Rhopalosiphum insertum (Walker)

Aphis crataegi Buckton
Aphis fitchii Sanderson
Rhopalosiphum prunifoliae (Fitch)
Rhopalosiphum crataegellum (Buckton)
Rhopalosiphum fitchii (Sanderson)
Rhopalosiphum viridis Richards

Rhopalosiphum padi (L.)

Aphis padi Linnaeus
Aphis prunifoliae Fitch
Aphis avenae Theobald
Aphis annuae Oestlund
Siphocoryne avenae Fabricius
Aphis psuedoavenae Patch

REFERENCES

1. Burnett PA: Barley yellow dwarf. A Proceedings of the Workshop. CIMMYT, 1984.
2. Jedlinski H: Plant disease 65, 975-978, 1981.
3. Rochow WF and JE Duffus: Luteoviruses and yellows diseases. In, E Kurstak (ed.), Handbook of Plant Virus Infections and Comparative Diagnosis. Amsterdam : Elsevier/North Holland, 1981.
4. Bruehl GW: Barley yellow dwarf, a virus disease of cereals and grasses. Monograph No. 1. The American Phytopathological Society, 1961.
5. Orlob GB: Entomology Experimental and Applied 4, 62-72, 1961.
6. Paliwal YC: Can. J. Plant Path. 2, 90-92, 1980.
7. Doncaster JP: Bulletine Entomological Research 47, 741-747, 1956.
8. Fitch A: Transactions of New York Agricultural Society 14, 826-828, 1855.
9. Baker AC and WF Turner: J. Agric. Res. 18, 311-324, 1919.
10. Theobald FV: The plant lice or aphididae of Great Britain. Headley Brothers, London, 1927.
11. Borner C: Anzeiger Schadlingskunde 7, 42-43, 1931.

12. Rogerson JP: Bulletin of Entomological Research 38, 157-176, 1947.
13. Palmer MA: Aphids of the Rocky Mountain region. The Thomas Say Foundation 5, 1952.
14. Hille Ris Lambers D: Virology 12, 487-488, 1960.
15. Robinson AG and YH Chen: Can. Entomologist 101, 110-112, 1969.
16. Sun RY and AG Robinson. Can. J. Zoology 44, 649-653, 1966.
17. Richards WR: A synopsis of the genus Rhopalosiphum in Canada (Homoptera:Aphididae). Canadian Entomologist Supplement 13, 1960.
18. Fargette D, RM Lister and EL Hood: Plant Disease 66, 1041-1045, 1982.

OAT CROWN RUST (PUCCINIA CORONATA AVENAE) RACES IDENTIFIED IN THE
WEST REGION OF THE USSR

B. P. NAMAJŪNAS and M. T. STRUKČINSKAS

Institute of Botany of the Academy of Sciences of the Lithuanian SSR,
Vilnius, Turistu 47, Lithuanian SSR, USSR 232021.

1. INTRODUCTION

Oat crown rust (Puccinia coronata f.sp. avenae Fr. et Led.) can be one
of the causes of reductions in grain yield and quality of oat crops in
many parts of the world, and this is the case in the west region of the
USSR.

In natural environments, the oat crown rust pathogen is widely
distributed as mixed populations consisting of a number of races which
infect different oat genotypes. Therefore, in this country, the
development of new oat cultivars is closely associated with studies of
pathogen race structure and its geographical spread, so as to enable the
successful development of disease resistant cultivars.

The objective of this work has been to investigate the oat crown rust
races prevalent in Lithuania, Latvia, Estonia, Bielorussia and the
Kaliningrad Region (the RSFSR).

2. MATERIALS AND METHODS

Characterisation of the pathogen race structure of oat crown rust was
carried out on a set of monogenous lines in the variety Pendek(1). More
than 300 pathogen isolates were tested. The type of reaction was
determined according to the well known scale of Stakman et al. as modified
for studies of oat rust by Murphy(2).

Crown rust samples were taken from various varieties and commercial
crops in 57 administrative regions in Lithuania, Latvia, Estonia,
Bielorussia and the Kaliningrad Region, over the period 1981-84.

3. RESULTS

Preliminary data for the period 1981-1984, indicate that on the set of
monogenous lines in the variety Pendek, 53 combinations of virulences have
been identified. The races are: c1, c2, c3, c9, c14, c26, c74, c153,
c258, c259, c265, c266, c268, c269, c274, c276, c281, c297, c299, c301,
c313, c315, c521, c522, c524, c525, c528, c537, c586, c587, c769, c770,
c771, c772, c777, c778, c780, c784, c793, c796, c800, c801, c802, c809,
c811, c812, c843, c844, c859, c874, c905, c908 and c910.

This result demonstrates the high differentiating quality of the test
set. The populations were dominated by races having 0-3 virulence genes,
which made up more than 50% of the populations investigated. The most
aggressive races, having 6-7 virulence genes, occurred rarely and
comprised only 10% of the pathogen populations.

Analysis of the data indicates that the most widely distributed races
were with the virulence genes Pc 35, Pc 38 and Pc 47.

Up to the present time, the rust resistant genes present in the monogenous lines have not been used in cultivated oat varieties. However, virulence genes in the rust pathogen population have been found which correspond to nearly all the resistance genes in the differential set(3). Therefore, various virulence genes are already present in populations of the pathogen before contact is made with the corresponding resistance genes in the host. By this means the preservation of the pathogen species is ensured.

REFERENCES

1. Fleischman G and PJ Baker: Can. J. Bot. 49, 1433-1437, 1971.
2. Murphy HC: Tech. Bull. US. Dep. Agric. N433, 1-48, 1935.
3. Chyptsova OE: Author's Precis, Candidate Dissertation, All-Union Scientific Research Institute of Plant Protection, Leningrad, 1981.

SESSION II. PESTS AND DISEASES OF OATS
CHAIRMAN'S COMMENTS AND SUMMING UP

J. SEBESTA
Research Institute for Crop Production, Prague, Czechoslovakia

Oats are attacked by a number of diseases and pests which may limit oat grain and forage production. In addition oat quality may be influenced by pathogens.

The chief diseases are crown and stem rusts, powdery mildew, smuts, Septoria blight, Helminthosporium leaf blotch, BYDV, root rots and blast. Of the pests, frit fly (Oscinella frit) and cereal cyst nematodes are reported to be of greatest importance.

Following the comprehensive introductory paper of Simons, which described the incidence and possible control of a wide range of oat diseases, other papers dealt with specific topics which can be broadly grouped into (a) prevalence of rusts and mildew, mainly in Europe and America, virulence spectra and sources of resistance, (b) the nature of horizontal (slow rusting) resistance of oats to crown rust (c) the relationship between phytoalexin production and resistance to crown rust (d) transgressive segregation for enhanced levels of adult plant resistance to mildew (e) transmission of BYDV isolates by Rhopalosiphum insertum and (f) resistance/tolerance in oats to cereal cyst nematodes.

Most of the discussion was concerned with the following aspects:

1. Problems of virulence in crown and stem rusts, methods of analysis, and the effectiveness of rust resistant sources.

2. Use of partial resistance to mildew with major gene or systemic fungicide seed treatments.

3. The nature of horizontal resistance (slow rusting) to crown rust and the possibility of detecting mechanisms responsible for its expression.

4. Problems in detecting tolerance to disease.

Clifford pointed out that the UK Virulence Survey has shown that crown rust occurs very sporadically in the British Isles, seriously affecting crops only in South West England and South Wales in perhaps one year in ten. The data on changes of virulence patterns are consequently very limited. Rhamnus cathartica does not play any role in the production of new virulence combinations as it has been eradicated for almost half a century in Britain. Also the varieties grown in the UK probably did not play an important role in changes of crown rust virulence.

The problems of stem rust virulence on the European continent and the methods of analysis were given some attention. The basis for the annual surveys in oats was created through the establishment by Sebesta and Zwatz of the European Oat Disease Nursery (EODN), for the rusts in 1969, and for mildew in 1978. This nursery is now grown at 28 localities in 10 countries of the continent, namely, Austria, Czechoslovakia,

German Democratic Republic, Federal Republic of Germany, Greece, Hungary, Poland, Spain, USSR, Yugoslavia and the United Kingdom. The EODN sets consist of stem and crown rust differentials and powdery mildew sources of resistance. However, other oat diseases have been recorded when present and these, according to Sebesta, have chiefly been <u>Septoria</u> blight, Helminthosporium leaf blotch and Barley Yellow Dwarf Virus.

Virulence analysis of crown and stem rusts is carried out at the seedling stage. On the Pc- and Pg-lines sets, the rust isolates are differentiated into avirulence/virulence combinations using the formula method described by Green. The percentage of isolates from each country, which show virulence on each Pc- and Pg-line is taken as a criterion for breeding importance. Evaluation of severity is according to Cobb's scale for the rusts and the James' scale for mildew.

Virulence on the Pg-13 and Pg-16 resistances was found in one stem rust isolate from Poland in 1978.

The origin of the two major genes identified in PC 50-2 and PC 50-4 is not clear. It may be that both originate from <u>A.sterilis</u> L. CW 486-1, but the second gene could not be detected owing to virulence to it. Another possible explanation is that the second gene appeared in the PC 50 population through spontaneous hybridization.

Jones explained that in seasons when the incidence of mildew is relatively low, the higher levels of partial, adult plant resistance now becoming available can prove sufficient to protect oats with its comparatively low hectarage. However, to ensure protection at the earlier stages, necessary in high mildew years, major gene resistances are being incorporated in backgrounds with the durable type partial resistance. This would also avoid the 'vertifolia' effect that can occur when corresponding virulence becomes prevalent to the major gene resistance. Field experimentation has shown that another alternative is to use certain new systemic fungicide seed treatments, which protect the early growth stages until the adult plant resistance becomes effective. The latter reduces the need for further spray treatments, which add to input costs and the problems of pollution.

With reference to horizontal resistance (slow rusting) to crown rust, Luke reported that resistance in the variety Red Rustproof consists of morphological components, such as exclusion or resistance to penetration but also some physiological attributes. Both types could be combined to form very effective barriers to the pathogen.

The problem of recognising 'true' tolerance assayed on the basis of yield reduction under disease attack compared with disease free plots is a continuing one. The difficulty of distinguishing between losses due to tolerance and those resulting from some degree of resistance was pointed out, although Clifford referred to some earlier experiments which indicated that slow rusting and tolerance can be differentiated. Satisfactory identification, by Barr and Dube, of genotypes with both resistance and tolerance to cereal cyst nematode appears very encouraging provided their yield potential can be raised to acceptable levels.

Hypersensitive, major gene controlled resistance is still being extensively utilized as defence against disease, especially in the rusts. It would appear, however, that the trend is for more complex forms of resistance produced through combination of individual specific resistances or the latter combined with minor or modifying genes. These, as Rothman reported, are providing adequate protection for all the known North American oat stem rust races. With mildew there is a paucity of major gene resistances and those used have given only ephemeral protection, hence emphasis has been placed on identifying types of adult plant resistance which have a history of durability.

Further research efforts should be aimed at elucidating the underlying biochemical and physiological mechanisms that collectively give stable resistance. Various interesting techniques like fluorescence microscopy (Luke) and phytoalexin production (Morikawa and Mayama) may in the meantime not only give important basic information but also useful resistance to use against specific oat diseases.

Nevertheless, monitoring of changes in virulence and effectiveness of horizontal partial resistance still requires the use of standard disease nurseries grown under field conditions at numerous sites. These provide fairly reliable data, bring about the exchange of germplasm between workers in different countries and foster international cooperation in the fight against diseases. Control of pests, was not given as much attention in this session but work such as that reported by Jedlinski on transmission rates and multiplication of various BYDV isolates is of great practical importance in understanding the epidemiology of this serious pest.

The 2nd International Oat Conference has been a unique opportunity to exchange and discuss the latest knowledge about the present problems of disease and pest control and how best they can be solved.

SESSION III

Efficiencies of breeding methods for oats

Chairman: V.D. Burrows

DEVELOPMENT OF OAT BREEDING METHODS

DEON D. STUTHMAN
Department of Agronomy and Plant Genetics,
University of Minnesota, St. Paul, MN 55108, USA

My topic is the development of oat and plant breeding methods. I will start with a short history of important ideas and reports, laying the ground work for current oat breeding approaches. The bulk of my report will describe and discuss, in a critical way, the philosophies and approaches currently in use and the technology that may become available to the breeder in the future.

1. INTRODUCTION

Oats have been cultivated for nearly 2000 years(1). Presumably, during most of that time oat breeding efforts were limited to mass selection. Commonly, after harvest and threshing, the grain was cleaned and the heaviest grains saved to plant the next season's crop. This process produced what are referred to as land races, i.e. varieties with local adaptation. This procedure has many limitations for long term improvement, especially for self-pollinated crops. However, considering the lack of knowledge of genetics before 1900, these attempts represented a reasonable approach to plant improvement.

With the report on progeny testing by Vilmorin in 1850 oat breeders acquired another option, that of plant selection within the land races(2). This approach was commonly used in the last part of the nineteenth and the early part of the twentieth century. It produced several widely grown varieties including Swedish Select and Sixty Day(3). However this procedure also has limitations, especially for crops which have insufficient outcrossing to produce new combinations of genes. Once the plant breeder identified the most productive individuals within the land race, additional selection within the progeny of a superior plant selection would not result in additional gain.

2. QUALITATIVE APPROACH

Three discoveries, all reported in the first decade of the twentieth century(2), transformed oat breeding efforts dramatically. First, the rediscovery of Mendel's work provided much needed information on how traits are passed from one generation to the next. The second, Johansen's "pure line theory", provided the explanation as to why continued plant selection would not produce additional improvement. The third was Biffen's report that genes, as described by Mendel, controlled resistance to some diseases, specifically the rusts.

The first two reports soon led oat breeders to realise that hybridisation between pure lines would produce progeny with the needed variability to allow selection to be effective. The report on genes controlling disease resistance, combined with the newly rediscovered Mendelian genetics, enabled oat breeders to devise a reliable and efficient procedure for handling the segregating populations resulting from the crosses they were making. For nearly a half century hybridisation followed by rust screening of seedlings in the early segregating generations dominated oat breeding everywhere that rusts were a problem.

One other paper that added to this approach was the report of Barrus on the physiological races within a pathogen(2). He explained why genotypes previously resistant could, over time, lose their ability to provide protection against certain forms of a pathogen. For the first time breeders began to understand the complexities in breeding resistant varieties. The race between the pathogen and the breeder had started.

The emphasis on selection for disease resistance in the early segregating generations worked well during this time. Some disease epidemics occurred, but many others were prevented by growing resistant varieties. Thus, this defensive approach resulted in valuable contributions to crop production mainly by reducing or eliminating potential losses. However, productivity increases in spring oats due to genetic improvement in the absence of diseases, at least in the midwestern United States, were minimal. This lack of gain in productivity among new cultivars is evidenced by the fact that until recently 'Gopher', the long term check in the American Uniform Midseason Oat Performance Nursery coordinated by the USDA, was exceeded in yield by few newer selections(4).

3. QUANTITATIVE APPROACH

Complementing the discoveries on simply inherited traits such as disease resistance were important early contributions to the understanding of the inheritance of traits controlled by multiple loci(2). Following a report by Nilsson-Ehle on multiple loci affecting oat kernel colour was the classical report by C.E.M. East in 1916 on the genetics of corolla length in tobacco. Several years later, R.A. Fischer wrote two paper(5,6) which also had considerable impact on plant research, genetics and breeding. One of these described correlations among relatives and the other dealt with statistical methods. The first of these two Fischer papers, along with the East paper, was the beginning of quantitative and population genetics on which oat breeders rely today. The second Fischer paper was the forerunner of the Analysis of Variance and various other experimental design concepts. Together these papers helped form the beginnings of the quantitative approach to oat breeding. This approach ultimately would encourage emphasis on the productive potential of the crop rather than merely protecting the crop from disease, i.e. an offensive as well as a defensive effort.

Another major contribution to the quantitative approach to plant breeding was Hull's (1945) idea of cyclic or recurrent selection(7). Initially this procedure was confined to cross-pollinated crops, but recently it has also been more commonly used in self-pollinated species including oats. Further, many oat breeders use cyclic crossing in their breeding programmes even if they do not use recurrent selection systematically.

The classical work of Dr J.E. Grafius made another important contribution to the quantitative approach to oat breeding. His delineation of the components of yield, number of panicles, seeds per panicle, and seed weight as well as their interrelationships provided much needed information on how best to manipulate the oat plant at this component level(8). A related proposal, that of the ideal ideotype by C.M. Donald, went beyond yield components(9). Based on known or presumed relationships between individual plant parts and grain yield, Donald described in morphological and physiological terms what he believed would constitute a high yielding genotype. The ideas on enhancing yield put forward by Grafius, Donald and others, plus the need to increase yield to make oats more competitive with other crops, have been primarily responsible for a major change in attitude among some oat breeders who feel the need to take a more aggressive (offensive) approach to yield improvement.

4. BREEDING STRATEGY ISSUES

Today's oat breeder has two basic sets of decisions to make when designing a breeding programme. The first involves selection of parents. The second concerns the choice of procedures for selection among progeny. Effective parental selection requires priority rankings among traits that need improvement. Obviously, needed genes must exist in one of the two parents for there to be a chance of finding the desired trait in the progeny. Because genetic resources have been discussed in previous papers at this conference, parental selection will not be discussed further here.

The choice of selection procedures may be considered by addressing four basic issues: 1) type of selection in early segregating generations; 2) sequential vs simultaneous treatment of the several characteristics important in variety development; 3) indirect vs direct vs partial approaches, especially for complex traits; and 4) the nature and extent of the selection environment.

Although the topic of early generation selection in self-pollinated crops has been the subject of numerous investigations, little, if any, consensus has developed. Three distinct options for handling the progeny immediately after hybridisation exist: no selection, natural selection and artificial selection by the breeder. These three choices relate to single seed descent(SSD), bulk, and pedigree and/or backcross breeding methods, respectively. As indicated earlier most oat breeding programmes traditionally have utilised either the pedigree or the backcross approach to select for disease (rust) resistance. Resistance to one or more diseases was usually given priority emphasis in the development of new cultivars. Recently, for various reasons oat breeders have increasingly adopted other approaches with SSD becoming a more common procedure. A successful bulk/mass selection effort as an alternative approach is described elsewhere in this session.

Because there are several traits which are important to the success of any oat variety programme, the issue of how to handle multiple characters is always a critical one. Selection which emphasises a single trait will usually result in changes in related traits. Many times these changes are in undesired directions and thus cannot be ignored. A second consequence of single trait selection might be reduced variability for one or more of the unselected, but important, traits. Some oat breeders suggest that early selection for disease resistance removes potentially high yielding genotypes prematurely. Thus a sequential approach requires careful consideration of the priority of traits(10).

Many different selection indices, some simple and others sophisticated, have been proposed to simultaneously deal with two or more important characters. In their most complex forms these indices consider relative economic value, heritability of individual traits, and relationships among traits collectively(11). Considerable effort is required to collect the necessary data and, until availability of computers, calculations were also time consuming. Even with ready availability of computers oat breeders are seemingly not using the complex selection indices, suggesting that they are not viewed to be cost effective. Further, breeders who use some form of index selection are opting for the simplest approaches such as obtaining an index value by simply adding or multiplying their different phenotypic values together.

The third issue, that of direct vs either indirect or partial (or component) approach, is particularly relevant for complex traits such as grain yield(12). As a further complicating consideration of this issue, it may be effective to use a combination of methods such as direct approach when selecting among progeny but a component approach when selecting parents. In particular, the apparent simplicity of the component of yield approach is attractive; however, component compensation usually reduces the rate of gain for grain yield when selection for only a single component among progeny is emphasised.

Every plant breeder, and oat breeders are no exception, would like to find a simple character which when manipulated would cause desired gain in a more complex character such as yield. Simple characters studied include ones which appear to contribute directly to the more complex character as well as those for which a relationship, albeit not a cause and effect one, seems to exist. The theory and/or hope is that modification of a more simply inherited trait, one that is easier to manipulate, will concurrently produce a desired change in the more complex trait. It seems reasonable to work with simpler traits which make a direct contribution to the more complex trait. However, the best example of such an approach, the ideal plant type, has not generally provided the desired results that its advocates had hoped. The ideotype concept has made a contribution, nonetheless, because it has focussed more attention on breeding for yield per se, or what I referred to earlier as an aggressive approach.

The fourth issue, that of the nature and the diversity of the selection environment is, by circumstance, situation specific. A breeder has to choose between using the environments as they exist or modifying them to effect a desired selection pressure. Examples of modifications include 1) irrigation, 2) soil nutrient levels either greater or lesser than those used for commercial production, 3) introduction of plant pests, and 4) using lower or higher population densities from those used to produce the crop. Such modifications may facilitate the identification of genotypes which will perform better in production systems by providing selective advantages to the desired genotypes. Regardless, it is essential that the selection environment predict the production environment(13).

Related to the issue of selection environment is that of the diversity within the target environment. With increasing heterogeneity it may well be advisable to divide the breeding programme into distinct parts at least during the selection phase. This decision should be based on the relative size of the genetic variance and that of the genotype by environment interaction. When those two values approach equality, the breeder will be forced to divide the programme to maintain genetic gain.

5. OGLE - A CASE STUDY

Although I have not yet revealed my personal specific choices on the four selection issues raised, an examination of the development of the cultivar 'Ogle' will be instructive. In the short time since it was released in 1981 by C.M. Brown and colleagues at the University of Illinois(14), Ogle has become the dominant spring oat variety in the U.S. and in parts of Canada. It does have good tolerance to barley yellow dwarf virus(BYDV) but is susceptible to leaf and stem rust and to many races of smut. Its bushel weight also is not high nor is its kernel especially plump. Clearly the principle reason Ogle is so widely grown across the eastern and central portions of the U.S. is its productivity. No oat cultivar and few cultivars in any other field crop have ever yielded so well over such a wide area of production. All of these performance records by a cultivar with a disease resistance profile which is questionable according to historical standards are noteworthy.

The procedure used to select Ogle is as unusual as its performance. Ogle is from a cross of Brave, Tyler and Egdolon No.23, none of which has been widely grown. According to Dr Brown(15), yield is one of the first traits evaluated in their system. Their approach differs from the traditional approach which I described earlier in which selection for yield is delayed. The Illinois system begins with planting single seeds from F_2 plants in 6 inch-clay glasshouse pots using a density such that each plant usually produces only one seed. The seeds are bulk harvested per pot and the process repeated one or two generations. All of the seed from the last harvest for a cross is planted in one or more ten foot rows. Up to several hundred random panicles are taken from these rows and threshed individually. Seed samples are evaluated and those not meeting minimum quality standards are eliminated. The following season panicle rows are grown with bulk harvesting of individual rows which are uniform. Three or four panicles are resampled from those rows which are not uniform. The following year the bulk sample is planted into an unreplicated yield plot consisting of four 10-foot rows. Smaller seed samples are also planted for crown rust and BYDV evaluations. If there is sufficient seed, a second replicate is planted for yield at a second location. All plots are planted in blocks of 20 with three common high yielding check varieties randomly distributed within each block. Less than one-half of these plots are harvested, with lodging resistance receiving considerable emphasis. Selection among those harvested is based primarily on plot yield and bushel weight while knowing the crown rust and BYDV reaction. A minimum of two additional years of two location testing is done before selections are entered into the USDA regional trials.

I would challenge you to think about whether another Ogle is more likely to be identified in your programme or the one used at the University of Illinois. Because of declining oat acreages more highly productive varieties like Ogle are essential if oats is to compete effectively with other crops.

6. MINNESOTA EFFORTS

Our programme at Minnesota is also what I have described earlier as more aggressive for yield improvement. We use SSD on crosses which have high potential for variety development: however, we do not use the high density plantings used at Illinois. By crossing in the glasshouse in November, growing F_2s in the field the following summer, and using SSD for two generations during the next fall and winter, we are growing F_5 seed increase rows the following summer (less than two years after the cross is made). Yield testing of F_4-derived F_6 lines in replicated hill plots is initiated the following summer (less than three years after making the cross). We eliminate low yielding lines (up to one-half of a population) using yield data from the hill plots and visual ratings of seed samples. All disease evaluations (crown rust and smut in particular) are initiated when remaining lines are first grown in multiple row yield plots (no more than five years after making the cross). These evaluations are done in the field on adult plants. Any genotype which is not fully susceptible would meet the minimum disease criteria for additional testing in the programme.

7. FUTURE CONSIDERATIONS

What about the future? I will finish with a few words about biotechnology and the potential assistance in four specific areas that appear most promising to aid in the improvement of crop plants. An obvious possibility of biotechnology is that of producing new gene combinations that are not possible by sexual means(16), e.g. the transfer of a maize gene to an oat genotype. A second related possibility is that of directed mutation, again in effect producing variability not now readily available. Some breeders argue that more variability exists now than can be fully utilised; however, if perfected, either of these suggestions would produce new genotypes which would better fit particular needs of the crop.

A third potential use would be to skew the distribution of progeny to something more favourable than random selection through some cellular or molecular manipulation. The possibility, for example, of the doubled haploid method for instantaneous homozygosity does not provide much of an advantage over SSD whenever two or three generations between field seasons in a glasshouse are available. If, however, after doubling the haploids, the frequency of desired genotypes was double that expected from random probability, oat breeders would anxiously await the details of how best to produce haploids in Avena.

A final useful outcome of current molecular and cellular efforts would be an improved understanding of genetic concepts critical to the success of plant breeding. Two examples are provided for illustrative purposes: 1) nature and function of repetitive DNA; 2) distinctive molecular markers on chromosomes so the parental chromosome contribution in progeny from wide crosses could be determined. In the first example classical genetics is not sufficient to explain repetitive DNA and a better appreciation of it might help explain a number of phenomenon which are now poorly understood. In the second, in a genus such as Avena where chromosome mapping is almost non-existent, molecular markers might well provide useful information about the connection between physical and genetic details of oat chromosomes. Only time will tell whether any of the items mentioned above will come to fruition and become useful in the improvement of the oat crop.

8. SUMMARY

In conclusion, ultimately the successful oat breeder will: 1) choose parents capable of producing progeny (using either sexual means or biotechnology) that meet programme objectives which are based on the needs of the crop in the production area of concern; and 2) match the appropriate selection method with the trait(s) to be improved and the breeding nursery circumstances. Both kinds of choices will obviously have to take into account the available resources and the timetable desired by producers.

REFERENCES

1. Coffman FA: In, FA Coffman (ed.), Oat and Oat Improvement. American Society of Agronomy, Madison, WI, pp.15-17, 1961.
2. Allard RW: Principles of Plant Breeding. John Wiley & Sons, Inc., New York, 1966.
3. Carleton MA: In, LH Bailey (ed.), Small Grains. MacMillan Company, New York, 1916.
4. Rodgers DM, JP Murphy and KJ Frey: Crop Science 23, 737-740, 1983.
5. Fischer RA: Trans. Royal Soc. Edinburgh 52, 399-433, 1918.

6. Fischer RA: Phil. Trans. Royal Soc. 222, 309-368, 1922.
7. Hull FH: J. Amer. Soc. Agron. 37, 134-145, 1945.
8. Grafius JE: Agron. J. 48, 419-423, 1956.
9. Donald CM: Euphytica 17, 395-403, 1968.
10. Habgood RM and BC Clifford, In, JF Jenkyn and RT Plumb (eds), Strategies for the Control of Cereal Diseases. Blackwell Scientific, Oxford, pp.15-25, 1981.
11. Hazel LN: Genetics 28, 476-490, 1943.
12. Helsel DB and RK Skrdla: Z. Pflanzenzuchtg. 90, 316-323, 1983.
13. Pederson DG and AJ Rathjen: Eighth Annual Australian Plant Breeding Conference, Adelaide, South Australia, 1983.
14. Brown CM and H Jedlinski: Crop Science 23, p.1012, 1983.
15. Brown CM: Personal communication, 1985.
16. Peacock WJ: Eighth Annual Australian Plant Breeding Conference, Adelaide, South Australia, 1983.

INHERITANCE OF PRIMARY:SECONDARY SEED WEIGHT RATIOS AND SECONDARY SEED WEIGHT IN OATS

A.C. TIBELIUS and H.R. KLINCK

Plant Science Department, Macdonald College of McGill University, Ste-Anne-de-Bellevue, Quebec, H9X 1C0, Canada.

1. INTRODUCTION

A low primary:secondary seed weight ratio in oats is desirable in that it reflects uniformity of seed size. Combined with high grain weights, this has advantages for all aspects of the oat industry. The potential for improving the ratio can be investigated using segregation patterns and sub-division of hereditary variance, both being useful tools for providing information on the inheritance of quantitative characters.

Improvement of the primary:secondary seed weight ratio implies a relative increase in secondary grain weight. For selection purposes, measurement of the weight of secondary seeds alone would be simpler than measurement of the ratio and, for this reason, its inheritance is also considered.

2. MATERIALS AND METHODS

Six genotypes providing a range of secondary grain weights and primary:secondary seed weight ratios were studied (Table 1). The genotypes were crossed in a half diallel pattern, utilising the controlled environment growth chambers at Macdonald College. The F_2 generation was grown in the glasshouse, and the F_3 in the field at Macdonald College during the summer of 1982.

Randomised complete block experiments with four replications, comprising the parents, F_2 and F_3, were seeded at Macdonald College and at Joliette, a regional test site 120 km to the northeast. Plots were hand-seeded and consisted of a single row, two meters in length, with seeds spaced at 6.5 cm intervals within the row.

TABLE 1. Thousand-grain weights of secondary seeds and primary:secondary seed weight ratios for parental genotypes of oats at two locations

Genotype	1000-grain weight (g)		Primary:secondary seed weight ratio	
	Macdonald	Joliette	Macdonald	Joliette
Ajax	22.9[e]	23.8[e]	1.50[b]	1.58[b]
Forward	32.4[a]	33.3[a]	1.51[b]	1.60[ab]
Nelson	26.7[d]	25.9[d]	1.41[c]	1.49[c]
Rodney	28.3[c]	28.2[c]	1.56[a]	1.64[a]
Roxton	32.0[ab]	30.8[b]	1.59[ab]	1.65[a]
X2078-1	30.7[b]	28.2[c]	1.52[ab]	1.58[b]

[a-e]Values within a column followed by the same letter are not significantly different (P < 0.05) by Duncan's mutiple range test

At maturity, single culms from each of ten plants per plot were hand-threshed. Primary, secondary and tertiary seeds were separated, counted and weighed. Mean values for the primary:secondary seed weight ratios and secondary grain weights were calculated for each plot.

3. RESULTS AND DISCUSSION

The analysis of variance at each location indicated significant effects of genotype on the primary:secondary seed weight ratio (Table 1). At both locations, Nelson had the lowest ratio. Rodney and Roxton tended to have high ratios. Values for the other parental genotypes were intermediate.

The mean values for the F_2 and F_3 generations relative to their parents were used to get a picture of whether additive or non-additive genetic effects were more important in the inheritance of the ratio. In some crosses the progeny means approached the mid-parent values, indicating additive inheritance, while in others, the progeny means approached one or other of the parental means, indicating non-additive inheritance. While both additive and non-additive genetic effects were evident, the general pattern indicative of dominance, where the F_2 generation is similar to one parent and the F_3 is closer to the mid-parent, was not common.

F_2 and F_3 progeny means did not differ significantly in any of the crosses. This pattern is generally associated with additive gene action, since dominance decreases with inbreeding[1]. Epistatic variation, while it may decrease with inbreeding, is partially fixable. The non-additive effects evident in the segregation patterns of this material are likely due to epistasis. Epistatic gene action is dependent on specific gene combinations and not only is influenced by the environment but may be positive or negative in effect[2]. This may explain some of the variation that was observed between locations and crosses.

Transgressive segregation was considered to have occurred when a progeny had a primary:secondary seed weight ratio 5% lower than that of the low parental extreme. Transgressive segregates with low ratios were found in six crosses, indicating that selections may be made to lower the ratio.

General combining ability (GCA) and specific combining ability (SCA) effects were calculated using the analysis described by Griffing[3]. Parents were not included in the analysis and the model was considered fixed (i.e. results apply only to the six genotypes studied). GCA is used to denote the average performance of a parent in hybrid combination and it is generally felt that additive gene action accounts for most of the GCA variance[4,5]. SCA applies to the interaction of two specified parents in hybrid combination, and non-additive gene action (dominance and epistasis) is felt to account for most of the SCA variance. However, the presence of epistasis can contribute to the estimates of additive and dominance effects such that the GCA variance does not represent merely additive gene action[6,7].

GCA mean squares for the ratio were significant for the F_2 and F_3 generations at both locations, indicating significant differences among the six parents (Table 2). In contrast, the SCA mean square was significant only in F_3 at Macdonald, suggesting that the major part of the genetic variability for the ratio was associated with GCA or additive gene action. That the estimates of GCA variance are invariably higher than those for SCA variance and that the ratio of GCA sums of squares to total genetic sums of squares is always high (0.77-0.97) reinforces the relative predominance of GCA in the establishment of the ratio (Table 3).

Broad and narrow sense heritability values were estimated using the GCA and SCA variance components[3]. If the mode of gene action is largely

TABLE 2. Analysis of variance of GCA and SCA effects for
primary:secondary seed weight ratios in oats

Source	d.f.	MS	F value
Macdonald - F_2 - GCA	5	3.40×10^{-3}	15.47**
SCA	9	6.67×10^{-5}	0.30
F_3 - GCA	5	4.68×10^{-3}	17.02**
SCA	9	6.16×10^{-4}	2.24*
Joliette - F_2 - GCA	5	4.98×10^{-3}	15.31**
⋰ SCA	9	2.28×10^{-4}	0.70
F_3 - GCA	5	4.80×10^{-3}	10.69**
- SCA	9	8.00×10^{-4}	1.73

*, **Significant at the 0.05 and 0.01 levels, respectively

additive, broad and narrow sense estimates will be similar(8).
Heritability estimates for the ratio ranged from 0.71 to 0.91, with the
differences between broad and narrow sense estimates ranging from 0.20 to
0.00 (Table 3). The value of zero was found for the F_2 generation at
Joliette where the SCA variance was negative and therefore equated to
zero. Narrow sense estimates were not found to increase in F_3 at Joliette
as would be expected if dominance was important(9). The increase at
Macdonald was not substantial.

TABLE 3. GCA and SCA variances, ratio of GCA SS to total SS and heritability
for primary:secondary seed weight ratios in oats

Location	Generation	GCA variance	SCA variance	GCA SS Total genetic SS	Heritability Narrow sense	Broad sense
Macdonald	F_2	8.33×10^{-4}	4.47×10^{-4}	0.97	0.71	0.91
	F_3	1.02×10^{-3}	3.41×10^{-4}	0.81	0.77	0.90
Joliette	F_2	1.19×10^{-3}	0.00	0.92	0.88	0.88
	F_3	1.00×10^{-3}	3.50×10^{-4}	0.77	0.71	0.84

The analysis of variance for 1000-grain weight of secondary seeds had
significant effects of genotype at both locations. Forward exhibited the
highest secondary grain weight, Ajax the lowest (Table 1).

As was found for the primary:secondary seed weight ratio, both additive
and non-additive genetic effects are involved in the inheritance of
secondary grain weight. Dominant gene action was more apparent than for
the ratio, however, since in four of the 15 crosses the F_2 progeny had
secondary grain weights similar to that of the higher parent, while the F_3
progeny had secondary grain weights which approached the mid-parent
values. In all but three crosses, F_2 and F_3 progeny means did not differ
significantly, which is indicative of additive gene action.

Transgressive segregation for both high and low secondary grain weight
was found. Each cross, with the exception of Ajax x Rodney, had high
transgressive segregates in at least one generation at one location. The
genetic material from which lines with higher secondary grain weights can

be selected therefore exists. Transgressive segregates for primary grain weight(10) and grain weight in general(11) have been reported in other studies.

GCA mean squares were significant at Macdonald and Joliette for both generations, indicating significant differences among the parental genotypes (Table 4). SCA mean squares were significant only for the F_3 generation at Macdonald, as was also found for the ratio. The estimates of the variance components due to GCA effects were invariably larger than those due to SCA effects (Table 5). Similarly, the ratio of GCA sums of squares to the total genetic sums of squares was always high (0.80-0.94), indicating that genetic variation for secondary grain weight was predominantly additive, to the extent that GCA reflects additive variation. Sampson and Tarumoto(9) reported significant GCA and SCA for 1000-grain weight in F_2 and F_3 of an 8-parent diallel without distinguishing between primary and secondary seeds. They found that the GCA variance was over five times as large as the SCA variance.

TABLE 4. Analysis of variance of GCA and SCA effects for
secondary grain weight in oats

Source	d.f.	MS	F value
Macdonald - F_2 - GCA	5	19.77	32.58**
SCA	9	0.72	1.19
F_3 - GCA	5	12.48	23.40**
SCA	9	1.78	3.24*
Joliette - F_2 - GCA	5	13.30	28.85**
- SCA	9	0.19	0.40
F_3 - GCA	5	15.00	20.95**
- SCA	9	1.43	2.00

*, **Significant at the 0.05 and 0.01 levels, respectively

Broad and narrow sense heritability estimates for secondary grain weight were similar, confirming that additive gene action was predominant (Table 5). Heritability estimates were high, indicating that genetic variation accounted for much of the variability and that selection for secondary grain weight may produce successful results. Narrow sense estimates did not increase in the F_3 generation at either location, as would be expected if dominance was important, reinforcing the predominance of additive gene action in the inheritance of secondary grain weight.

TABLE 5. GCA and SCA variances, ratio of GCA SS to total SS and heritability
for secondary grain weight ratios in oats

Location	Generation	GCA variance	SCA variance	GCA SS Total genetic SS	Heritability Narrow sense	Heritability Broad sense
Macdonald	F_2	4.76	0.11	0.94	0.93	0.93
	F_3	2.69	1.20	0.80	0.76	0.92
Joliette	F_2	3.28	0.00	0.93	0.93	0.93
	F_3	3.39	0.71	0.85	0.83	0.91

116

While no other studies examining the heritability of secondary seed weight specifically could be found, several have considered the heritability of seed weight in general. Sampson and Tarumoto(9) calculated narrow sense estimates of 0.77 and 0.88 for F_2 and F_3, respectively; Wesenberg and Shands(10) found that broad sense estimates for primary grain weight, based on variance components, ranged from 0.71 to 0.89 in F_3. Results in both cases are similar to what we found for secondary grain weight.

A low primary:secondary seed weight ratio combined with a high 1000 grain weight represents the ideal combination for selection. One cross, Nelson x X2078-1, had transgressive segregates with both a low ratio and a high secondary grain weight, indicating that it is possible for desirable combinations of these two traits to co-exist. Appearance of transgressive segregation in a cross involving Nelson is especially encouraging since this genotype had the lowest primary:secondary seed weight ratio of the parents included in the study.

4. CONCLUSIONS

Segregation patterns and combining ability analyses clearly indicate the presence of additive gene action in the inheritance of both the primary:secondary seed weight ratio and secondary grain weight. Large amounts of additive genetic variance allow for substantial progress using standard selection schemes in the development of pure-line cultivars.

Heritability estimates for secondary grain weight were generally slightly higher than those for the ratio. In addition, transgressive segregates with high secondary seed weights were more prevalent than those with low ratios. Selection for high secondary seed weight, in addition to being a simpler process, may therefore be more successful than selection for low ratios. Selection for higher grain weight in oats has indeed proved fairly successful when no distinction was made between the primary and secondary components(12,13). However, selection for secondary seed weight alone may not necessarily result in lower ratios because of strong correlations between primary and secondary grain weights (0.83-0.98).

REFERENCES

1. Brim CM and CC Cockerham: Crop Sci. 1, 187-190, 1961.
2. Sun PLF, HL Shands and RA Forsberg: Crop Sci. 12, 1-5, 1972.
3. Griffing B: Aust. J. Biol. Sci. 9, 463-493, 1956.
4. Sampson DR: Can. J. Genet. Cytol. 13, 864-872, 1971.
5. Singh M and RK Singh: Theor. Appl. Genet. 67, 323-326, 1984.
6. Falconer DS: Introduction to Quantitative Genetics. The Ronald Press Co., New York, 1960.
7. Gardner CO and SA Eberhart: Biometrics 22, 439-452, 1966.
8. Brown CM, AN Aryeetey and SN Dubey: Crop Sci. 14, 67-69. 1974.
9. Sampson DR and I Tarumoto: Can. J. Genet. Cytol. 18, 419-427, 1976.
10. Wesenberg DM and HL Shands: Crop Sci. 13, 481-484, 1973.
11. Murphy CF and KJ Frey: Crop Sci. 2, 509-512, 1962.
12. Chandhanamutta P and KJ Frey: Crop Sci. 13, 470-473, 1973.
13. Geadelmann JL and KJ Frey: Crop Sci. 15, 490-494, 1975.

STABILITY OF GERMAN OAT CULTIVARS (AVENA SATIVA L.): SPECIAL CONSIDERATION IS GIVEN TO NON-TYPICAL OAT FLORETS

U. BICKELMANN and N. LEIST
LUFA Augustenberg, Nesslerstr. 23, Postfach 430230,
D-7500 Karlsruhe 41, FRG

1. INTRODUCTION

In plant breeding, genetic variability, available or even artificially increased, is used to improve qualitative or quantitative characteristics. On the other hand, for commercialization of new cultivars, homogeneity and stability are the important factors. Therefore in seed identification and classification emphasis must be placed on typical oat florets. During recent years special attention has been given to non-typical oat florets which combine the characters of cultivated oats (A.sativa) and wild oats (A.fatua): the so called fatuoids.

In the literature two theories(1) have been presented for the origin of these types of florets: i.e. a) mutation or b) crossing with wild oats. Either of these possibilities produce problems for certification of seed. However, mutation should not lead to rejection, whereas outcrosses would inevitably lead to disapprobation. Depending on the year, region and cultivar, fatuoids are found in 1-20% of German oat seed samples and this just refers to the best selling cultivars.

2. MATERIALS AND METHODS

The florets of the seven cultivars: Erich, Flaemingsstern, Flamo, Platin, Ponta, Selma and Tiger were radiated with gamma-rays. Two hundred kernels of each variety were treated with 15 kr and 300 kernels with 30 kr. Diallel crosses were made between A.fatua and the nine cultivars: Erich, Flaemingsstern, Flaemingskrone, Flamo, Leanda, Moyencourt, Ponta, Selma and Tiger. Four successive generations of the radiated and crossed material and of naturally found fatuoids were grown under controlled pollination conditions.

The potential for crossing under relatively natural conditions was tested in an emasculation experiment. For this purpose plots of eight cultivars (Alfred, Borka, Erbgraf, Fabian, Flaemingsnova, Ponta, Selma, Tiger), were sown. Each plot was surrounded by mixture of brown (Siréne), naked (Caesar) and wild oats (A.fatua). On five successive days, 200 flowers of each of the plots were emasculated and then exposed to natural pollen dispersal without any protection.

The viability of pollen was examined not only by in vitro germination on a medium containing 2.5% gelatine and 10% sucrose but also by colour reaction with 1% triphenyltetrazolium-chloride at pH 7.0.

Isoelectric focussing was carried out with 5% polyacrylamide gels containing 4 M urea and 2% Servalyt pH 5-8. The prolamins of single seeds were extracted by 30% chloroethanol.

3. RESULTS

If the tendency to form fatuoids is of genetic origin, it is important to know in which cultivars fatuoids are likely to occur and whether there are any common traits in their genealogy.

118

Fatuoids have been found in 16 out of 28 oat cultivars licenced in Germany. Twelve of them have the cultivar Minor as a common ancestor (see Fig. 1). The other four cultivars have either completely different progenitors or we do not know the source of their parental lines. In any case, it is impossible to assess the capability of cultivars to generate fatuoids just on the basis of their pedigrees.

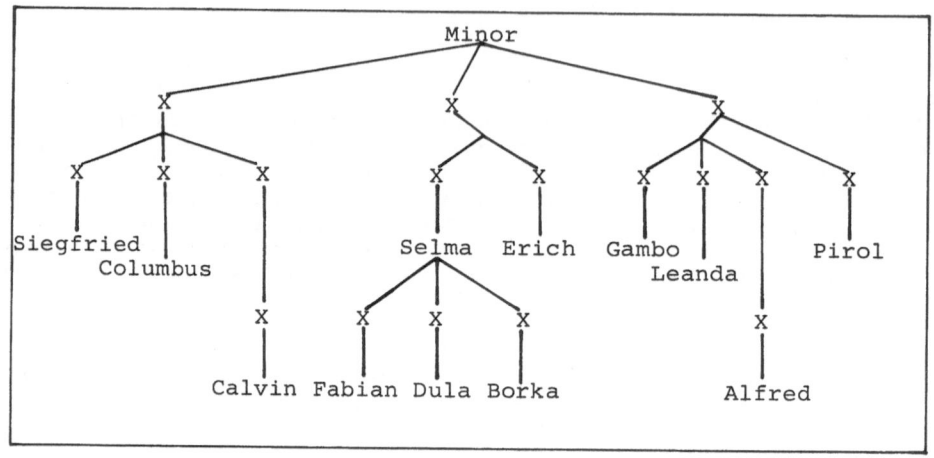

FIGURE 1. Pedigree of German oat cultivars which generate fatuoids

As the frequency of fatuoids varies from year to year and from region to region, it is possible that environmental conditions have a strong influence on the frequencies of mutations or crossing. An experiment was therefore carried out to investigate whether fatuoids develop by mutation and whether the result of mutation is dependent on the genotype of the cultivar. For this purpose oat cultivars were irradiated at different intensities.

After treatment fatuoids developed, but only in those cultivars that also produce fatuoids naturally. Overall the radiated material, homozygous fatuoids developed to an extent of on average, 0.02% after the 15 kr treatment and 0.10% after 30 kr (see Table 1).

Homozygous fatuoids have glabrous lemmas, of the colour of the cultivar in which they occur, awns are strong, twisted and geniculate, floret bases are formed like sucker-mouths and there are dense hairs on the callus and rachilla. Primary and secondary florets of homozygous fatuoids have identical morphology. Heterozygous fatuoids also exist but these may be easily overlooked as they are intermediate in character between homozygous fatuoids and normal cultivated oats. They have twisted, geniculate awns on the primary grain of each spikelet only, and the sucker-mouth and hairiness are much reduced (see Fig. 2).

Another possibility is that naturally occurring crosses of A.sativa with A.fatua give rise to fatuoids. To check this possibility, appropriate diallel crosses were carried out. Progeny showed large variability for floret morphology, comprising of fatuoids even in those cultivars in which

TABLE 1. Frequency of homozygous fatuoids occurring after
gamma-radiation and crossing between A.sativa x A.fatua

| Varieties | % hom. fatuoids in M1 and M2 | | % hom. fatuoids in F$_2$ |
	15 kr	30 kr	A.sativa x A.fatua
Erich*	-	-	1.67
Flaemingsstern	-	-	0.42
Flamo	-	-	-
Platin	-	-	-
Ponta*	0.05	1.16	1.47
Selma*	0.06	-	-
Tiger	-	-	0.10
Mean	0.02	0.10	0.33

*Varieties in which fatuoids are naturally found

fatuoids are not naturally found. The progenies of crossing segregated,
on average, giving 0.33% fatuoids. In the following generations
homozygous and heterozygous fatuoids behaved in the same way, no matter
whether they were of natural, mutagen or crossing origin. The homozygous
fatuoids were true breeding, whereas the heterozygous segregated into
homozygous, heterozygous and sativa types in the ratio of 1:2:1.

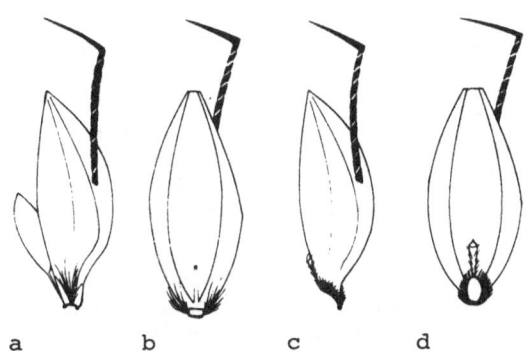

FIGURE 2. a, b) heterozygous fatuoid - c,d) homozygous fatuoid

The question now arises, as to how effective crossing would be under
natural conditions. Observations of flowering behaviour showed that all
cultivars of oats are capable of flowering, with widely opened lemma and
palea, spread hilum and extruded anthers, if certain environmental
conditions are provided: i.e. about 20°C temperature and 40% relative
humidity in the afternoon. If the conditions are less favourable (warmer
and drier), the lemma and palea open at least a little and any gentle
breeze may spread the pollen. Pollen is 90-100% viable at the time of

120

anthesis and open flowering. Under these conditions natural crossing of oats should be possible.

To test this supposition under relatively natural conditions, an emasculation experiment was carried out. On average, 18% of emasculated flowers set seed. Comparing seed setting and climatic conditions at flowering time, a clear dependence of fructification on climatic factors can be observed. Cold, rainly weather enforces selfing. If emasculated flowers are not too severely damaged by these conditions, fertilisation may occur during a following period of fine weather, when outcrossing is favoured. We are now growing F_1 out of last year's emasculation experiment, and these show a very heterogenous picture; it can be recognised that among the yellow oats - brown, naked and wild oats pollinated the emasculated flowers. According to the results of these tests, it is evident that natural crossing may occur depending on the climatic conditions.

As fatuoids may be generated from either crossing or mutations, the question arises as to whether the mutants or hybrids can be distinguished by biochemical methods. No distinction is possible by morphological characteristics. To study this the prolamins of single seeds were analysed by isoelectric focussing. Cultivars and their fatuoids, of natural or mutagen origin, and the fatuoids from A.sativa x A.fatua crosses show electrophoregrams with identical banding patterns. A.fatua and all the other multishaped progenies of the diallel crosses with A.sativa and A.fatua produce very heterogenous banding patterns. With regard to thousand grain weight and protein and lipid content, there is a clear difference between A.sativa and the fatuoids on the one hand and off-types and A.fatua on the other.

From the results of these experiments, it is clear that the fatuoids have morphological and biochemical characteristics which are independent of their mode of origin.

Now it should be discussed, how important mutation is as a contributor to the origin of fatuoids, when compared to out-crossings as contributor in a propagation for certified seed of oats. That fatuoids occur frequently in closely related varieties may support the mutation theory, if these cultivars have certain genetical instabilities in common. Concerning flowering behaviour, our results show that crossing may be important. The frequency of fatuoids which occur under natural conditions, and among the descendants of our artificial crosses may confirm the crossing theory. In oats there is a predominance of self-pollination, but certain environmental conditions encourage open-flowering and outcrossing. In contrast to our experiments with wild and cultivated oats, wild oats rarely grow in a normal crop of oats. The descendants of our crosses between wild and cultivated oats produced fatuoids, but most of the segregates combined more of the other morphological characteristics of wild and cultivated oats than the fatuoids. Now the multishaped descendants of crosses are expected to be found more frequently in purity tests. It appears that crossing between plants of the same variety probably frequently occurs depending on the region of growing and the climatic conditions. Such crossing within or between varieties, or mutation events, are responsible for the frequency of fatuoids in the seed of oats.

REFERENCE

1. Huskins CL: Bot. Rev. 12, 457-514, 1946.

MASS (GRAVIMETRIC) SELECTION IN DISEASED NURSERIES

URIEL MALDONADO*, CARLOS JIMENEZ* and DEON D. STUTHMAN**
*Instituto Nacional de Investigaciones Agricolas, Chapingo, Mexico
**University of Minnesota, St. Paul, Minnesota, MN 55108, USA

1. INTRODUCTION

Since its beginning more than 25 years ago, the Mexican oat breeding programme has utilised pedigree selection. The programme began by importing germplasm from the USDA and testing it at three sites, Chapingo, Celeya (winter crop) and Chihuahua. Priority traits include stem rust resistance, earliness, kernel quality and grain yield. As oat acreage increased in the Mexico City area (near Chapingo), stem rust became a serious problem because inoculum is always present and climatic conditions in the area are usually very favourable for stem rust development.

In 1968 the oat programme at the University of Minnesota and the USDA Cereal Rust Laboratory joined forces with the programme at Chapingo to develop stem rust resistant varieties. CI 3034 was chosen as the principle source of resistance. The programme was mainly pedigree selection and backcrossing to incorporate that resistance into Mexican varieties.

About 10 years ago the programme began using a form of mass or bulk selection, attempting to take advantage of a heavy and rather uniform stem rust infection occurring annually at the main breeding nursery at Chapingo. In this report we are presenting data from evaluations of lines so identified. Both agronomic and pathological traits are included.

2. LITERATURE REVIEW

Since early man, mass selection has been used for crop plant domestication and improvement(1,2). The method can be based on subjective or objective criteria. Another important feature of this method, as pointed out by Frey(3), is that it permits inexpensive propagation of both a large number of crosses and a large number of plants within crosses. Today this method is still often used by both novices and masters of plant breeding.

In selecting for disease resistance to several diseases, breeders have frequently selected inherently poorer yielding lines. In early generation selection for disease resistance, in many breeding procedures, lines with genes that would improve yields in disease free conditions are discarded because of their apparent susceptibility(4). However in other situations plants that appear to be heavily infested with a disease organism may still develop grain nearly normally with no appreciable reduction in yield(5,6). This type of plant protection, designated as 'tolerance', is perhaps most easily identified using a form of mass or gravimetric selection.

Tiyawalee and Frey(7) used mass selection to increase gene frequency of oat crown rust resistance alleles in genetically heterogeneous populations without loss of agronomic performance. An epiphytotic condition was created and plants were exposed as uniformly as possible. Harvested seed was divided into high- and low-density classes. Supposedly crown rust susceptible plants would produce seed in the low-density fraction. By eliminating this fraction the frequency of resistance genes in the remnant population would be increased. Using this approach, they were able to demonstrate the increased frequency of a desirable gene(s) in genetically heterogeneous populations.

Kernel weight itself can be used as an objective criteria for mass selection regardless of the amount of disease infection. This methodology apparently offers a good opportunity for selecting high yielding lines. Studying the relation of seed weight to grain yields in oats, Frey and Huang(8) were able to demonstrate the association of these two traits and explain how it is possible to obtain gains in grain yield when selecting on the basis of seed weight. Others have also found and/or taken advantage of the relationship between seed weight and grain yield. Among the three yield components, kernel weight is probably the most likely to show a positive correlation with grain yield(9).

Indirect mass selection has been used extensively to select soybean populations for high seed protein or high seed oil by using density separation of the seed(1). Indirect mass selection for grain yield was highly effective when selecting for heavy primary panicles in bulk oat populations(10).

3. MATERIALS AND METHODS

The 1976 F_2 nursery of the INIA oat breeding programme at Chapingo, Mexico was used to generate the selections evaluated in this report. The nursery consisted of about 100,000 plants from 84 different crosses from combinations of parents selected for high yield, high protein or resistance to stem rust (Puccinia graminis f.sp. avenae). Each F_2 family was planted in rows 11 m long, with 30 cm between rows and 10 cm between plants within a row.

Visual selection was practiced, as normally done, to initiate the pedigree selection (Table 1). A total of 1905 individual F_2 plants were selected with emphasis on stem rust reaction, earliness, acceptable height and general plant type. In the F_3 and the F_4 generation, visual selection was first on a row and then on a plant within selected row basis. In the

TABLE 1. Procedure for line development using the pedigree system

Generation	Season-location	No. of plants or lines selected[a]
F_2	S 76 Ch	1905
F_3	W 76-77 C	1030
F_4	S 77 Ch	314
F_5	W 77-78 C	199 bulk
F_6	S 78 Ch	118
F_7	W 78-79 C	97
F_8	S 79 Ch	15

[a]Visual selection for stem rust resistance, early maturity, general plant type, and against extremes in plant height.
S = Summer at Chapingo (Ch); W = Winter at Celeya (C)

F_5 generation the entire selected row was bulk harvested and all subsequent selection was on a line basis only, ultimately producing 15 lines.

Following the harvest of the selected F_2 plants, the remaining plants in the entire nursery were combine harvested (Table 2). The seed obtained

TABLE 2. Procedure for line development using mass selection

Generation	Season-location	Treatment[a]
F_2	S 76 Ch	MS
F_3	S 77 Ch	MS + EPE
F_4	W 77-78 C	Same as F_2
F_5	S 78 Ch	Same as F_3 + WF
F_6	S 79 Ch	1000 space plants
F_7	W 79-80 C	500 random lines SI

[a]MS = Mass selection; WF = Water flotation; SI = Seed increased;
EPE = Extreme Plant Elimination-stem rust susceptibility,
late maturity and tall plants.
S = Summer at Chapingo (Ch); W = Winter at Celeya (C)

was rethreshed in a head thresher, fanned and screened to remove the lighter and less dense seed. A sample of the remaining seed was planted at Chapingo in the summer of 1977. Prior to combine harvest of these plants, extremely undesirable plants which were badly rusted, extremely late or very tall were manually eliminated. The F_4 generation was grown at Celeya (no rust but considerable BYDV) during the 1977-78 winter season. This planting was harvested in the same manner as the F_2 generation. The F_5 generation was grown in summer 1978 at Chapingo and was handled similar to the F_3 generation. In addition , the remaining seeds were floated on water to further differentiate between light and dense seeds. The following summer 1000 spaced plants were grown at Chapingo and a random one-half of these was increased in Celeya during the 1979-80 winter season.

A preliminary yield comparison of lines from the two selection methods was made by growing 377 mass selected lines, 15 pedigree selections and 8 check cultivars. These trials were grown in four environments, two summer tests at Chapingo with heavy stem rust and two winter tests at Celeya, which were rust free.

Seed from the first three 'cycles', MI (F_2), MII a and b (F_3), and MIII (F_5) were sent to St. Paul, Minnesota to determine the effect of multiple cycles of selection. The two versions of MII resulted because the seed lot was divided and slightly different procedures were used to produce advanced generations. One hundred lines were extracted randomly from each of the four populations and exposed to races NA 26 and NA 27 in the field. Individual plant progeny from resistant lines were further evaluated as seedlings in the glasshouse to determine the origin of the resistance gene(s), i.e. CI 3034 or some other source. Eight F_8 lines from MIII and two from MII were also entered into the 1983 and 1984 International Oat Rust Nursery to obtain information on their stem rust reactions at other diverse locations.

Subsequent to the summer of 1976, F_2 nurseries were harvested in a similar manner as in 1976. The following generations were also handled similarly to those derived from the initial combine harvest in 1976. Thus, distinct mass selected populations originating from each summer have also been created.

4. RESULTS AND DISCUSSION

The performance of the mass selected lines, the pedigree selected lines and the check cultivars in the preliminary yield comparison is summarised in Table 3. The range for each of the included traits is greatest for the

TABLE 3. Ranges for stem rust(SR), grains/panicle(G/P), groat weight(GW), and yield/plot(Y) for mass selection lines(MS), pedigree selections(P), and check cultivars(C) grown in four environments, consisting of two summers with rust and two winters without rust

Trait	MS (377)	P (15)	C (8)
SR (% infection)	7 - 90	7 - 35	12 - 85
G/P (n)	31 - 78	44 - 71	47 - 73
GW (g/1000)	16 - 26	18 - 23	18 - 22
Y (g/plot)	229 - 578	324 - 437	338 - 443
	(397)[a]	(388)	(405)

[a]Value is mean yield for lines in this group

mass selected lines. The mean yield for each group was similar, 397 g/plot for mass selected lines, 388 g/plot for pedigree selections and 405 g/plot for the check cultivars. Thus there seems to be little advantage in terms of productivity for either selection system.

The values for the best yielding lines from each group are given in Table 4. The five highest yielding mass selected lines were more than 10% greater yielding than either the highest yielding pedigree selected line or the highest yielding check cutivar. Each of these five mass selected

TABLE 4. Performance of five best mass selection lines, compared with the best pedigree selection and best check cultivar for grain yield, groat weight and stem rust grown in four environments, two with rust and two rust free

Entry	Yield (kg/ha)	Groat weight (g/100)	Stem rust (% infection)
219	2404	22.32	75
212	2262	20.75	17
114	2237	21.66	65
80	2200	21.32	30
109	2183	21.84	75
Pedigree	1816	17.68	27
Check	1845	20.14	45

lines also had a greater groat weight than did the highest yielding pedigree selected line. This latter superiority is not surprising since the mass selection procedure should give selective advantage to genotypes with heavier groats. It is of interest to see high yielding, mass selected lines that are rated quite susceptible when using a qualitative stem rust reaction scale. However, it should be remembered that stem rust infection occurred at only two of the four locations producing the yield values.

Populations with more cycles of selections had a greater proportion of stem rust resistant plants (Table 5). The percentage of resistant plants increased from 4 to 18 from cycle I (F_2) compared to cycle III (F_5). This evaluation utilised inoculation with both races NA 26 and NA 27. The observed resistance to race NA 26 is interesting since only race NA 27 has been isolated from the nursery at Chapingo during the time of population development. The resistant reaction to race NA 26 adds support to the suggestion that the resistance gene in CI 3034 effectively protects against several races of stem rust(11).

TABLE 5. Field and glasshouse stem rust reactions at St. Paul, Minnesota, for 100 random lines from each of four mass selected populations

Cycles of selection	Percentage resistant lines[a]	Seedling reaction of field resistant lines
3	18	Susceptible
2	9	Mixed
2	11	Mixed
1	4	Mixed

[a]Field and glasshouse inoculation, race NA 26 and NA 27

The results from the seedling screening of plants identified as resistant in the field confirmed the hypothesis that sources of resistance other than CI 3034 were also selected with the mass selection procedure. Because CI 3034 has only adult plant resistance, plants with seedling resistance must contain another stem rust resistance gene(s). One likely source is CI 7114 which does contain such genes and was a part of parental pedigrees.

The performance of ten lines selected from populations with either two or three cycles of mass selection with stem rust exposure in the IORN further characterises the stem rust resistance of this material. In 1983 these ten lines were among the 34 best lines (154 total) for coefficient of stem rust infection. The ten lines with the lowest coefficient of infection averaged 1.6, the ten Mexican lines averaged 2.1 while the ten most susceptible lines averaged 48.5. Their performance in the 1984 IORN was similar. Of a total of 155 lines tested, the most susceptible Mexican entry ranked 51st. The 10 most resistant entries, which included four Mexican entries, averaged 2.1, the Mexican entries averaged 17.0 and the ten most susceptible lines had an average coefficient of 80.0. Thus, although there was more stem rust infection in 1984, the mass selected lines from Mexico continued to exhibit useful stem rust resistance in many parts of the world.

5. CONCLUSIONS
The mass selection procedure employed under conditions of heavy stem rust infestation produced lines which were at least as productive as those from pedigree selection and check cultivars. Because this mass selection procedure requires fewer resources than the pedigree system used, it is a cost effective method for early generation selection in this or a similar set of conditions.

126

A sample of 10 lines from the mass selection procedure had good resistance to stem rust in the 1983 and 1984 International Oat Rust Nursery. Thus the procedure was effective in identifying stem rust resistance which will be useful in locations other than the selection site. Because some of the selected lines also possess seedling resistance, an additional source of resistance other than CI 3034 was also recovered.

REFERENCES

1. Wood DR: Crop breeding. American Society of Agronomy, Crop Science Society of America, 294 pp., 1983.
2. Chandhanamutta P and KJ Frey: Crop Sci. 13, 470-473, 1973.
3. Frey KJ: Euphytica 16, 341-349, 1967.
4. Schafer JF: Ann. Review of Phytopathology 9, 235-252, 1971.
5. Caldwell RM: Breeding for general and/or specific plant disease resistance. Proc. 3rd Int. Wheat Genet. Symp., Canberra, Aust., 1968.
6. Kerr EA: Hort. Sci. 18, 27-30, 1983.
7. Tiyawalee D and KJ Frey: Iowa State J. of Sci. 45, 217-231, 1970.
8. Frey KJ and TF Huang: Euphytica 18, 417-424, 1969.
9. Williams SE: Utilizing yield components to breed for high yield in semidwarf barley, Hordeum vulgare L. MS Thesis, University of Minnesota, 65pp., 1983.
10. Geadelmann JL and KJ Frey: Crop Sci. 15, 490-494, 1975.
11. McKenzie RIH and JW Martens: Crop Sci. 60, 625-627, 1968.

MASS SELECTION FOR IMPROVED MILLING PERFORMANCE

E. SOUZA and M.E. SORRELLS
Cornell University, Ithaca, NY 14853, USA

1. INTRODUCTION

Milling yield is one of the most important quality characteristics for millers of oats; it is defined as the weight of oats required to produce 100 lbs of milled groats. Milling percentage can be described as the groat percentage of the fraction of a seed left after discarding slim, pin and double or bosom oats(1). Seed characteristics of oats have been studied extensively. The conclusion of several studies is that groat percentage is one of the most important factors determining milling yield and is more tightly correlated to milling yield than other traditionally used traits such as test weight(1,2). Groat percentage is a quantitatively inherited trait. Wesenberg and Shands determined that the trait had a broad sense heritability ranging from 36% to 93%(2). Stuthman and Granger obtained a similar range of heritabilities in their later paper and were able to realise a gain from selection by selecting for groat percentage among F_4 headrows(3). Groat percentage, however, is a time consuming trait to measure as it requires dehulling and cleaning samples.

Mass selection has often been used by breeders to improve a population for a given trait. For example, Frey increased the seed size in oats through five cycles of mass selection for seed width(4). The density of the oat caryopsis in the absence of disease remains fairly constant(5). Variation in seed density therefore should be largely due to variation in hull percentage and the associated air-space. Selection of seed for high specific gravity should improve the groat percentage of a seed lot and increase the frequency of high groat percentage genotypes.

2. MATERIALS AND METHODS

Three mass selection techniques were compared in the 1984 growing season: aspiration (ASP) of seed samples using an office sized air-screen cleaner; selection on a gravity table (GT); and selection with the gravity table using seed polished in a paint shaker (PS). We selected a heterogeneous population of F_5 seed that had been advanced in bulk from the cross Orbit x Porter. The seed lots were divided into three seed sizes using slotted screens: small (<2mm), middle (2-2.4 mm) and large (>2.4 mm). The seed was sized to limit the confounding effect of selection for seed size by the density selection and to evaluate the rate of progress in each size fraction. The size lots were selected for high and low specific gravity with a selection intensity of 25%. The process of selection for density was replicated twice. Samples of 2000 seeds were planted in 5 m six-row plots that were later trimmed to 3 m. The experimental design was a factorial of selection technique, seed size and density. The plots were planted in a two block randomised complete block design with the high and low density plots of a given treatment paired:

unselected checks of the size lots and the original population were also included in the experiment. The plots were planted on 20 April and harvested 15 August.

Since seed source effects may have biased some of the data from last summer's trial, vigour evaluations of remnant ASP selected seed of the 1984 planting were conducted similarly to Perry(6). Eighteen seeds from each treatment were placed in a rolled-up 'doll' of germination paper as one replicate. Each treatment was replicated three times and in two time blocks. The seeds were germinated for seven days at 24°C in a dark growth cabinet. On the seventh day the average length of the coleoptiles in the dolls was measured and used as the index of seed vigour.

3. RESULTS AND DISCUSSION

Remnant seed from all three selection methods showed significant separation between high and low density treatments for groat percentage (Table 1). There was a significant interaction between the selection for density and seed size; the large and small seed size fractions showed larger differences between the high and low density fractions. The middle size fraction however had the highest groat percentage initially and would be expected to show the least improvement. There were significant shifts in the seed width with the ASP and PS selection for density; the mean of the high density fraction was significantly (95% confidence by LSD) greater than the low density fraction. This shift was most prominent in the small seed size fraction, probably because it had the greatest variance in size of the three fractions. The selection for size could be limited by truncating the small seed size by discarding the smallest seed.

TABLE 1. Groat percentage of seed selected for density

Treatments		Seed sizes			Mean of treatments	
		Small	Medium	Large		
ASP	High density	70.45*	71.02	70.69**+	70.72**	
	Low density	66.79	68.68	57.91	64.46	
GT	High density	69.76**	70.28*	70.98**+	70.34**	
	Low density	62.67	66.75	57.08	63.32	
PS	High density	73.34**	70.66**	72.13**	72.05**	
	Low density	65.28	65.55	57.45	65.80	
Unselected seed		69.78	70.67	66.55	69.00	MSE=2.757

*Mean is significantly greater than the paired density treatment at the 95% confidence level by the LSD method of pairwise comparison.

**Mean is significantly greater than the paired density treatment at the 99% confidence level.

+Mean is significantly different from the unselected seed lot at the 95% confidence level.

The results of the field trial comparing the selection techniques showed significant changes in the population in response to the selections. Using the groat percentage of the harvested seed as the primary evaluation, the PS selection showed significant increase in the groat percentage in the high density selection over the low density selection (Table 2). The GT gave significant changes, but in the small

TABLE 2. Groat percentage of harvested plots

Treatments		Seed sizes			Mean of treatments	
		Small	Medium	Large		
ASP	High density	69.48	70.00	70.19	68.89	
	Low density	69.19	69.54	69.82	69.64	
GT	High density	69.19	70.21	69.06	69.48	
	Low density	70.21*	69.46	69.91*	69.86	
PS	High density	69.89*	70.05	69.87*	69.94*	
	Low density	69.18	69.89	69.38	69.38	
Unselected seed		69.68	69.80	69.91	69.54	MSE=1.040

Note: Symbols for significance as for Table 1

and large seed sizes the changes were a reduction in groat percentages of the high density selections. The ASP and PS selections also showed significant increase in test weight of the high density selections over the low density selections; the gravity table alone showed no change (Table 3). The poor performance of the GT selection may be due to irregular seed flow. The binding caused by awns and ragged ends of hulls is eliminated by polishing the seed.

TABLE 3. Test weight of harvested plots
(values in kg/hectalitre)

Treatments		Seed sizes			Mean of treatments	
		Small	Medium	Large		
ASP	High density	42.24*	42.01	42.52*	42.26**	
	Low density	41.21	41.58	41.62	41.47	
GT	High density	41.68	42.22	41.94	41.95	
	Low density	41.16	41.99	41.85	41.67	
PS	High density	42.30*	43.11**	42.04	42.48**	
	Low density	41.38	41.91	41.30	41.53	
Unselected seed		42.08	42.52	42.32	42.31	MSE=0.5069

Note: Symbols for significance as for Table 1

The selection for density in essence is a selection for a larger groat in a given hull size. This selection may cause non-heritable seed source effects due to different caryopsis sizes; the seed source would have most pronounced effects in early stand vigour. Early vigour may be a factor in some of the changes in agronomic traits with selection. Larger seed size and higher density treatments had better germination (Table 4), increased

130

TABLE 4. Seedling emergence, 1984 trial (Seedlings per meter of row,
average of replications and seed size)

| Density | Selection method | | | Mean of treatments | |
	ASP	GT	PS		
High density	59.83*	62.67	61.00*	61.16**	
Low density	51.75+	56.58	52.58+	53.37	
Unselected plots				60.17	MSE=96.49

Note: Symbols for significance as for Table 1

tillering, and earlier heading dates. The high density selection also
showed a significantly higher yield than the low density selection (Table
5).   Seed source effects may also bias the groat percentage either

TABLE 5. Yield of harvested plots (data in grams per plot,
average of replications and seed sizes)

| Density | Selection method | | | Mean of treatments | |
	ASP	GT	PS		
High density	1416	1413	1440*	1423**	
Low density	1353	1380	1345	1359	
Unselected plots				1378	MSE=9803.6

Note: Symbols for significance as for Table 1

directly or through interaction with the other yield components. The seed
vigour test was conducted to evaluate the significance of the change in
seed vigour due to selection. In the standardised test, larger and denser
seed had greater coleoptile elongation and thus higher vigour ratings.
The exception was the large seed size which had a reduced rating due to a
higher frequency of bosom oats (Table 6). For valid testing of genetic
components future trials should be one generation removed from the
selection to limit seed source effects.

TABLE 6. Standardised vigour test of seed lots
(average coleoptile length in mm)

| Density | Seed size | | | Mean of treatments | |
	Small	Medium	Large		
High density	108.23*	118.96*	107.98**	111.72**	
Low density	90.93	94.31	53.09+	79.44+	
Unselected seed lots	100.99	111.41	98.94	103.78	MSE=210.99

Note: Symbols for significance as for Table 1

There may be a heritable component to the increased seed vigour due to
density selection, as indicated by a trial currently in the field. Seed
of a composite of crosses between OA 338 and New York cultivars was
selected for high density using ASP in the F_3 and F_4 generations; cycle 0,
cycle 1 and cycle 2 populations were grown out in the glasshouse last

winter and thirty to thirty-five headrows of F_5 lines were taken from each cycle. These lines were replicated twice in a randomised complete block design at two locations. The cycle 2 lines showed significantly higher germination percentages at sixteen and twenty-six days after planting than the cycle 0 and cycle 1 lines. The harvest data from this population and two other populations selected for two cycles with both ASP and PS will be used to evaluate gain from selection and to make a second comparison of the two selection techniques.

REFERENCES

1. Peek JM and JM Poehlman: Agron. J. 41, 462-466, 1949.
2. Wesenberg DM and HL Shands: Crop Sci. 13, 481-484, 1973.
3. Stuthman DD and RM Granger: Crop Sci. 17, 411-414, 1977.
4. Frey KJ: Euphytica 16, 341-349, 1967.
5. Murphy CF and KJ Frey: Crop Sci. 2, 509-512, 1962.
6. Perry DA: Seed Sci. and Technol. 5, 709-719, 1977.

PRESENT STATUS OF RESEARCH TO DEVELOP USEFUL SEMIDWARF OAT GERMPLASM

H. G. MARSHALL

Agricultural Research Service, US Dept of Agriculture and
The Pennsylvania State Univ., University Park, PA 16802, USA

1. INTRODUCTION

Lodging of oats (Avena sativa L.) is likely to cause severe grain yield losses in humid regions when the crop is grown at high levels of nitrogen (N). Even when conservative amounts of N fertilizer are applied to an oat crop in the Northeastern USA, lodging may be severe because manure is regularly applied to most fields, legumes are used in most crop rotations, and high levels of N fertilizer are applied to the corn grown in most rotations. Several modern, conventional height cultivars possess good lodging resistance, but additional improvement in that trait is needed to consistently prevent lodging when oats are grown under high levels of fertilizer in an intensive management system.

The value of semidwarf cultivars to prevent or reduce lodging of wheat (Triticum aestivum L. em Thell.) and rice (Oryza sativa L.) has been demonstrated(1,2,3,4). Similar benefits should result from the use of very short or demidwarf cultivars of oats, but useful dwarf or semidwarf germplasm was not available until recently(5,6).

A programme to develop lodging resistant spring oat germplasm and cultivars was initiated in 1974 by the Agricultural Research Service of the US Department of Agriculture in cooperation with the Pennsylvania Agricultural Experiment Station. The progress in that research is summarized in this paper.

2. MATERIALS AND METHODS

Adapted cultivars were crossed to several lodging resistant germplasm lines or cultivars, but the primary effort has been devoted to populations derived from crosses involving the germplasm lines Egdolon 26 (from Cornell Agric. Exp. Station), OT 207 (from Agriculture Canada, Winnipeg), and CI 8447 (from North Carolina Agric. Exp. Station). The first line is of medium height but possesses elite lodging resistance. OT 207 is a semidwarf line carrying the Dw 6 dwarfing gene(5). It has excellent lodging resistance but maturity is too late under Pennsylvania conditions, and the panicle does not completely emerge from the boot. CI 8447 is a dwarf winter oat that carries the Dw 7 dwarfing gene(6) and has a compact, club-type panicle.

All populations were handled by using the bulk breeding method. After the F_2 generation each population was grown in a single, five row plot that was 5.5 m long. The row spacing was 17.8 cm, and the seeding rate was only 50 kg ha^{-1}. No mass selection pressure for reduced plant height was applied. Plots were harvested with a plot combine.

Selection normally was begun in the F_4 generation although a few unusual plants were selected in the F_2 and F_3 generations. Lines were extracted by selecting panicles and growing the resulting seed in a short row. The best lines were advanced to replicated tests at two locations in

Pennsylvania. All lines with short plant height or conventional height and stiff stems were grown with N applications of 90 kg ha^{-1}.

Pennlo, the most promising semidwarf developed in this programme(7), was compared to Ogle during 1982 and 1983 for responsiveness to row spacing, seeding rate and N fertilizer rate. Those cultivars were grown in a three replicate, factorial experiment at row spacings of 12.7 and 17.8 cm, seeding rates of 67, 101 and 134 kg ha^{-1} and N rates of 67, 101 and 134 kg ha^{-1}. Plots were 7.3 m long and had either seven or nine rows depending on the row spacing.

3. RESULTS AND DISCUSSION
3.1 CI 8447 derivatives

Numerous lines were isolated from various populations having CI 8447 as a parent and compared to recommended cultivars. Essentially all of the short lines had a compact, club-type panicle and were very late in maturity. The best semidwarf line of this type to date is Pennline 116 which was released as a germplasm line in 1982(8). It was selected from a James (hull-less)/CI 8447 population and carries the hull-less trait. Most spring oat lines and populations tracing to CI 8447 have been dropped. Several very short lines with open-type panicles still are being tested in the winter oat programme. These may prove to be useful as parents in the spring oat programme.

3.2 OT 207 derivatives

Lines with the Dw 6 gene have shown excellent lodging resistance, but all have been late in maturity and low yielding because of incomplete grain filling. A few lines with panicles clear of the boot were selected and crossed to very early maturing cultivars. Some of the progeny are earlier than the OT 207 parent, but all are too late for use under Pennsylvania conditions.

One problem with the Dw 6 gene is complete dominance for semidwarf plant height with little modification by other genes. This severely reduces the probability of developing cultivars somewhat taller than OT 207 in order to have satisfactory yields of straw.

3.3 Egdolon 26 and Astro derivatives

Unexpectedly, the best semidwarf lines were selected from certain populations involving either Egdolon 26 or Astro (Egdolon 26 was derived from an Astro/PI 193027 cross). The semidwarf plant height apparently has resulted from transgressive segregation and genetic control is complex. Two lines, Pennlo and Pennline 6571, were released as germplasm lines in 1982(7), and they have been used extensively as parents in the current breeding programme. They are about 28% shorter in plant height than Ogle, the highest yielding and most lodging resistant recommended cultivar. Both lines have been nearly free of lodging when grown under severe stress. Pennlo is higher yielding than Pennline 6751 and, with the exception of Ogle, has equalled the yields of recommended cultivars grown under conventional management in Pennsylvania. Compared to Ogle over a 2 year period in the management study described earlier, Pennlo averaged 13.5% lower yield over all treatment combinations, but lodging was not a limiting factor. The relative characteristics of Pennlo and Ogle in that study are given in Table 1.

Unexpectedly, there were no major differences between Pennlo and Ogle for responsiveness to the variations in row spacing, seeding rate, and N fertilizer rate. The harvest index was high and identical for both cultivars.

TABLE 1.   Relative characteristics of Pennlo (semidwarf) and
Ogle spring oats in Pennsylvania, 1982-83

Characteristics	Cultivar	
	Pennlo	Ogle
Plant height (cm)	69.4**	96.8
Lodging (%)	0**	29.0
Grain yield (kg ha^{-1})	4757	5502**
Test weight (kg m^{-3})	430**	418
Harvest index	0.49	0.49

**Cultivars significantly different at the P = 0.01 level

Populations derived from crosses involving Pennlo and Pennline 6571 have a wide range of genetic variability for plant height, and many of the short types have superior lodging resistance.   In order to provide breeders with populations for selection, two bulk composites have been released.   Penncomp 29 was made up by compositing 31 bulk populations that have Pennlo as a parent.   Penncomp 30 was made up by compositing 29 bulk populations that have Pennline 6571 as a parent.   Because of the conventional height parents used, both composites include a wide range of genetic variability for plant height, lodging resistance, maturity, kernel size, protein content, and resistance to barley yellow dwarf virus.   Seed is available from the Cereal Crops Research Unit, USDA-ARS, Department of Agronomy, The Pennsylvania State University, University Park, PA 16802.

In conclusion, Pennlo and several related semidwarf lines are highly resistant to lodging, even under severe stress, and have a high yield potential.   Primary weaknesses are poor disease resistance and relatively small kernel size.   Populations with genetic variability for these traits and semidwarf plant height currently are being subjected to selection and shared with other breeders.

REFERENCES

1.   Athwal DS: The Quarterly Rev. of Biol. 46, 1-34, 1971.
2.   Borlaug NE: In, KW Finlay and KW Shepherd (eds), Wheat breeding and its impact on world food supply.   3rd Int. Wheat Genet. Symp. Butterworth, London, pp.1-36, 1968.
3.   Vogel OA, RE Allan and CJ Peterson: Agron. J. 55, 397-398, 1963.
4.   Chandler RF: In, JD Easton, FA Haskins, CY Sullivan and CHM van Banvel (eds), Physiological Aspects of Crop Yield.   Amer. Soc. of Agron., Madison, Wis., pp.265-289, 1969.
5.   Brown PD, RIH McKenzie and K Mickaelsen.   Crop Sci. 20, 303-306, 1980.
6.   Marshall HG and CF Murphy: Crop Sci. 21, 335-338, 1981.
7.   Marshall HG, FL Kolb and JA Frank: Crop Sci. 23, p.404, 1983.
8.   Marshall HG and FL Kolb: Crop Sci. 23, p.190, 1982.

HYBRIDISING OF OATS UTILISING CHEMICAL HYBRIDISING AGENTS

M. E. McDANIEL
Texas A&M University, College Station, Texas 77843, USA

1. INTRODUCTION

Data pertaining to the performance of F_1 oat (Avena spp.) hybrids compared to commercial cultivars or parental lines are limited. Almost all oat heterosis studies have been based on small numbers of space-planted F_1 hybrid plants grown from hand-crossed seeds. Space-planted experiments probably do not give realistic estimates of heterosis, since parents and hybrids may not perform as they would in competitive stands. It is likely that heterosis estimates will be somewhat 'inflated' in space-planted tests, since environmental constraints to high yield on a per-plant basis are minimised. It also is likely that the relative contribution of various grain-yield components would be very different for space-planted and 'full-stand' trials. For example, it is certain that 'spaced' plants of winter-type oats produce a much higher number of functional tillers than plants grown in full stands. Hybrids having outstanding tillering capacity might exhibit most of their grain yield heterosis in terms of this yield component in space-planted tests. However in full-stand trials, this yield component likely would be expressed more moderately, and the other yield components might account for a larger proportion of heterosis for grain yield.

2. MATERIALS AND METHODS - HETEROSIS STUDY

When we discovered that the 'approach' crossing method[1] could be used to produce a relatively large number of hybrid seeds in oats, we conducted a study[2,3] of oat heterosis in more nearly 'normal' stands than those used in previous studies. Ten pure-line oat cultivars and selections were used as parents to produce seven F_1 hybrids for this study. Seven of the parents ('Houston', 'Ora', 'Alamo-X', 'Coronado', 'Florida 500', 'Cimarron', and 'Norwin') were cultivars grown commercially in Texas at the time this study was conducted. The remaining parents were experimental lines from complex crosses. The 10 parents represented a range of genotypes, as only Coronado and ABD 555 (sibs) were closely related. All parents except Wis. X941-4 were 'winter' oats typical of the type grown in the southern oat producing region of the USA. These 'winter' parents differed widely in winterhardiness and growth type. Norwin and Cimarron were the most prostrate and winterhardy of this group, and they did not perform as well as the other winter-type parents at the two Central Texas test locations during the season the test was conducted. The 'spring' parent, Wis. X941-4, also was not well adapted, although the relatively mild winter allowed full survival of this entry.

The approach cross method was used to obtain more than 500 F_1 seeds for each of six crosses, and approximately 250 F_1 seeds of a seventh cross. Seed set ranged from 38% to 73% for individual crosses, and averaged 61%

for the 7606 florets emasculated. Although some crosses were made specifically because of wide differences in agronomic type of the parents, most hybrid combinations were decided by coincidence of flowering time. An effort was made to minimise differential growth and vigour due to differences in seed size by planting only seeds from primary florets. It was obvious that the F_1 seeds produced in the 'clipped' emasculated florets were smaller than either parental or F_2 seeds; smaller seed size may have placed the F_1 plants at some disadvantage, at least in terms of initial growth. All parent, F_1, and F_2 plants were individually established in the glasshouse in 4.4 cm square peat pots. At approximately the three-leaf stage, plants were transplanted to the field locations at Temple and McGregor, Texas, during the first and second weeks of November, 1967.

Plots of parent, F_1 and F_2 entries were single rows 140 cm long containing 30 plants spaced approximately 4.7 cm apart; the between-row spacing was 30 cm. For each cross, the parents, F_1, and two F_2 populations derived from different F_1 plants were included at each location. Five replicates of six crosses were planted at each location, while the seventh cross was planted in only two replications at each test site. Since parent-hybrid comparisons within crosses were of greatest interest, a split-plot experimental design was chosen. Crosses assigned as main plots were randomised within each block, and the parent and hybrid populations within crosses assigned as sub-plots were randomised within main plots. Traits measured were grain yield and the yield components tillers per plant, seeds per panicle, and weight per seed. Individual plant data were collected for these characters at Temple, but only whole-plot data were collected at McGregor due to severe lodging and some culm breakage at harvest.

3. RESULTS - HETEROSIS STUDY

In these results, the term 'heterosis' is used to denote the superiority of the F_1 or F_2 over the higher parent. In some cases, research workers have used 'high-parent heterosis' or 'heterobeltiosis' to denote this comparison.

F_1 hybrids at Temple showed grain yield heterosis of 6 to 105% (Figs. 1-7). Six of the seven F_1s yielded significantly more grain than their better parents, and four F_1s exceeded their better parents by more than 40%. At McGregor, grain yields of the F_1s ranged from 94 to 230% of the high-parent yields (-6% to 130% 'heterosis'), with three of the yield increases being statistically significant. To conserve space, yield data for the McGregor test are not given; tabular data for this location have been published(3).

Grain yields of the hybrids also were compared to that of the highest-yielding pure line entry at each test location, since this yield comparison would determine the 'practical' competitiveness of any hybrid. Six of the seven F_1s at Temple had higher average yields than Coronado, with three of the yield advantages being statistically significant (Fig. 8). At McGregor, five of the seven hybrids produced higher average grain yields than the highest-yielding pure line entry, Florida 500. One hybrid outyielded Florida 500 by 40%, and two others exhibited appreciable (although statistically non-significant) yield advantages of 15 and 22% over this pure-line entry. Inbreeding depression was relatively severe in most crosses. However, two F_2s at Temple and two F_2s at McGregor yielded significantly more grain than their better-yielding pure-line parents.

The F_1 of CI 8262 x Norwin surpassed the best cultivar at Temple by 36% and the best cultivar at McGregor by 40%; it was the highest yielding entry at both locations. The agronomically 'diverse' parents of this F_1 were not outstanding at either location, demonstrating that superior hybrid combinations may be obtained from crosses of relatively unadapted or low-yielding parents. Similarly, the Wis. X941-4 x Cimarron hybrid (a cross between a spring oat and a 'strong' winter type with the parents having very different agronomic traits) had low yielding parents, but produced a hybrid that yielded about the same as the best pure-line entry at each location.

Heterosis generally was greater for yield than for yield components. Yield component data for individual hybrids at Temple are presented in Figures 1-7. In each of these figures, the grain yield and grain yield component data for the parent having the highest grain yield are equated to 100%, with these values represented by the line labelled 'H.P.'. The yield and yield components for the lower grain yielding parent are identified by a 'P', while the F_1 and F_2 yields and yield components are identified by '1' and '2' designations. This scaling system facilitated visual comparison of the yield and yield components of the two parents and of their F_1 and F_2 progenies. These figures show that seeds per panicle had the greatest and most consistent heterotic response and contributed most to grain yield in the hybrids studied. Four F_1s at Temple and three F_1s at McGregor exhibited significant heterosis for seeds per panicle. In contrast, only one cross showed any heterosis for number of tillers. This cross, CI 8262 x Norwin had a slightly higher tiller number than either parent at Temple, and a significantly higher tiller number than either parent at McGregor. In most other crosses, tiller number for hybrids was somewhere between values for the two parents. No significant heterosis for seed weight was observed. However, seed weight tended to be considerably above the mid-parent value for most crosses.

Although some of the hybrids exhibiting high grain-yield heterosis also showed significant increases for one or more of the grain yield components, it should be emphasised that the product of these yield components determines ultimate yield(4). Substantial yield increases can occur without high-parent heterosis for any yield component if the hybrid combines the better features of the parents. The F_1 of Coronado x Florida 500 provides an example of this type. Coronado has larger seed and more seeds per panicle than Florida 500, but has a much lower number of tillers. Although the F_1 did not significantly exceed the better parent for any of the individual yield components at either Temple or McGregor, the superior combination of the yield components of the F_1 produced significant yield increases of 31 and 25%, respectively, at these locations.

F_1 yield increases of the magnitude we observed indicate that if hybrid seed could be economically produced, grain yield of oats could be substantially increased. The production of larger quantities of hybrid seed would allow more critical testing of hybrids in standard nursery plots, and also would allow testing the hybrids for forage-yielding ability. Since the production of larger quantities of hybrid seed by hand crossing is not practical, and no practical genetic or cytoplasmic/genetic sterility system currently is available in oats, the use of chemical hybridising agents to 'sterilise' oats and allow field production of hybrid seed in crossing-blocks represents the only viable option at present. The high degree of heterosis we observed in our hybrid oat research makes pursuit of a successful hybridisation method worthwhile,

138

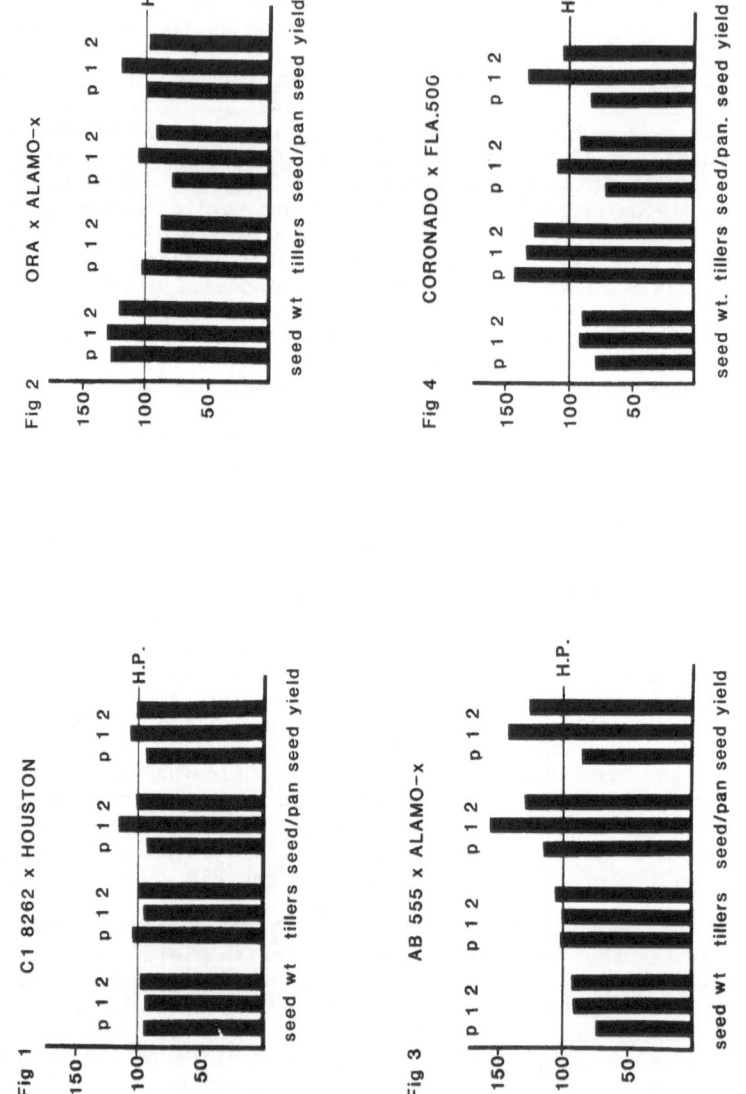

FIGURES 1 - 7. Grain yield components of parents and F_1 and F_2 oat hybrids at Temple, Texas. Yield and components of high parent (H.P.) = 100%. P = Low parent; 1 = F_1; 2 = F_2.

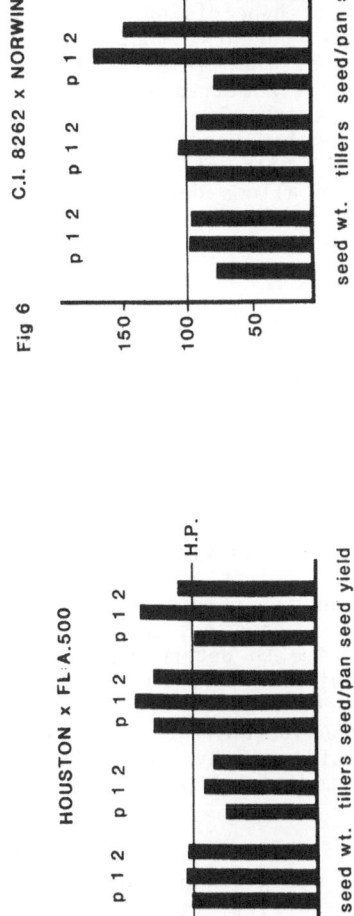

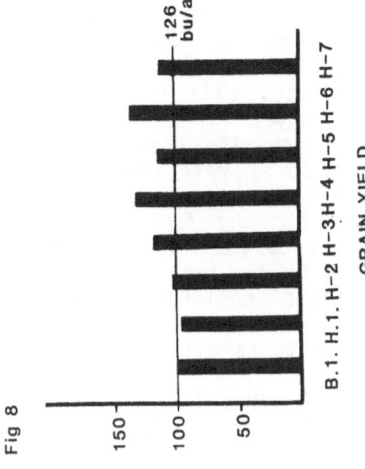

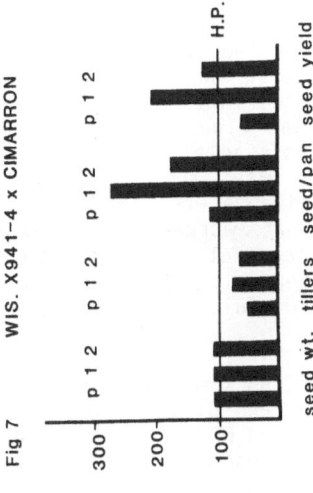

FIGURES 1-7. Continued

FIGURE 8.   Comparison of grain yield of best inbred parent (B.l.) to yields
of the seven F₁ hybrids in Figures 1-7 (H – 1 to H – 7)

140

and prompted our interest in the possible use of 'wheat' chemical
hybridising agents in producing experimental quantities of hybrid oat
seed.   Our initial results with two of these hybridising agents is
reported in the sections which follow.

4. MATERIALS AND METHODS - HYBRIDISING AGENTS

We have conducted preliminary field-hybridisation experiments with two
wheat chemical hybridising agents.   The chemicals utilised were SD84811
(Shell Development Company) and 0007 (Rohm and Haas).   Work with these
chemicals has been conducted under terms of Memorandi of Agreement with
the companies, and involves some confidentiality.   Details of application
timing, rates etc. will not be given here but the 'general' research
results will be presented.   Each of the chemical hybridising agents was
used in field crossing blocks in which 'female' parents (those treated
with the chemicals) were completely surrounded by the 'male' pollinator
parent.   Standard nursery plots of approximately 1.2 metres x 3.4 metres
were utilised for the females, so the pollen parent always was in very
close proximity to the 'sterilised' female.   Females used in each study
were susceptible to race 264B of crown rust, while all male parents
carried 'dominant' resistance to this race.   This allows determination of
the percentage of hybrid plants in the population from seed produced by
the female parent, as all resistant F_1 plants are hybrids.

5. RESULTS-FIELD HYBRIDISATION

Both of the 'wheat' hybridising agents are effective in inducing male
sterility in oats.   Female fertility appears to be reasonably good based
on limited attempts to pollinate sterilised plants in the glasshouse
(approach crossing method).   However, field seed set percentages have been
quite low, ranging from less than 1% to 7% in most cases.   The percentage
of hybrid seed in progenies from 'female' parents has ranged from 30 to
70% (rust test results).   These results are not very encouraging, and it
appears that the production of reasonably large quantities of hybrid seed
with these chemicals will not be easy.

The reason for the low percentage of field outcrossing is not known.   We
have observed that the flower 'gaping' or opening is not as pronounced as
one might like to see.   We have not yet determined the stigma longevity of
'sterilised' oat flowers.   In some cases we also have not been able to
observe the crossing blocks at the appropriate time to determine that the
'nick' between male and female parents was suitable and that adequate
pollen was available when female florets were receptive.

However, wheat workers also experienced considerable difficulty in
developing successful treatments to allow production of hybrid wheat seed
with these chemicals, so there is hope that additional work with oats
would prove fruitful.

REFERENCES
1. McDaniel ME, HB Kim and BR Hathcock: Crop Sci. 7, 538-540, 1967.
2. Hathcock BR: PhD Dissertation, Texas A & M University, 1970.
3. Hathcock BR and ME McDaniel: Crop Sci. 13, 8-10, 1973.
4. Grafius JE: Crop Sci. 4, 241-246, 1964.

BREEDING WINTER OATS - SPECIAL CONSIDERATIONS

J. Valentine and B.T. Middleton
Welsh Plant Breeding Station, Aberystwyth, Dyfed, Wales, UK

> "If you can look into the seeds of time,
> And say which grain will grow, and which will not,
> Speak, then, to me ...."
>
> (Macbeth, Act 1, Scene III)

1. INTRODUCTION

Some aspects of winter oat breeding in the UK have already been dealt with in a paper presented at the EUCARPIA cereal section meeting in Weihenstephan in 1984(1). General areas covered included an outline of the evolution of the Aberystwyth programme, recent varieties released, and the encouraging progress being made in breeding for appreciably higher winter-hardiness and in the development of naked or huskless oats. The present paper restricts itself to presenting an account of our work on selection for yield in the early generations of the pedigree breeding method with special reference to the Aberystwyth winter oat breeding programme.

2. VISUAL SELECTION OF YIELD AND ITS COMPONENTS

In the early generations, the eyes are the breeder's most valuable asset. Direct measurement of yield or yield components are impractical while the consensus of evidence in oats and other cereals is that visual assessment in rows in the F_3 generation and later is worthwhile. In winter oat rows, Valentine and Ismail(2) found visual assessment of yield and its components to be moderately to highly effective. Spacing rows 0.3 m apart to ensure clear separation of families between rows was considered to favour the high efficiency of visual assessment.

The value of visual assessment, or indeed any other selection procedure, in the earlier F_2 generation is more questionable. This is undoubtedly due to the high environmental component of variation between F_2 plants or ears rather than to inefficient visual assessment per se.

In the winter oat breeding programme, we find that spacing F_2 plants even only 5 cm apart predisposes the whole generation to virtually complete winter kill in cold winters. We now grow F_2s at more or less crop density, in which individual plants cannot readily be identified. Selection is therefore practised on an individual ear rather than on a plant basis.

2.1 Visual selection for grains/panicle in the F_2 generation

A recent experiment was undertaken to investigate whether visual selection for grains/panicle in two F_2 populations of winter oats improved grains/panicle or yield/row in the subsequent F_3 generation. This experiment took a different approach from Valentine and Ismail(2) in oats, and Ismail and Valentine(3,4) in barley, in which the efficiency of visual assessment within a single season and selection unit was considered. The term 'visual assessment' is used to indicate assessment of lines within a generation and season, while visual selection is used to indicate selection between generation and seasons.

2.1.1 Results

Results are presented in Table 1. While F_2 ears judged to have the highest number of grains ('high' selections) did in fact have about 80% more grains than randomly chosen ears ('random' selections), the effect was not carried over to the next generation. 'High' selections in F_3 had similar values for grains/panicle as 'random' selections.

TABLE 1. Effect of visual assessment of grains/panicle in the F_2 on F_3 performance

			F_2 1983			F_3 1984			
	Type of selection	n	Grains/ panicle	TGW (g)	Yield (g)	Grains/ panicle	TGW (g)	Tillers/ row	Yield (g)
Pop. 1	High	31	119.4	37.2	4.44	57.4	38.0	122.2	263
	Random	31	70.1	37.1	2.59	54.8	38.6	106.1	223
Pop. 2	High	22	129.8	32.9	4.27	59.7	32.6	125.8	245
	Random	22	68.6	33.6	2.29	62.9	32.9	93.4	192

Population 1 = 07198 Cn I/3 x 76-17 Cn 24
Population 2 = Bulwark x 76-17 Cn 26

'High' selections in F_3 had 25% more tillers/row than 'random' selections. This can be attributed to the fact that following our normal procedures we did not standardize the number of grains sown in each row but sowed the total number of grains in the F_2 panicle. This did not appear to be responsible for diminishing grains/panicle of 'high' selections as there were no strong negative correlations between tillers/row and grains/panicle.

2.1.2 Discussion

Looking at the associations between values of individual genotypes (Table 2), grains/panicle in the F_2 were not correlated with grains/panicle in the F_3 in either 'high' or 'random' selections.

TABLE 2. Correlations between F_2 ears in 1983 and F_3 rows in 1984

	Type of selection	n	Grains/panicle	TGW
Population 1	High	31	-0.04 NS	0.67***
	Random	31	-0.11 NS	0.69***
Population 2	High	22	-0.04 NS	0.15 NS
	Random	22	-0.36 NS	0.26 NS

Population 1 = 07198 Cn I/3 x 76-17 Cn 24
Population 2 = Bulwark x 76-17 Cn 26
NS = $P > 0.05$; *** = $P < 0.001$

Therefore, in terms of the overall means and individual values, we can conclude that visual selection of grains/panicle practised between individual ears in the two F_2 populations examined did not have a recognisable effect on grains/panicle in the subsequent F_3 generation. Since 'high' selections did have significantly more grains in the F_2 than 'random' selections (Table 1), the ineffectiveness of visual selection for grains/panicle appears to lie not in visual assessment of grains/panicle per se, but in the interactions of genotypes with selection unit, generation and season.

2.2 Correlations between F_2 and F_3 for thousand grain weight

This however, is not the end of the story. Although visual selection for high thousand grain weight was not a part of this experiment, it is interesting to note from Table 2 that, in one cross, values of thousand grain weight in the F_2 were correlated with values of thousand grain weight in the F_3. From the regression of F_2 values against F_3 values of thousand grain weight in population 1, a broad sense heritability of over 70% has been found. This strongly suggests that visual assessment for high thousand grain weight in the F_2 (derived from 07198 Cn I/3) would have been successful in selecting lines with improved thousand grain weight in the F_3 generation.

2.3 An association between high thousand grain weight and lemma non-fluorescence

Moreover, high thousand grain weight was associated with lemma non-fluorescence under UV light, and lower thousand grain weight was associated with lemma fluorescence (Table 3). (Fluorescence and non-fluorescence are also associated with white and yellow lemma colour respectively but is a more reliable character for distinguishing intermediate types and for categorising individual grains). Fluorescence in the F_1 and preponderance of fluorescent grains in the segregating F_3 families indicates non-fluorescence to be recessive to fluorescence, but it is not possible to tell from this experiment whether there are 1 or 2 genes identical to those now designated Lf-1 and Lf-2(5,6). From these results and other observations, including the fact that we have obtained a 100% fluorescent genotype with large grains following several rounds of gametogenesis, it appears that we may have identified an incompletely recessive gene fairly closely linked to a gene for non-fluorescence. If this is the case - and further genetic studies are under way - the identification of a major gene or gene block conferring 20% larger grains than normal, and its close linkage to a reasonably easily identifiable marker gene may be a major step towards breeding oats with appreciably higher yields and milling quality.

TABLE 3. Association between non-fluorescence and thousand grain weight in non-segregating families of Population 1 (07198 Cn I/3 x 76-17 Cn 24) and controls

	n	Non-fluorescence (%)	Thousand grain weight (g)
76-17 Cn 24	8	0	36.6
F_3 families	34	0	35.0
07198 Cn I/3/2	8	100	43.8
F_3 families	15	100	42.5

144

3. ACCELERATED PEDIGREE SELECTION - AN ALTERNATIVE APPROACH

Apart from this favourable situation, visual selection in the F_2 may also be useful in crosses involving unadapted genotypes, where many plants can be readily discarded. For very good breeding material, on the other hand, indirect selection using visual assessment of yield components in the F_2 generation, as illustrated for grains/panicle in two crosses, and thousand grain weight in one cross in the absence of major gene effects, is likely to have relatively low efficiency. In these circumstances, we adopt an alternative approach for some desirable crosses, which we call 'accelerated pedigree selection' (APS). This bases initial selection not on individual plants but on the assessment of lines in rows derived from unselected individual plants. Accelerated generation procedures can be used to derive these lines so that they are available for assessment in the same period as individual plants. In fact there is a shortening of the overall length of the breeding cycle(7).

We have used APS to produce a winter naked oat, Kynon, which is in National List trials in 1985 from a cross only made in 1979.

APS differs from single seed descent (SSD) in that pedigree selection begins at an earlier stage of the breeding cycle. APS involves only 1 or 2 accelerated generations in which the breeder can afford to grow plants at a wider spacing compared to SSD. It has two main advantages over SSD. First, APS results in an earlier assessment of breeding material in rows under field conditions (2 years after crossing compared with three or more in SSD). Secondly, APS avoids the considerable risk of mortality and sterility which results from growing plants at very close spacing(8,9).

For both reasons, we consider APS more suitable than SSD, particularly for winter cereals, such as winter oats, where there is segregation for vernalisation requirement as well as for maturity and height. The provision in APS of unselected single plant progenies may also allow greater emphasis to be placed on breeding for characteristics for which special early generation tests are required, such as high nutritional quality or high winter-hardiness, for which individual plant selection is not possible. The routine provision in APS of random F_2 plants and progenies may also allow closer genetic examination of breeding material.

REFERENCES
1. Valentine J: Progress in breeding winter oats in the UK. Proceedings of the Eucarpia Cereal Section Meeting, Weinhenstephan 1984, Vortrage für Pflanzenzuchtung 6, 239-248, 1984.
2. Valentine J and AB Ismail: Ann. Appl. Biol. 102, 551-556, 1983.
3. Ismail AB and J Valentine: Ann. Appl. Biol. 102, 539-549. 1983.
4. Ismail AB and J Valentine: Ann. Appl. Biol. 104, 367-373, 1984.
5. Finkner RE, HC Murphy, RE Atkins and DW West: Agron. J. 46, 270-274, 1954.
6. Simons MD, JW Martens, RIH McKenzie, I Nishiyama, K Sadanaga, J Sebesta and H Thomas: USDA, Agric. Handbook No. 509, 1978.
7. Valentine J.: Euphytica 33, 943-951, 1984.
8. Riggs TJ and AM Hayter: Practical aspects of the single seed descent methods in barley breeding. In, Barley Genetics III. Proceedings of the Third International Barley Genetics Symposium, Garching 1975, 708-717, 1976.
9. Jenkins G: Breeding winter barley and the exploitation of spring x winter crosses. In, Proceedings of the 4th Regional Winter Barley Workshop. Vol. II, Amman, Jordan 1977, 70-81, 1978.

PEDIGREE OF GERMAN OAT CULTIVARS AND THEIR IDENTIFICATION BY ISOELECTRIC FOCUSSING

U. BICKELMANN
LUFA Augustenberg, Nesslerstr.23, Postfach 430230,
D-7500 Kaslsruhe 41, FRG

On establishing the pedigree of all oat cultivars licensed in Germany since 1900, we find a very narrow relationship between them. Nearly all cultivars trace back to the crossing of two landraces: i.e. a yellow and a white oat. The early breeding work included crossings between yellow and white oats on the one hand and brown oats on the other.

Twelve out of the 27 existent oat cultivars (Avena sativa) have the Danish variety Minor as a common ancestor. Although the relationship is very close, it is possible to distinguish 23 out of 27 oat cultivars. The method used is isoelectric focussing of the prolamins of single seeds at pH 5-8. Some varieties develop two or three different types of banding patterns. The banding patterns of the varieties often combine the banding patterns of their parents. But three varieties deriving from Selma have the same type of banding pattern as Selma. These have to be distinguished by another electrophoretic method.

SESSION IV

Crop physiology and production methods of oats

Chairman: C.M. Brown

CROP PHYSIOLOGICAL APPROACHES TO INCREASED PRODUCTIVITY IN OATS

J. B. BROUWER
Victorian Crops Research Institute, Horsham, Victoria, Australia.

1. INTRODUCTION

Physiological approaches to increased productivity in oats focus on either crop growth or crop development or both. These phenomena need separate consideration as they are affected differently by genetic and environmental factors. Plant breeders may increase crop productivity by improving the crop as a photosynthetic system, i.e. improved harvest index(HI) or total dry matter(DM) production, or by dissecting economic yield into its separate components, for example, individual grain weight or grain number per unit area in oat grain production, and by concentrating on one or more of these yield determinants.

2. DEVELOPMENT

2.1 Developmental phases

Crop development can be divided into several phases. The period from emergence to floral initiation is referred to as the vegetative phase. It is followed by the reproductive phase (jointing, shooting or elongation) which finishes at heading or anthesis. The period from anthesis to maturity is called the grain development phase. Interactions between genotype and environment during each phase determine the time-scale of the crop.

Oat cultivars are commonly classified by their respective maturity type as being either early, mid-season or late. This classification is based on the usual period of time taken from emergence to either heading or final grain maturity in certain environments. It is important to know to what extent photoperiod and temperature control this development and whether there is scope for partitioning the life cycle of the crop into phases of differential sensitivity.

Differences in days from emergence to heading were found to be largely attributable to variation in days from emergence to floral initiation or jointing in spring cerals(1,2,3). This is despite the fact that for some combinations of cultivar and sowing date the reproductive phase may be equal or even longer than the vegetative phase.

The duration of the grain development phase can also vary substantially. In field sowings of a range of different oat cultivars, varietal differences were recorded of up to 4 days(4) and 11 days(5) for the duration of the period from heading to maturity, and from anthesis to maturity, respectively. By manipulating controlled environmental conditions, Klinck and Sim(6) created a difference of at least 12 days in the period between anthesis and maturity.

2.2 Photoperiod and temperature

Oats in general have been regarded as quantitative long-day plants(7) with increasing photoperiods hastening the onset of flowering. Some oats have been observed not to flower at photoperiods of less than 12 hours, unless the dark period was interrupted by a short interval of light(8). No further responses are obtained in oats by increasing photoperiods above 18-19 hours(9).

Considerable diversity in responses to photoperiod has been found in several _Avena_ species, including the cultivated oat(1,10). Greater responsiveness to daylength occurs generally among oats orginating from higher latitudes. Griffiths(11) found a definite relationship between geographic origin and the limits of tolerance to short day. Response to photoperiod appeared under genetic control(9,11).

The strong influence of photoperiod, even on adaptation of comparatively recent introductions of the wild oat species A.fatua and A.barbata, is evident in Western Australia(12). Van Dobben and Wiersema(13) reported that two European oat cultivars were not responsive to photoperiod until the plants reached the tillering stage.

In hexaploid wheat, the adaptability to diverse climates depends, to a large extent, on its vernalisation and photoperiod responses(14). In addition, the term 'basic development rate' has been introduced(15) to indicate the existence of differences in rate of development in the absence of influences of vernalisation and photoperiod. This phenomenon appears to be influenced by temperature and may exhibit close interaction with photoperiod response(15). It is likely that a similar adaptive mechanism exists in oats, but most studies of the effects of temperature on crop development in oats have been conducted in the field without control over temperature.

Aitken(1) reported that floral initiation was hastened by decreasing temperature and that temperature above 15°C delayed initiation, and Wiggans(16) observed varietal differences in number of heat units required from planting to maturity. Variation caused by other factors such as disease, rainfall, soil type and cultural technique was of only minor significance.

Klinck and Sim(6) grew two oat cultivars in controlled environments and showed that warm temperatures (21°C day, 15°C night) and long days (18 h) shortened each of the three phases delineated by sowing, floral initiation, anthesis and maturity, irrespective of cultivar or environment during the preceding phase. Daylength was the major factor influencing the rate of development in the second phase, and temperature had the greatest effect in the grain development phase. Development during this final phase was especially hastened by warm conditions, when the preceding phase had been under cool temperatures. Conversely, maturity was delayed when the grain development phase was subjected to cool temperatures after a reproductive period under warm conditions. Temperature treatments caused greater variation in rate of development in one cultivar, whereas the other cultivar was more influenced by daylength.

2.3 Vernalisation

According to Kirby(7), oats have no vernalisation requirements and in contrast to rye, wheat or barley, the winter habit is less distinct(9). So-called winter cultivars of oats are considered to have quantitative rather than obligatory requirements for low temperatures as they will eventually flower without such treatment, if given a sufficiently long growing season(10,11).

Nevertheless, there are numerous reports of responses to natural and induced vernalisation in cultivated oats. Coffman and Frey(17), quote, inter alia, results published by Borodin who observed that even 50 days was insufficient to vernalise seed of Winter Turf oats at 2°C and 50% seed moisture content.

Sampson and Burrows(10) were the first to point out that response to cold vernalisation is widespread among the wild Avena species from the Mediterranean, and even demonstrated that responses to short-day vernalisation occur. Cold vernalisation responses have since been reported for A.barbata and A.sterilis in Australia(12,18) with evidence of a relationship with geographic distribution(12). Inheritance studies for vernalisation responses in oats are limited(19).

2.4 Dormancy
Dormancy is another trait which can be manipulated in adapting crop development to specific environments.

In many parts of the world with severe winters, autumn-sown oat crops often fail because of insufficient cold-hardiness for overwintering being available during the vegetative stage. As an alternative management system 'dormoats'(20) have been synthesised in Canada which can be sown in autumn but, because of the survival mechanisms of seed dormancy and overwintering capacity in the seed stage, introduced from A.fatua, they tend to delay germination until spring, thus avoiding the winter-kill problem. Problems of low spring emergence from autumn sowings still remain to be solved before commercial application is possible.

3. YIELD, DEVELOPMENT AND GROWTH UNDER OPTIMAL CONDITIONS
Grain yield can be regarded as a function of growth rate, growth duration and HI.

Final yield may not always be associated with duration of developmental phases(6) and where variation in growth period is restricted by environmental limits, manipulation of growth rates or improved partitioning of assimilate products must be considered. Donald and Hamblin(21) suggested that for grain yield to increase, plant breeders should combine high HI and high biomass productivity.

It remains important, however, to relate such responses to their initiation during specific and different developmental phases as many successful manipulations of the crop during early stages may be altered or even nullified by compensatory physiological changes later in its development.

3.1 Harvest index
The genetic improvement of oat cultivars has been accompanied largely by increased HI which may or may not be associated with higher biomass productivity(22,23,24,25).

A high degree of variability is found for HI in all cereals, with a range in oats of 40-50%(26,27). Prospects for further improvement in oats beyond this level have been questioned(28) but the possibility of a rise to 60% has been suggested for winter wheat(29). Rosielle and Frey(30) showed that HI would be most efficiently used as a secondary trait in selection for grain yield with further restrictions put on heading date and plant height. Inheritance of HI was found to be based primarily on additive gene action, but high HI x high HI crosses tended to produce limited variation(30).

In using HI as a selection criterion, relationships with other plant attributes, such as competitive ability, must be considered. HI tends to show a negative association with dry weight of all vegetative parts of the plant(26). Improved HI values have often been accompanied by shorter straw, but as the period during which a high proportion of assimilates is needed for stem growth, coincides with the period of maximum growth of the inflorescence(21), improvement in favour of the panicle may also occur because of increased competitiveness within the plant.

Population density and application of nitrogen affect HI values and in dryland cropping there may be an optimum level of vegetative growth imposed by water availability(31).

Plant ideotype is of relevance to contrasting stages in a plant breeding programme, which are typified by the use of spaced plants and mixed populations in early generations and dense homogeneous stands in later generations(21), and Bertram and Stuthman(32) have expressed doubts about the validity of HI measurements on spaced oat plants as criteria for selecting genotypes which are high yielding in a crop situation.

3.2 Growth

The grain yield improvement associated with introgression of A.sterilis germplasm has been attributed primarily to increased vegetative growth rate(VGR) by Takeda and Frey(27), but high and positive correlations were reported for VGR with grain yield in both interspecific crosses (sterilis x sativa) and in intraspecific (sativa) crosses(27,28). VGR was suggested as an efficient criterion for indirect selection for grain yield(33,34). This trait was found to be quantitatively inherited, with heritability values similar to or slightly higher than those of grain yield and HI(27,28,35). Additive and non-additive gene action were considered to be involved in different cross combinations(33). Positive transgressive segregation occurred for VGR at a higher rate than for biomass or for grain yield(36). Cox and Frey(37) reported percentages of transgressive segregants which were two to three times higher in populations from interspecific than from intraspecific crosses, although 2.5 times more transgressive segregants for grain yield were found in intra-sativa crosses compared to interspecific crosses.

There is ample evidence for specific combining ability for grain yield, biomass and VGR in both types of crosses (35,36,38), but broad-based A.sativa crosses may constitute a more efficient short-term strategy for improvement of both grain yield and VGR than introgression from a limited number of randomly chosen accessions of A.sterilis germplasm(39).

When the genetic relationship between HI and VGR was examined in progenies of A.sativa crosses and of sterilis x sativa(27,28,37), the two traits were found to be nearly independent, indicating that grain yield gains could be made by selecting for either separately or by simultaneous selection for both traits. Up to 95% of increases in grain yield have been attributed to variation in VGR and HI(28).

VGR values in the above studies have been obtained by dividing straw yield at maturity by days to heading. The method is based on the findings of Frey et al.(40) that straw yield does not change between heading and maturity. The results of Jennings and Shibles(41) support this view whereas Wych et al.(5) reported that straw dry weight may increase or decline after heading and suggested that changes in post-heading 'phytomass' are subject to genotype x environment interactions. Evans(42) and Stoy(43) considered that remobilisation or assimilate reserves stored in vegetative parts before anthesis contributes only a relatively small proportion to final grain yield in temperate cereals (ca. 10%) under non-stress conditions. However, greater redistribution of mobilised reserves from stems and roots to panicles has been found by Gmelig Meyling(44) for more modern oat cultivars, with 30-50% of the dry matter accumulation in the panicle attributable to such reserves. Peterson et al.(45) calculated that the net loss of non-structural carbohydrates shown by stems and leaves in the six oat cultivars studied, was equal to 15-24% of their accumulation in the panicle. Substantial differences (up to 124%) have been recorded between oat cultivars in respiration and remobilisation to the kernel during kernel development(46).

Generally, cooler temperatures are associated with increased dry matter production and growth rate in oats during several developmental phases(6,47), with some cultivars being more sensitive to temperature than others(48). A correlation has been reported between net assimilation rate(NAR) and the difference between day and night temperatures(49).

Van Dobben(50) regarded the shoot:root ratio as an important growth regulating factor, which changes from the vegetative to the reproductive phase, with more productive cultivars having a higher shoot:root ratio during the grain-filling phase. Greater root development was shown to be a feature of modern oat cultivars adapted to spring-sowing(51).

3.3 Photosynthetic efficiency and canopy structure

Differences in photosynthetic activity per unit leaf area exist among oat genotypes(52), but no correlation with final grain yield or total dry weight at anthesis has been established. NAR is known to differ between cultivars(53,54) and interspecific differences have been demonstrated during tillering(55). However, Pazos(56) did not observe a noticeable change in rate of dry weight accumulation in A.sativa lines during the transition of pre- to post-anthesis growth.

Investigations into sink-source relationships have identifed the importance of the contributions of the panicle(41,57) and the two top leaves(58,59).

The importance of crop canopy duration and structure for crop growth has been pointed out for many crops(42,60). Greater leaf area and leaf-area duration or both, for either flag leaf or lower leaves, were found to be associated with substantial yield increases in oats(61,62). The relationship between leaf angle and grain yield has received ample attention in other crops(63) but only a few studies have been conducted in oats. Tanner et al.(64) noted that oat cultivars with small erect leaves were the highest yielding ones. Plant tissue below the flag leaf contributed more towards grain yield in oat cultivars with sparse leaf canopy and erect leaves, than in cultivars with moderately dense canopy and droopy leaves(41,58).

3.4 Grain yield components

There are numerous analyses attributing grain yields of certain cultivars in particular environments to individual yield components(9,17). These components may be inherited independently, but intra-crop competition often causes yield component compensation in plants(65). That components determined in earlier developmental phases have a great influence on components determined later in the developmental sequence, has also been demonstrated for oats(66,67).

Tillering capacity is a relatively stable plant trait(17). Although winter cultivars often have a higher mean tiller number than spring cultivars, indicating a positive relationship with the duration of the vegetative period, opinions differ about the desirability of profuse tillering in cereals(68,69). The contribution of panicles to grain yield has received comparatively more attention.

3.4.1 Panicle development. Panicle development can be partitioned into a spikelet addition phase, followed by panicle internode elongation, floral morphogenesis and finally, grain development.

Processes governing ear development in cereals may offer possibilities for increasing the number of potential grain sites(70). Higher rate and duration of spikelet initiation, increased apex size, and more leaves on the seed embryo could be potential selection criteria. Cisar and Shands(71) detected genotypic differences for duration of spikelet addition and for growth rate during this period.

Opportunities for increasing panicle yield can also be sought in its structure and development pattern. The number of spikelets per panicle in oats depends to a great extent on the length of rachis, the number of whorls, and the number of primary branches per whorl(9). The length of the rachis is genetically fixed(9,72), but there is no correlation with straw height(9). Up to nine panicle nodes or whorls have been observed(9). Finkner et al.(73) found sufficient genetic variation for rachis node number to conclude that selection based on F3 progeny means would be successful in selecting for increased numbers.

Bonnett(74) described the branching habit of the oat panicle with the number and length of branches at each successive node increasing from apical to basal node. Since the latter node produces the greatest number of branches and possesses the longest branch of comparable order, the spikelet sites there are the most numerous. Some cultivars have one-third or more of all spikelets on branches at the basal node(75).

Spikelet development within an oat panicle proceeds basipetally, but floret development within a spikelet occurs acropetally(74). Initially, there may be up to 12 potential florets per spikelet, which only completely develop in naked oats. The final number of florets per spikelet decreases basipetally within the panicle(9). A high correlation between number of rachis nodes and number of florets per panicle has been reported(76). Spikelets with one grain were more frequent at the base whorl, whereas three-grain spikelets occurred more often at the top.

Development of cultivars with tertiary seed set may be another way of increasing sink size. Takeda and Frey(77) observed varietal differences for this trait, which appears to be under control of multiple alleles at two loci(78).

The capacity of vascular tissue for transport of assimilates has to be considered when attempting to increase sink size within the panicle. A strong positive relationship was observed between vascular measurements of peduncle and panicle internodes, and number of spikelets and kernel, but not with grain weight(79,80). Number and size of vascular bundles appear to develop in concord with the number of spikelets initiated. Some cultivars seem to respond more than others to environment in altering their vascular morphology.

3.4.2. Grain development. In studies of the effects of rate of grain filling and its duration in cereals, the former trait is often positively correlated with grain yield(4,5,81). Extension of grain filling periods tends to be associated with earlier heading. Higher temperatures towards the end of the growing season shorten the grain filling phase, but enhance grain filling rate. Genotypic differences in heading date or synchrony of flowering can further confound their relative importance as yield determinants. In spite of so many factors influencing individual grain weight, thousand grain weight appears to be one of the more stable yield components(23), but opinions differ as to its use as a selection criterion for improving grain yield(82,83).

Murphy and Frey(84) studied ways of improving grain weight by means of its components, viz., kernel weight, size and shape. Kernel width required nearly three weeks for complete development and was no more heritable than kernel weight. However, variation in kernel width had a greater effect on kernel weight than kernel length, which was largely determined within eight days after anthesis, but had a higher heritability. The independent inheritance of kernel length and kernel width offers possibilities for kernel weight improvement through simultaneous selection for both components.

The position of the grain in the panicle has an effect on grain characteristics. Youngs and Shands(85) found highly significant and positive associations between distance of the grain from the basal panicle node, and both grain weight and kernel weight. This agrees with the data of Frimmel and Von Koppelow(76), showing that grains in the lowest whorl where most grains were located (ca. 40%), tended to have lower kernel percentage and kernel weight. While the number of vascular bundles decreased in each successive higher internode of the panicle main axis(74), grains developing in the upper whorls were served by more vascular tissue per grain than those lower down(79). Most tertiary grains appeared at the top of the panicle. Whilst these findings may point to limitations in assimilate supply, primary and secondary groats from three-grain spikelets had a higher average weight than those of two-grain spikelets. This indicates that assimilate supply may not be limiting, even where local sink size is increased.

Although synchronous development of flowering and of grains within panicles has been suggested(76,86) as a means of increasing uniformity in grain size, since later developing grains may receive progressively less photosynthates or may even abort, the ability to adjust the number of developing grains in response to prevailing conditions, may improve adaptability(87).

4. RELATIONSHIPS DEVELOPMENT-GROWTH UNDER STRESS

Knowledge of sensitivity to environmental stress during certain developmental phases may assist in avoiding or minimising such problems.

4.1 Drought tolerance

Oats are generally regarded as being the least efficient of all small grains with the exception of rice, in their use of water to produce a unit of dry matter(17,88), although contradictory data are available(89).

Drought stress just prior to or during heading causes the greatest reduction in final grain yields, whereas drought during the earlier tillering stage results in little damage(17,88). Chinnici and Peterson(90) subjected three oat cultivars to stress treatments of heat, cold and drought, at four stages of panicle development. Drought caused the greatest yield reduction at all stages of application, in particular at the booting stage. Sandhu and Horton(91) reported, however, that the cultivar Jaycee showed the greatest sensitivity to drought at the anthesis to early grain development stage.

Percentage blast or abortive floret development has been suggested as a possible selection criterion in oats for tolerance to drought stress during the reproductive phase(90).

Genotypic variability for root volume and distribution system in oats has been demonstrated(92,93) with prospects of a four-fold increase in root volume. In oat growing regions where transient drought problems occur, increased root depth may be required and a well developed seminal root system may be desirable(94).

The highest percentage of dry matter in the roots of some spring oats occur at the start of panicle formation(88). Stucker and Frey(93) found that the root dry weight to a soil depth of 60 cm did not change between the boot stage and maturity for five oat cultivars. A later study using nutrient solution(92), showed that root volumes of certain oat lines can continue to increase right up to maturity.

In regions of Sweden, where drought conditions occur in early summer before heading, cultivars in the 'black oat' group showed the highest adaptation(95). They exhibit drought avoidance, which is associated with

156

a high root:shoot ratio in the seedling stage, and an extended period of tillering as compared to other varieties. A rapid, non-destructive screening method using measurements of seedling length and width, has been developed for such environments(95).

4.2 Waterlogging tolerance

Research into variability for tolerance to transient or short-term waterlogging has been somewhat neglected in oats, and comparisons of oats with other cereals for sensitivity to waterlogging are scarce(96,97,98).

The extent of the damage caused by waterlogging depends on the developmental stage during which the plant is under stress. Although waterlogging depresses growth at all stages, the most sensitive period appears to be just prior to flowering or during early flower development(99).

Post-stress growth rate and recovery time is of major consideration in assessing waterlogging tolerance where transient waterlogging problems occur. The oat cultivars studied in waterlogging experiments(97,98) were more affected by excess water than other cereals as indicated by degree of reduction in tillering and in number of grains per panicle. However, they compensated better for such losses later in development, in particular, by increased grain weight.

5. GENERAL CONCLUSIONS

Partitioning of determinants of yield whether they be developmental or morphological factors, is a potentially powerful tool in plant breeding, and deserves more attention in oat improvement. There is little evidence of purposeful development of oat lines with specific physiological characteristics in both uniform and diverse genetic backgrounds. Such material would allow more precise evaluation of the effects of physiological traits, and would also lead to a better understanding of genotype x environment interactions affecting their expression.

Greater variability is now becoming available through introgression of wild oat germplasm, in particular A.sterilis, and is opening exciting vistas for oat improvement. The findings of specific combining ability among oat genotypes for various physiological attributes, with evidence of transgressive segregation giving rise to even greater variation, focusses on the necessity of gaining knowledge about the genetic basis of variation in physiological traits.

As Murphy and Frey(39) have pointed out, there is a need for efficiently identifying superior wild parents 'among the myriads'. A rapid but systematic evaluation of genetic diversity for useful adaptive traits, is probably more urgent than in-depth studies of metabolic processes, which by their nature, are restricted to limited samples of variability.

REFERENCES

1. Aitken Y: Aust. J. Agric. Res. 12, 371-388, 1961.
2. Pinthus MJ: Bull. Res. Council Israel D7, pp.71-80, 1959.
3. Ross WM: Agron. J. 47, 453-457, 1955.
4. McKee GW, HJ Lee, DP Knievel and LD Hoffman: Agron. J. 71, 1029-1034, 1979.
5. Wych RD, RL McGraw and DD Stuthman: Crop Sci. 22, 1025-1028, 1982.
6. Klinck HR and SL Sim: Can. J. Bot. 55, 96-106, 1977.
7. Kirby EJM: Field Crop Abstr. 22, 1-7, 1969.
8. Wiggans SC and KJ Frey: Proc. Iowa Acad. Sci. 62, 125-130, 1955.
9. Coffman FA and J MacKey: In, H Kappert & W Rudorf (eds). Handbuch für Pflanzenzüchtung. Paul Parey, Vol. 2, pp.425-531, 1959.

10. Sampson DR and VD Burrows: Can. J. Plant Sci. 52, 471-482, 1972.
11. Griffiths DJ: J. Agric. Sci. 57, 279-288, 1961.
12. Paterson JG, WJR Boyd and NA Goodchild: J. Appl. Ecol. 13, 265-272, 1976.
13. Dobben WH van and AC Wiersema: Verslag CILO, Wageningen, pp.164-166, 1954.
14. Evans LT, IF Wardlaw and RA Fisher: In, LT Evans (ed.). Crop Physiology, Some Case Histories. Cambridge Univ. Press, pp.101-149, 1980.
15. Flood RG and GM Halloran: Adv. Agron. 39, 1986 (in press).
16. Wiggans SC: J. Agron. 48, 21-24, 1956.
17. Coffman FA and KJ Frey: In, FA Coffman (ed.). Oats and Oat Improvement. Am. Soc. Agron., pp.420-464, 1961.
18. Whalley RDB and JM Burfitt: Aust. J. Agric. Res. 23, 799-810, 1972.
19. Rose JL: Proc. Aust. Cereal Pasture Breed. Conf., Brisbane, pp.15-16, 1966.
20. Burrows VD: Can. J. Plant Sci. 50, 371-377, 1970.
21. Donald CM and J Hamblin: Adv. Agron. 28, 361-405, 1976.
22. Gmelig Meyling HD and WH van Dobben: Jaarboek IBS 1965, Wageningen, pp.81-88, 1965.
23. Lawes DA: J. Agric. Sci. 89, 751-757, 1977.
24. Sims HJ: Aust. J. Exp. Agric. Anim. Husb. 3, 198-202, 1963.
25. Wych RD and DD Stuthman: Crop Sci. 23, 879-881, 1983.
26. Singh ID and NC Stoskopf: Agron. J. 63, 224-226, 1971.
27. Takeda K and KJ Frey: Crop Sci. 16, 817-821, 1976.
28. Takeda K, KJ Frey and TB Bailey: Can. J. Plant Sci. 60, 379-384, 1980.
29. Austin RB, J Bingham, RD Blackwell, LT Evans, MA Ford, CL Morgan and M Taylor: J. Agric. Sci. 94, 675-689, 1980.
30. Rosielle AA and KJ Frey: Crop Sci. 15, 544-547, 1975.
31. Fisher RA and GD Kohn: Aust. J. Agric. Res. 17, 281-295, 1966.
32. Bertram RB and DD Stuthman: Agron. Abstr. p.47, 1978.
33. Helsel DB and KJ Frey: Theor. Appl. Genet. 65, 219-223, 1983.
34. Helsel DB and RK Skdla: Zeitschr. Pflanzenzücht. 90, 316-323, 1983.
35. Takeda K, KJ Frey and DB Helsel: Zeitschr. Pflanzenzücht. 82, 237-249, 1979.
36. Cox DJ and KJ Frey: Crop Sci. 24, 963-967, 1984.
37. Cox DJ and KJ Frey: Theor. Appl. Genet. 68, 239-245, 1984.
38. Takeda K and KJ Frey: Euphytica 26, 309-317, 1977.
39. Murphy JP and KJ Frey: Crop Sci. 25, 531-536, 1984.
40. Frey KJ, PL Rodgers, WF Wedin, L Walter, WJ Moline and JC Burns: Iowa State J. Sci. 42, 9-18, 1967.
41. Jennings VM and RM Shibles: Crop Sci. 8, 173-175, 1968.
42. Evans LT: In, LT Evans (ed). Crop Physiology, Some Case Histories, Cambridge Univ. Press, pp.327-355, 1980.
43. Stoy V: In, JHJ Spiertz & T Kramer (eds). Crop Physiology and Cereal Breeding. Proc. Eucarpia Workshop, Wageningen, Pudoc, pp.55-59, 1978.
44. Gmelig Meyling HD: Meded. IBS, Wageningen, 415, p.1-7, 1969.
45. Peterson DM, LE Schrader, DA Cataldo, VL Youngs and D Smith: Can. J. Plant Sci. 55, 19-28, 1975.
46. Cataldo DA, LE Schrader, DM Peterson and D Smith: Crop Sci. 15, 19-23, 1975.
47. Smith D: Can. J. Plant Sci. 54, 725-730, 1974.
48. Peterson DM and LE Schrader: Crop Sci. 14, 857-861, 1974.
49. Stoskopf NC, HR Klinck and HA Steppler: Can. J. Plant Sci. 46, 397-404, 1966.

158

50. Dobben WH van: Neth. J. Agric. Sci. 10, 377-389, 1962.
51. Murphy CF, RC Long and LA Nelson: Crop Sci. 22, 1005-1009, 1982.
52. Lawes DA and KJ Treharne: Euphytica 20, 86-92, 1971.
53. Criswell JG and RM Shibles: Crop Sci. 11, 550-553, 1971.
54. Jenkins G: Ann. Rep. Plant Breed. Inst., Cambridge for 1962-63, p.59, 1964.
55. Thurston JM: Ann. Appl. Biol 47, 716-739, 1959.
56. Pazos DB: MS Thesis Iowa State Univ., Ames (cited in ref. 27).
57. Takeda G: Bull. Nat. Inst. Agric. Sci., Japan, D, 29, pp.1-65, 1978.
58. Frey KJ: Iowa State J. Sci. 37, 17-22, 1962.
59. Klinck HR and SL Sim: Ann. Bot. 40, 785-793, 1976.
60. Wilson D: In, KJ Frey (ed.). Plant Breeding II. Iowa State Univ. Press, pp.233-290, 1981.
61. Brinkman MA and KJ Frey: Crop Sci. 17, 426-430, 1977.
62. Pazos DB and KJ Frey: Agron. Abstr. p.75, 1976.
63. Duncan WG: Crop Sci. 11, 482-485, 1971.
64. Tanner JW, CJ Gardener, NC Stoskopf and E Reinbergs: Can. J. Plant Sci. 46, p.690, 1966.
65. Adams MW: Crop Sci. 7, 505-510, 1967.
66. Barnard J: Diss. Abstr. Int. B. 39,7,3074B, 1979.
67. Brinkman MA and KJ Frey: Crop Sci. 17, 165-168, 1977.
68. Donald CM and J Hamblin: Adv. Agron. 36, 97-145, 1983.
69. Evans LT and IF Wardlaw: Adv. Agron. 28, 301-359, 1976.
70. Gallagher JN: In, JHJ Spiertz & T Kramer (eds.). Crop Physiology and Cereal Breeding. Proc. Eucarpia Workshop, Wageningen, Pudoc, pp.3-9, 1979.
71. Cisar G and HL Shands: Crop Sci. 18, 461-464, 1978.
72. Young-am Chae and RA Forsberg: Crop Sci. 15, 457-460, 1975.
73. Finkner VC, CG Ponoleit and DL Davis: Crop Sci. 13, 84-85, 1973.
74. Bonnett OT: Univ. Ill. Agric. Exp. Stn Bull. 672, p.112, 1961.
75. Bonnett OT: Univ. Ill. Agric. Exp. Stn Bull. 721, p.105, 1966.
76. Frimmel G and E von Koppelow: Bodenkultur 26, 254-260, 1975.
77. Takeda K and KJ Frey: Crop Sci. 20, 771-774, 1980.
78. McBratney BD and KJ Frey: Cereal Res. Commun. 11, 91-97, 1983.
79. Housley TL and DM Peterson: Crop Sci. 22, 259-263, 1982.
80. Peterson DM, TL Housley and TM Luk: Crop Sci. 22, 274-278, 1982.
81. Metzger DD, SJ Czaplewski and DC Rasmusson: Crop Sci. 24, 1101-1105, 1984.
82. Frey KJ and TF Huang: Euphytica 18, 417-424, 1969.
83. Sampson DR: Can. J. Genet. Cytol. 13, 864-872, 1971.
84. Murphy CF and KJ Frey: Crop Sci. 2, 509-512, 1962.
85. Youngs VL and HL Shands: Crop Sci. 14, 578-580, 1974.
86. Klinck HR and C Deslauriers: Can. J. Plant Sci. 60, 316-317, 1980.
87. Pinthus MJ: World Crops 32, 47-48, 1980.
88. Salter PJ and JE Goode: Crop responses to water at different stages of growth. Comm. Agric. Bur., Res. Rev. No. 2, p.246, 1967.
89. Hobbs EH and KK Krogman: Can. J. Plant Sci. 54, 23-27, 1974.
90. Chinnici MF and DM Peterson: Crop Sci. 19, 893-897, 1979.
91. Sandhu HS and ML Horton: Agron. J. 69, 361-364, 1977.
92. Carrigan L and KJ Frey: Crop Sci. 20, 407-408, 1980.
93. Stucker R and KJ Frey: Proc. Iowa Acad. Sci. 61, 98-102, 1960.
94. Salim MH, GW Todd and AM Schlehuber: Agron. J. 57, 603-607, 1965.
95. Larsson S: Zeitschr. Pflanzenzücht. 89, 206-221, 1982.
96. Hoorn JW van: Neth. J. Agric. Sci. 6, 1-10, 1958.

97. Mann AP: In, Drainage of Agricultural Lands Seminar. Aust. Nat. Comm. and Int. Comm. Irrig. Drainage, Melbourne, pp.16-22, 1981.
98. Watson ER, P Lapins and RJW Barron: Aust. J. Exp. Agric. Anim. Husb. 16, 114-122, 1976.
99. Krizek DT: In, MN Christiansen & CF Lewis, (eds). Breeding Plants for Less Favourable Environments. John Wiley and Sons, pp.293-334, 1982.

SOME METABOLIC CONSTRAINTS TO OAT PRODUCTIVITY

DAVID M. PETERSON

Cereal Crops Research Unit, Agricultural Research Service, United States Department of Agriculture and Department of Agronomy, University of Wisconsin, Madison, WI 53706, USA

1. INTRODUCTION

The yield of oats and other cereals has been increased considerably over the years through plant breeding. To a large extent this has been accomplished through the elimination of defects, such as insect and disease susceptibility and weak straw. However, yield per se has been increased, largely through the breeding of cultivars that have a higher proportion of their dry matter at maturity in the grains. This has been called the harvest index, and has been defined as: grain wt/(grain wt + straw wt). This trend was demonstrated by Wych and Stuthman[1] who compared nine oat cultivars released over six decades in Minnesota. They found a correlation coefficient of 0.90 between grain yield and harvest index (Table 1). Biological yield was also positively correlated with grain yield, but the correlation coeffecient was lower (0.67). Lawes[2] found that grain yield of European cultivars introduced since 1908 was also highly correlated with harvest index, although the harvest indices of the European cultivars averaged considerably higher than the American ones.

TABLE 1. Grain yield and harvest index (HI) of Minnesota adapted oat cultivars released between 1923 and 1979[a]

Cultivar	Year released	Grain yield (q/ha)	HI
Gopher	1923	22.4	0.27
Anthony	1929	20.0	0.23
Andrew	1949	25.9	0.32
Rodney	1954	25.1	0.27
Minhafer	1957	26.9	0.32
Lodi	1963	27.9	0.30
Lyon	1977	29.9	0.32
Benson	1979	30.6	0.32
Moore	1979	31.3	0.34

[a]Wych and Stuthman[1]

There has also been some attempt to breed oat and other cereal cultivars with increased protein percentage in their grain, or increased protein yield/hectare. The success rate for this endeavour has been less than for yield improvement, and often increased protein percentage has come at the expense of total dry matter yield.

To date, plant breeders have placed little consideration on the metabolic and physiological components of yield. They have, in some instances, looked at morphology to guide them in their selections. I would not argue that their strategy has been in error. Physiologists have not been able to provide plant breeders with reliable criteria for selection of metabolic yield components, since yield is a complex phenomenon derived from many metabolic reactions. Through many centuries of natural selection and, more recently, decades of plant breeding, our best cultivars are probably well balanced in terms of their metabolic pathways. It is unlikely that our best cultivars contain any severely rate limiting steps, which, if improved, could markedly increase yield.

However, I do believe that it is more appropriate now than ever before for breeders and physiologists to work closely together towards the goal of yield improvement. This is timely for two reasons. First, as yield is increased higher and higher by conventional techniques, each additional increment will become more difficult to achieve, and we will have to renew our efforts to identify and combine optimum characteristics into improved cultivars. Second, as genetic engineering through recombinant DNA becomes possible, it will be necessary to identify which gene products (enzymes) are to be modified. This effort will require a concerted effort by geneticists, physiologists and molecular biologists to be successful.

In this paper I will consider a number of possible limiting metabolic or physiological factors that are important for grain or protein yield in oats. Because there are very little data available for oats, as compared to wheat, barley and other species, I have drawn freely on examples from other species, assuming the same factors to be important in oats. The companion paper in this session considers the response of oats to various environmental stresses, and that will not be considered here. Rather, I will emphasise the genetic potential of oats under optimum growing conditions.

2. ENHANCEMENT OF TOTAL YIELD

The relative growth rate does not appear to have increased in the course of domestication in a number of species despite a remarkable advance in crop yield. Furthermore, there is no direct evidence for an advance in the carbon exchange rate per unit leaf area. Past improvements in yield of grain crops have been accompanied by an increased harvest index(1,2,3). On the basis of this evidence, one may argue that:

a) further increases are likely to come from increased photosynthesis, since we have nearly exhausted the possibilities for improvement in harvest index, or

b) further increases must come from improved harvest index, because photosynthesis has not proven to be related to yield.

Each of these conflicting hypotheses will be considered.

The first strategy concerns the possibility for improvement of photosynthetic efficiency. For example, oats and other temperate cereals fix carbon by the C_3 pathway, the primary enzyme being ribulose-1,5-bisphosphate carboxylase/oxygenase or rubisco. This enzyme combines one molecule of CO_2 with ribulose bisphosphate to form two molecules of phosphoglyceric acid for a net gain of one carbon. However, this enzyme can also combine with oxygen, a process called photorespiration, which is inefficient in two ways. First, if oxygen reacts with the enzyme, CO_2 cannot react at the same time; thus CO_2 fixation is slowed. Second, a molecule of CO_2 is lost in a subsequent decarboxylation reaction. Either an increase in the carboxylation reaction (photosynthesis) or a decrease in the oxygenation reaction (photorespiration) would potentially increase biological yield.

In a study of 21 lines of wheat representing wild diploid and tetraploid progenitors and cultivated hexaploids, it was shown that photosynthetic rate per unit leaf area actually decreased over the course of evolution, but this was more than compensated for by increased leaf area(4). These workers also noted a decrease in photorespiration, as measured by the rate of photosynthesis in the absence of oxygen.

Jordan and Ogren(5,6) measured the carboxylase and oxygenase activities of purified rubisco from a number of species, including C_3 and C_4 higher plants, green algae, cyanobacteria and photosynthetic bacteria. The relative rates of the two reactions varied considerably among species (Table 2). The specificity factor, which is a measure of the relative carboxylase:oxygenase rates, was highest for the C_3 and some of the C_4 plants, but was lower for more primitive species. The importance of these findings is that the balance of the carboxylation and oxygenation reactions by rubisco is not constant but has been modified over the course of evolution. These authors also found that the K_m for CO_2 decreased as the specificity factor increased (Table 2), indicating that lower CO_2 concentrations were needed for half maximal velocity in C_3 species than in most C_4 species and more primitive plants and bacteria. In a comparison among grass species, Yeoh et al.(7) also found lower K_m values for C_3 than for C_4 species, and among C_3 species, the K_m for oats was 20 uM compared to 13-15 uM for several Triticeae species. This indicates that further improvement for oats may be possible. The question remains as to whether further decreases in K_m can be obtained by selection pressure in combination with mutagenesis, or by other techniques such as enzyme modification by recombinant DNA.

TABLE 2. Specificity factors and Michaelis constants for CO_2 for RuBP carboxylase/oxygenase from several species[a]

Species	Specificity factor $(V_c K_o / V_o K_c)$	K_c (uM CO_2)
C_3 plants		
Glycine max	82 ± 5	9
Spinacea oleracea	80 ± 1	14
Lolium perenne	80 ± 1	16
Medicago sativa	77 ± 3	16
C_4 plants		
Amaranthus hybridus	82 ± 4	16
Zea mays	78 ± 3	34
Setaria italica	58 ± 2	32
Sorghum bicolor	70 ± 1	30
Green algae		
Chlamydomonas reinhardii	61 ± 5	29
Cyanobacteria		
Plectonema boryanum	54 ± 2	100
Photosynthetic bacteria		
Rhodospirillum rubrum	15 ± 1	89

[a]Jordan and Ogren(5,6)

Rubisco is a relatively inefficient enzyme, having a very high K_m for CO_2. Selection for a more efficient version of rubisco might be potentially useful. Evans and Seemann(8) compared the in vitro carboxylase activity per unit of leaf nitrogen between Triticum aestivum and the diploid T.monococcum and found it to be 30% higher in the hexaploid (Fig. 1). This was due to a higher specific activity of the enzyme from the hexaploid. In this case, however, T.aestivum apparently had a greater resistance to CO_2 transfer between the intercellular spaces and the site of carboxylation, which compensated for the difference in specific activity. The net result was an identical rate of CO_2 assimilation for a given intercellular $p(CO_2)$ (Fig. 1).

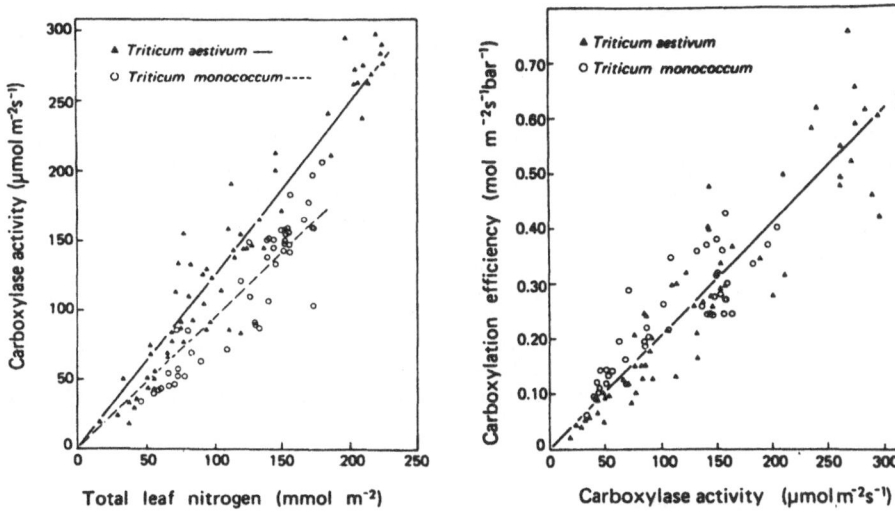

FIGURE 1. In vitro rubisco activity versus total leaf N and carboxylation efficiency versus in vitro rubisco activity for leaves of two Triticum species(8)

Another strategy to be considered is the reduction of dark respiration. Respiration is necessary, of course, for growth and maintenance, but it is possible that the level of respiration in oat cultivars exceeds that required for these processes. Lambers(9) has demonstrated the presence of cyanide-resistant respiration in plant leaves and roots. This so-called alternative pathway respiration does not produce ATP, and is apparently wasteful. The level of alternative pathway respiration appears to be related to the level of respiratory substrate(10), indicating that it does not operate until the regular cytochrome oxidase pathway is fully engaged. Selection for reduction of this pathway may increase yield.

This was demonstrated by Wilson(11), who succeeded in selecting ryegrass (Lolium perenne) genotypes that have low rates of respiration in fully expanded leaves. These low respiration genotypes display about a 10% yield advantage in field plots(12). The increased growth rate of slow versus fast respiration lines tended to be more evident during later cutting stages, when a relatively larger proportion of the respiration would be directed towards maintenance as opposed to growth. If similar low respiration genotypes of an annual cereal such as oats could be identified, the effect on either increased grain yield or total biomass could be determined.

A final strategy to be considered for increasing total yield of biomass is to lengthen the duration of growth. It is obvious that if growth duration could be increased without altering the rate of growth, total biomass yield would be increased. Indeed, Rosielle and Frey(13) noted that unrestricted selection for yield resulted in selection of later heading types. The problem with this approach is that late maturing oats may be undesirable for other reasons. For example, in the midwestern United States, early maturing cultivars are preferred because they are more likely to escape the heat and drought of the mid- to late summer months. Also, if oats are underseeded with legumes, as they usually are in Wisconsin, it is desirable to get the oats off early to allow growth of the legume for a fall cutting. In Europe, where the climate is milder, later maturing varieties may be more acceptable than in the United States.

3. INCREASED HARVEST INDEX

Harvest index can be increased by increasing the size of grains or the number of grains per unit area compared to total plant weight. With regard to the former, evidence with wheat indicates that potential grain size is closely related to the number of endosperm cells formed(14). Experiments have clearly shown that subjecting the plant to stress during the critical period shortly after anthesis reduces cell number and mature dry weight commensurately(15). Of course, if subsequent environmental conditions do not favour grain filling, the potential grain size may not be achieved, even if the maximum cell number was obtained. Treatments which alter the supply of photosynthetic assimilates, such as shading or removal of a portion of the spikelets, have been shown to affect single grain weights(16,17,18). An experiment with oats (Fig. 2) clearly shows

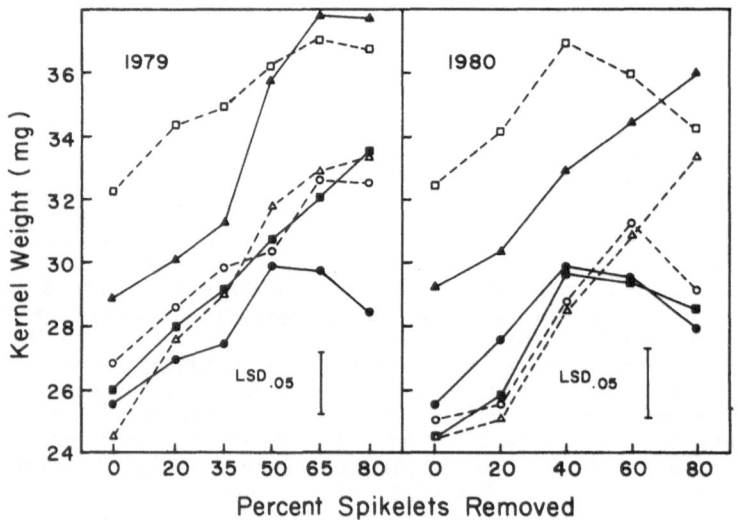

FIGURE 2. Effect of spikelet removal on kernel weight of remaining spikelets in six oat cultivars(19).

the effects of spikelet removal at anthesis on mature kernel weight(19). The implicit assumption in the interpretation of these results is that with fewer grains developing per inflorescence, there is more assimilate available for the remaining ones, allowing more growth. However, this assumption has been difficult to prove. Singh and Jenner(20), in experiments with shading and cultured ears, found no association between endosperm sucrose concentration and growth (in cell number or dry weight) of wheat. Because the free sucrose pool in the endosperm is in a dynamic state, dependent upon the rate of input and utilization, it is not too surprising that a correlation was not found. In addition, these authors expressed the sucrose concentrations on a dry weight basis. Expression of sucrose concentration as a molar concentration in the liquid phase of the cells might have been a better choice, although one still doesn't know how much is available for metabolism and how much may be sequestered into vacuoles.

Once the maximum grain potential is set, development and growth of starch grains and protein bodies to fill the cells is dependent on current photosynthesis and stored reserves in the plant. Generally, under normal environmental conditions photosynthesis is adequate to meet the sink demand. It is obvious that variation in the rate and duration of assimilate transfer to the endosperm will determine ultimate grain weight. There has been much research to determine the limiting steps of assimilate transfer, including phloem loading in the source leaf, transport in the sieve tubes, and unloading in the vicinity of the sink(21). There is some evidence that delayed senescence of the leaves may allow additional assimilates to be provided for the grain, thus enhancing yield(22).

Little is known about the control of starch granule formation or how the number and size of starch granules relates to grain filling. In barley and wheat, large starch granules are initiated early, and smaller ones later. Oats may be similar in this respect. Important questions are how is the movement of substrate across the envelope of the amyloplast regulated, and what regulates the activity of the enzymes of starch synthesis within the amyloplast? We are addressing these questions in my laboratory with isolated amyloplasts from developing barley endosperm. Currently, we have developed procedures that yield amyloplasts that are bounded by a double membrane envelope, as are chloroplasts. The outer membrane is permeable to sucrose, whereas the inner one is not. Our future plans are to characterise the selective nature of the inner envelope membrane and also to investigate the starch synthesis enzymes associated with the cytosol and with cytosol-free preparations of amyloplasts.

Another question of importance is what causes the synthesis of starch to cease as maturation approaches? Experiments with wheat(23) have shown that the sucrose concentration in the endosperm cells is not diminished at maturity. These data have been interpreted to indicate that lack of photosynthetic assimilates is not the cause of the termination of starch synthesis. These physiological experiments do not agree with the observations on delayed senescence genotypes(22) mentioned above. Other workers(24) have suggested that in wheat, the chalaza becomes plugged with lipid impeding the transport of assimilates from the vascular bundle to the nucellar projection and endosperm cavity. This is probably not the case in barley (25; Felker, Peterson and Nelson, unpublished data). At this time, there is no widely accepted theory on the cause of cessation of grain filling.

166

The other approach, inducing the plants to form additional florets and set more grains per plant, has received less attention than that of grain filling. Floral initiation and terminal spikelet initiation are known to be under photoperiodic control in wheat, and genetic differences are evident(26). Manipulation of photoperiod at critical stages can increase spikelet number. Klinck and Sim(27) demonstrated in controlled environments that cool temperatures and short daylengths at critical stages markedly increased the number of grains per panicle in the oat cultivar, Garry, but had little effect on Clintland 60 (Table 3). Treatments during the first phase, seeding to initiation of panicle development, were more effective than during panicle development (phase 2) or post anthesis (phase 3).

TABLE 3. Effect of temperature and daylength during different growth phases on grains per panicle and grain yield of oats[a]

Treatment/parameter	Garry			Clintland 60		
	Phase 1	Phase 2	Phase 3	Phase 1	Phase 2	Phase 3
Cool temperature[b]						
Grains/panicle	151[d]	52	37	-31	-5	-17
Grain yield	51	56	94	3	42	24
Short daylength[c]						
Grains/panicle	103	67	0	-3	8	-5
Grain yield	39	63	-8	7	16	-4

[a]Klinck and Sim(27); [b]15°C versus 21°C for the control; [c]13 h photoperiod versus 18 h for the control; [d]Percent increase or decrease.

4. STRATEGIES TO ENHANCE PROTEIN YIELD

Because of the often observed negative correlation between protein concentration and yield, many plant breeders have been striving to increase protein yield per hectare, rather than protein concentration. Protein yield is the product of grain yield and protein concentration. Any breeding programme designed to increase protein yield must select for high grain yield as well. Two primary factors may influence grain protein: the total amount of nitrogen assimilated from the soil and the efficiency of its mobilization into the grain.

Nitrate is the most commonly found form of N in the soil, and its assimilation proceeds by the nitrate reductase - nitrite reductase pathway. It is now believed that the NH_3 formed is incorporated into organic compounds via the glutamine synthetase-GOGAT pathway. There is a substantial body of evidence that points to nitrate reductase as the rate limiting enzyme in nitrogen assimilation. For example, ammonium does not accumulate in plants under normal conditions, nor does nitrite, indicating that once the first step of nitrate assimilation is accomplished, the reduced N is incorporated rapidly into amino acids. Also, the in vitro activity of nitrite reductase is always higher than nitrate reductase. However, it has been difficult to establish good correlations between nitrate reductase activity and yield of total reduced N despite many attempts in many species. More recently, barley mutants were induced which apparently lacked nitrate reductase but nevertheless grew and yielded almost normally with nitrate as the sole N source. These results have been explained by the discovery of another nitrate reductase enzyme which utilises either NADH or NADPH as the source of reductant(28). This enzyme apparently is induced in mutants lacking the common NADH-specific form. However, the activity of the bispecific form is low, indicating that nitrate reductase activity, as measured in field-grown plants, may not be a limiting factor to growth in most cases.

The question of whether the protein concentration of the grain is a function of the capacity of the grain itself to synthesise protein or of the plant to supply amino acid substrates was addressed by Lesar and Peterson(29). They excised panicles of oats shortly after anthesis and cultured them on artificial media. During a 12 day culture period, dry weight of the developing groats increased almost as much as in those developing on intact plants. However, by manipulation of the N to carbohydrate ratio in the media, increases of 30, 45 and 72% N in the groats were achieved with mixed amino acids, glutamine and NH_4NO_3, respectively. This increased N was predominantly protein. This experiment demonstrated that developing oat groats have excess capacity to synthesise protein and indicates that the supply of amino acids and amides from the plant may be limiting under normal circumstances.

Nitrate moves to the leaves in the xylem where it is reduced and incorporated into amino acids. Some of these amino acids are recycled to the roots or to other growing points where they are assembled into proteins. Most remain in the leaves where they form rubisco and other enzymes. Generally, by the time of anthesis, the roots are much less active, and the plant already contains most of the N it will contain at maturity. It is the remobilization of this N from leaves, stems and roots that accounts for the majority of the grain protein at maturity. Herein lies the paradox: to supply N to the grains, leaves must senesce, their rubisco and other proteins be degraded, and their amino acid residues be transported out. However, this very process diminishes photosynthetic capacity, perhaps reducing the potential for further yield. Perhaps a desirable plant trait would be to select for a more active root function during maturation, coupled with delayed leaf senescence. In this respect, it has recently been shown that treatment of wheat roots with solutions of cytokinin at anthesis and at 4-day intervals thereafter increased grain protein by 3 percentage points(30).

It has been demonstrated that oat genotypes vary in their efficiency of N remobilisation, as determined by the percentage of total plant N located in the grain at maturity(31,32,33). Wych and Stuthman (personal communication) found that the nitrogen harvest index (NHI) ranged from 0.64 to 0.80 in a field study of 20 genotypes in Minnesota. NHI was positively correlated with HI and negatively correlated with total N accumulation. Unless the straw is to be used for feed, it would seem desirable to select for genotypes that remobilise the highest percentage of N possible.

5. CONCLUSIONS

I have suggested a number of possible enzymes, reactions, pathways and physiological processes that may potentially enhance yield or protein yield if they could be optimised. It should be obvious that there are little data available for oats and much of what I have said is inferred from other species. As physiologists, we need to be selective in choosing those parameters for study which offer the most promise for oats improvement or for uncovering basic biological principles. In some cases, it will be most appropriate to select examples that have been demonstrated to be of vital importance in other species and verify these data for oats. In other instances, oat metabolism may be unique. For example, the storage proteins of oats are predominantly globulins, in contrast to the prolamins of other cereals. These instances should be exploited.

I began with the premise that we need to identify specific genes that can be altered or transferred by recombinant DNA techniques. Much information on specific gene structure and function will be obtained for other species more amenable to research than oats, such as pea and tobacco. We need to remain alert to this literature and adapt findings to oats wherever appropriate. With the hexaploid nature of oats, and the lack of a good working nullisomic series such as is available for wheat, gene localisation in oats will be a long time coming. The lesser economic importance of the crop and lack of research attention also suggests slow progress.

REFERENCES

1. Wych RD and DD Stuthman: Crop Sci. 23, 879-881, 1983.
2. Lawes DA: J. Agric. Sci. 89, 751-757, 1977.
3. Austin RB, J Bingham, RD Blackwell, LT Evans, MA Ford, CL Morgan and M Taylor: J. Agric. Sci. 94, 675-689, 1980.
4. Evans LT and RL Dunstone: Aust. J. Biol. Sci. 23, 725-741, 1970.
5. Jordan DB and WL Ogren: Nature 291, 513-515, 1981.
6. Jordan DB and WL Ogren: Arch. Biochem. Biophys. 227, 425-433, 1983.
7. Yeoh HH, MR Badger and L Watson: Plant Physiol. 66, 1110-1112, 1980.
8. Evans JR and JR Seemann: Plant Physiol. 74, 759-765, 1984.
9. Lambers H: Physiol. Plant. 55, 478-485, 1982.
10. Azcon-Bieto J, H Lambers and DA Day: Plant Physiol. 72, 598-603, 1983.
11. Wilson D: Ann. Bot. 49, 303-312, 1982.
12. Wilson D and JG Jones: Ann. Bot. 49, 313-320, 1982.
13. Rosielle AA and KJ Frey: Crop Sci. 15, 544-547, 1975.
14. Brocklehurst PA: Nature 266, 348-349, 1977.
15. Brocklehurst PA, JP Moss and W Williams: Ann. Appl. Biol. 90, 265-276, 1978.
16. Bingham J: J. Agric. Sci. 68, 411-422, 1967.
17 Fischer RA and D HilleRisLambers: Aust. J. Agric. Res. 29, 443-458, 1978.
18. Nosberger J and GN Thorne: Ann Bot. 29, 635-644, 1965.
19. Peterson DM: Field Crops Res. 7, 41-50, 1983.
20. Singh BK and CF Jenner: Aust. J. Plant Physiol. 11, 151-163, 1984.
21. Geiger DR: Bot. Gaz. 140, 241-248, 1979.
22. Helsel DB and KJ Frey: Crop Sci. 18, 765-769, 1978.
23. Jenner CF and AJ Rathjen: Aust. J. Plant Physiol. 2, 311-322, 1975.
24. Sofield I, IF Wardlaw, LT Evans and SY Zee: Aust. J. Plant Physiol. 4, 799-810, 1977.
25. Cochrane MP: Aust. J. Plant Physiol. 10, 473-491, 1983.
26. Pinthus MJ and H Nerson: Aust. J. Plant Physiol. 11, 17-22, 1984.
27. Klinck HR and SL Sim: Can. J. Bot. 55, 96-106, 1977.
28. Warner RL, A Kleinhofs and KR Narayanan: In, JE Harper et al. (eds). Exploitation of physiological and genetic variability to enhance crop productivity. Amer. Soc. Plant Physiologists, Rockville, Md. p.92, 1985.
29. Lesar LE and DM Peterson: Crop Sci. 21, 741-747, 1981.

30. Dalling MJ: In, JE Harper et al. (eds). Exploitation of physiological and genetic variability to enhance crop productivity. Amer. Soc. Plant Physiologists, Rockville, Md. p.92, 1985.
31. Eagles HA, RM Haslemore and CA Stewart: N.Z. J. Agric. Res. 21, 65-72, 1978.
32. Peterson DM, LE Schrader, DA Cataldo, VL Youngs and D Smith: Can. J. Plant Sci. 55, 19-28, 1975.
33. Welch RW and YY Yong: J. Sci. Food Agric. 31, 541-548, 1980.

EFFECTS OF GENOTYPE, SEEDING RATE AND THE GRAIN YIELD COMPONENTS
ON THE PRIMARY:SECONDARY SEED WEIGHT RATIO IN OATS

A. C. TIBELIUS and H. R. KLINCK
Plant Science Department, Macdonald College of McGill University,
Ste-Anne-de-Bellevue, Quebec, H9X 1C0, Canada

1. INTRODUCTION

If the primary:secondary seed weight ratio is to be considered as a selection criterion, it is important to recognise the sources of variation which affect it. Since the ratio is calculated from primary and secondary grain weights, the factors which effect these traits will determine the ratio, especially if they are affected differentially. In these experiments, the effects of genotype and seeding rate on the ratio and the relationship between the ratio and yield components were investigated.

2. MATERIALS AND METHODS

Two sets of six genotypes were chosen to represent a wide range of 1000-grain weights and primary:secondary seed weight ratios. The first set, consisting of Laurent, O.A.347.1, O.A.366, Q.0.151.58, Q.0.170.11 and Scott, was studied in 1980, 1981 and 1983 at Macdonald College. These experiments, designated 1A, 2A and 3A, were arranged in split-plot designs with genotypes as the main plots, and seeding rates, 16 and 100 seeds per 2 m row, as the sub-plot units. Plots were hand-sown and consisted of two 2 m rows, 20 cm apart. The second set of six genotypes was studied in four experiments and included Ajax, Forward, Nelson, Rodney, Roxton and X2078-1. Three experiments were conducted at Macdonald College: 1B in 1981, 2B and 3B in 1983. Experiment 4B was located at Joliette, a regional test site 120 km to the northeast, in 1983. Two seeding rates, as noted above, were used for 1B and 2B in a split-plot design. In 3B and 4B plots consisted of single 2 m rows, 40 cm apart, seeds being hand-sown at 31 seeds per row. A randomised complete blocks design was used. In all experiments, 10 plants were chosen randomly in each plot for data collection and statistical analysis. Seeds were extracted from the panicles by hand.

3. RESULTS AND DISCUSSION

In all experiments significant effects of genotype on the primary:secondary seed weight ratio were found. However, when the genotypes were ranked according to their ratio values from the different experiments, variation in their relative rankings was apparent. Scott consistenty had a lower ratio than the other genotypes (Table 1). Q.0.151.58 also tended to have a low ratio but the rankings for the remainder of the genotypes were variable. In the other set of genotypes (Table 2), Nelson had a consistently low ratio relative to the other genotypes, while Roxton had a relatively high ratio. The rankings of the intermediate genotypes were less stable. It would appear that some genotypes are more susceptible to environmental variation than others in terms of their primary:secondary seed weight ratios.

TABLE 1. Primary:secondary seed weight ratios in oats

| Genotype | Experiment | | | % variation |
	1A	2A	3A	
Laurent	1.62[b]	1.74[a]	1.54[ab]	13
O.A.347.1	1.66[a]	1.62[b]	1.47[c]	13
O.A.366	1.61[b]	1.59[bc]	1.56[a]	3
Q.0.151.58	1.54[c]	1.59[bc]	1.48[c]	7
Q.0.170.11	1.64[ab]	1.55[c]	1.51[bc]	9
Scott	1.45[d]	1.46[d]	1.37[d]	7

[a-d]Values within a column followed by the same letter are not significantly different at the 0.05 level

TABLE 2. Primary:secondary seed weight ratios in oats

| Genotype | Experiment | | | | % variation |
	1B	2B	3B	4B	
Ajax	1.63[c]	1.51[c]	1.56[b]	1.49[b]	9
Forward	1.67[bc]	1.54[bc]	1.56[ab]	1.50[b]	11
Nelson	1.54[d]	1.46[d]	1.46[c]	1.41[b]	9
Rodney	1.70[b]	1.52[bc]	1.60[a]	1.56[a]	12
Roxton	1.86[a]	1.56[ab]	1.62[a]	1.53[ab]	22
X2078-1	1.51[d]	1.57[a]	1.55[b]	1.52[ab]	4

[a-d]Values within a column followed by the same letter are not significantly different at the 0.05 level

Significant effects of genotype were found for both primary and secondary grain weight in all experiments. The coefficients of variability for secondary grain weight were greater than those for primary grain weight in all but experiment 2A (Table 3). Secondary grain weight appears to be more susceptible to environmental variation than primary grain weight and is a probable source of variability in the primary:secondary seed weight ratio. A limitation in the supply of assimilates for grain filling in wheat affects the distal florets proportionally more than those at the base of a spikelet(1,2,3). In oats, too, proximity to the peduncle seems to confer an advantage in grain weight. Deslauriers(4) found that even within the panicle, secondary grains tended to show more variation in weight than primary grains.

When selecting parents for crosses, it is important to consider both the value of the primary:secondary seed weight ratio and its stability. X2078-1 had a relatively constant ratio across experiments, varying only 4%. Thus, while its ranking relative to other genotypes varied considerably (Table 2), its actual mean value did not. Both the primary and secondary grain weights of X2078-1 were also fairly stable, each varying 11% across experiments (Table 4). Ajax exhibited more variation in its ratio across experiments than X2078-1, but less than a number of the other genotypes (Table 2). For Ajax, more than a 30% difference was found across experiments for both primary and secondary grain weights, but the two seed types did not vary greatly in their relative responses to the different environments, leading to a ratio which was relatively stable (Table 4).

172

TABLE 3. Coefficients of variability for primary and
secondary grain weights in oats

| | Coefficient of variability | |
Experiment	Primary grain weight	Secondary grain weight
1A	4.05	4.94
2A	5.67	5.50
3A	4.23	5.30
1B	3.17	4.85
2B	4.13	5.54
3B	3.13	3.42
4B	2.67	4.07

In contrast to X2078-1 and Ajax, Roxton showed consistency in its
ranking between experiments but varied considerably in its mean ratio
value 22% (Table 2).   High levels of variability were found for both
primary and secondary grain weights of Roxton, but the variability was
relatively greater for the secondary seeds, which would account for the
ratio differences. The ratios of Rodney and Roxton differed significantly
in only one experiment, a result not unexpected since Rodney is a
descendant of a cross involving Roxton.   Rodney, however, exhibited much
more stability in its ratio.   Variation in primary grain weight was
similar for the two cultivars, but Roxton showed much more variation in
its secondary grain weight, leading to its less stable ratio.

TABLE 4. Variation in primary and secondary grain weights
in oats across experiments

| | Percent variation | |
Genotype	Primary grain weight	Secondary grain weight
Laurent	29	45
O.A.347.1	32	50
O.A.366	26	28
Q.O.151.58	34	42
Q.O.170.11	24	35
Scott	21	28
Ajax	33	38
Forward	38	44
Nelson	40	50
Rodney	33	44
Roxton	39	66
X2078-1	11	11

In experiments 1A-3A, the genotypes O.A.366, Q.O.151.58 and Scott
exhibited the least variation in their primary:secondary seed weight
ratios (Table 1). Scott and Q.O.151.58 retained their relative rankings,
while O.A.366 was less consistent.   In comparison with the other
genotypes, primary and secondary grain weights were least variable for

Scott and O.A.366 and they showed similar responses to environmental variation (Table 4). While Scott and Laurent both carry the genotype Garry as a common parent, it is interesing to note that grain weights and ratios were much more stable across environments for Scott. In addition, grain weights are higher and ratios lower for Scott, both traits being desirable.

There was a significant effect of seeding rate on the primary:secondary seed weight ratio in experiments 2A and 1B (Table 5). The ratios were lower at the reduced seeding rate and both primary and secondary grain weights were higher. Larger grain weights at lower seeding rates is a trend confirmed in the literature for oats in general(5,6,7). However, the difference in grain weight between the two seeding rates was proportionately greater for the secondary grains (5-6%) than for the primary grains (2-3%), again emphasising the greater susceptibility of secondary grains to environmental variation. More intense competition between plants at the higher seeding rate could result in limiting levels of assimilates within the panicle. Because of their greater relative distance from the assimilate source, secondary grains may be more restricted in growth compared with primary grains. At the lower seeding rate where interplant competition is reduced, the secondary seeds may benefit proportionately more than the primary seeds.

TABLE 5.   Primary:secondary seed weight ratios and 1000-grain weights (g) of oat cultivars at two seeding rates, 16 and 100 seeds per 2 metre row

Experiment	Primary:secondary seed weight ratio		Primary grain weight		Secondary grain weight	
	16	100	16	100	16	100
1A#	-	-	-	-	-	-
2A	1.57	1.62*	36.0	35.0*	23.1	21.8**
3A	1.49	1.49	44.8	44.7	30.2	30.0
1B	1.63	1.67**	35.0	34.2*	21.6	20.5**
2B	1.52	1.54	44.2	43.7	29.1	28.4

#Genotype x seeding rate interaction effects noted in Experiment 1A
*, **Significant at the 0.05 and 0.01 levels, respectively

Primary and secondary grain weights were not always found to be higher at the lower seeding rate, however. In experiment 3A and 2B there were no seeding rate effects on the ratio or on grain weight (Table 5). In experiment 1A there were significant genotype x seeding rate effects on both grain weights and the ratio. When each genotype was analysed individually, only O.A.366 was found to be influenced by the seeding rate. The ratio was lower at the reduced seeding rate, as was found in experiments 2A and 1B. Primary grain weights of O.A.366 were unaffected by the seeding rate but secondary grain weights were increased by 5% at the lower rate, which would account for the seeding rate effect on the ratio.

Correlation and covariance analysis did not reveal a relationship between primary:secondary seed weight ratios and either number of panicles per plant or number of seeds per panicle. Correlations between 1000-grain weight and the ratio varied between experiments (Table 6). A significant correlation (-0.40**) was found only in experiment 2A, though a similar

174

correlation coefficient (-0.54**) was also found when 22 genotypes were
studied in 1981. The relationship between the ratio and 1000-grain weight
was substantiated by covariance anaysis. It would appear that, at least
in 1981, there was a trend toward genotypes with higher grain weights also
having lower ratios. However, further studies with a wider range of
genotypes is required.

TABLE 6. Correlations between primary:secondary seed weight ratios
and 1000-grain weights in oats

Experiment	Correlation coefficient
1B	-0.12
2B	0.35
3B	0.40
4B	0.39
1A	-0.12
2A	-0.40**
3A	-0.11

**Significant at the 0.01 level

4. CONCLUSIONS
Varying degrees of stability in both primary and secondary grain weights
and primary:secondary seed weight ratios were found for the different
genotypes across environments. Stability of both primary and secondary
grain weight or, alternatively, parallel responses of these two factors to
environmental variability, would be desirable, with resultant stability in
the primary:secondary seed weight ratio. Variation in secondary grain
weight across experiments and seeding rates appeared to be the cause of
variability in the ratio.

REFERENCES

1. Chowdhury SI and IF Wardlaw: Aust. J. Agric. Res. 29, 205-223, 1978.
2. Cook MG and LT Evans: Aust. J. Plant Physiol. 10, 313-327, 1983.
3. Walpole PR and DG Morgan: Ann. Bot. 34, 309-318, 1970.
4. Deslauriers C: M.Sc. Thesis, Macdonald College of McGill Univ.,
Ste-Anne-de-Bellevue, P.Q., 1982.
5. Jones IT and JD Hayes: J. Agric. Sci., Camb. 70, 405-410, 1968.
6. Pendleton JW and GM Dungan: Agron. J. 45, 442-444, 1953.
7. Woodward RW: Agron. J. 48, 160-162, 1956.

BREEDING OATS BY SELECTION OF PARENTAL PAIRS FROM DIFFERENT ECOLOGICAL
AND GEOGRAPHICAL BACKGROUNDS

E. LYSLOV* and H.D. KÜÜTS**
*Research Institute of Agricultural Management of Central District
 of the Nonchernozem Zone, Nemtchinovka, Moscow Region, USSR
**Jõgeva Plant Breeding Station, Estonia, USSR

1. INTRODUCTION
Yield is the product of two components - the potential productivity of
the plant and resistance to unfavourable factors in the environment in
which it is grown.
The oat varieties which have been developed in the Soviet Union in the
last ten years such as Hercules, Sinelnikovsky 21, Kirovsky, Mirny,
Horisont and Ruslan possess good yield potentials of 7,000 - 7,500 kg of
grain per ha.   However, these varieties have a tendency to lodge, are
susceptible to plant diseases and suffer yield reductions in dry summers.
The best West-European oat varieties on the other hand are more often
characterised by resistance to lodging, although they are also prone to
infection by plant diseases and to an even greater degree suffer severe
yield reductions as a result of drought in dry years.   As a result, big
fluctuations in oat yields are observed from year to year.   In the
temperate climate conditions of the Moscow Region the coefficient of
variation of the yield of the oat variety Hercules in the last ten years
in trials has been 34.6%, while that of the variety Astor has been 42.5%.
The problem of drought resistance is most critical for the steppe and
forest-steppe zones of the USSR.   Analyses have shown that the factors
causing the most common reductions in oat yields during the last ten years
have been lodging (4 years), drought (3 years), infection with crown rust
(1 year), bacterial diseases (2 years).   The development of new varieties,
which combine high productivity with resistance to lodging, drought and
plant diseases would contribute greatly to increased stability of the oat
crop and its harvest.   Considering the range of soil and climatic
conditions,   which   include   considerable   seasonal   fluctuations   in
precipitation, it is desirable to grow varieties of broad ecological
plasticity.   In the breeding of new varieties of oats, the main emphasis
must be therefore on widening adaptability to a range of physical and
biological environments.   The major objective is to improve the stability
of yields.   Hybridisation of genotypes from diverse ecological and
geographical backgrounds, along with the selection and evaluation of lines
under various soil and climatic conditions, is an effective method of
developing such varieties.

2. MATERIALS AND METHODS
The VIR world collection (located in Leningrad) served as starting
material and varieties to be used in the breeding programme were selected
on the basis of individual characters or groups of characters.   This basic
material represents a collection of samples collected in different
countries of Europe, North America and Latin America.   Hybridisations were
made with such domestic varieties as L'govsky, Hercules and Horisont;

176

highly productive West-European varieties resistant to lodging such as
Astor, Leanda, Sang, Borrus, WZ 437, Maris Tabard and Gambo 60;
early-ripening varieties resistant to drought and fungal diseases;
varieties from the American continent with high protein content, such as
Putnam 71, Diamante R-31, Fraser, Froker and Cayuse. When selecting pairs
for crossing the complementation of economically valuable characters was
taken into consideration. Selected lines were evaluated in different
soil-climatic zones near Moscow, in the Ulyanovsk Region, in West Siberia
and in the Far East.

3. RESULTS AND DISCUSSION
The results have shown the advantages of ecologically remote crosses. In
comparison with hybrids between closer types, such crosses have yielded
three times as many valuable hybrid populations, and some of these have
been entered for final strain testing. The most promising lines have been
obtained from crosses between the best Soviet varieties and the
lodging-resistant West-European varieties on the one hand and varieties
from Canada, USA, Mexico and other countries of the American continent on
the other; the latter, however, are more resistant to drought and fungal
diseases.
Over the last few years, several new varieties have been developed by
this method. These varieties combine economically valuable characters
with a broad ecological range of plasticity. Drug is one of these
varieties, it has been developed from the hybrid population resulting from
Fraser (Canada) x L'govski 78 (USSR) and has been tested in two regions:
near Moscow on sodpodzolic soils and in the Ulyanovsk Region on black
soils which suffer a shortage of moisture. The new variety has shown high
productivity, resistance to lodging, a mild degree of infection by loose
smut and crown rust, and good grain qualities in combination with
increased drought resistance and ecological plasticity (Table 1). Drug
has been recommended for cultivaton in several regions of the Volga
area.

TABLE 1. Results of final strain testing of the oat variety Drug at the
Ulyanovsk Experimental Station (averages for 1978-1980)

Variety	Grain yield (tonnes/ha)	Resistance to lodging	% Disease infection Crown rust	Loose smut	1000 grain wt (g)	Hull (%)	Grain protein content (%)
Astor	5.70	9.0	10 - 40	0.07	36.0	25.5	11.3
Drug	6.34	8.5	0	0	42.0	23.1	11.7

A new highly productive variety Alo has been developed at the Jõgeva
Plant Breeding Station (in Estonia) by crossing Leanda (Netherlands) and
Lodi (USA). This variety is resistant to lodging and is less infected
with rust diseases. By the hybridisation of various ecotypes, new and
valuable lines with sets of useful characters have been developed which
are characterised by a wide-ecological plasticity. Another new line,
8 H 288, has been developed from a hybrid population of Fraser x Astor,
and this has shown equally good results in trials near Moscow and in the
Ulyanovsk Region (Table 2). With its increased produtivity and by its
ability to respond to improved agronomy 8 H 288 exceeded the grain yield
of the standard variety Astor, even in dry years (1981 in the Ulyanovsk
Region and 1983 near Moscow) as shown by the data presented in Table 3.

TABLE 2. Results of final testing of the oat line 8 H 288
(averages for 1981-1983)

Character	Nemtchinovka 8 H 288 : Astor		Ulyanovsk 8 H 288 : Astor	
Grain yield (tonnes/ha)	4.67	4.23	5.62	5.17
Resistance to lodging	8.4	8.4	8.3	8.3
Plant height (cm)	88	87	100	95
Infection with crown rust (% and type of reaction)	12/2	80/4	0	15/4
Infection with loose smut (%)	-	-	0	0.57
1000 grain wt (g)	38.0	35.0	34.4	34.4
Hull (%)	23.6	25.6	26.9	29.0
Protein content (%)	12.5	12.2	12.1	11.8

Other promising lines have been developed from crosses between several Dutch, Swedish and USSR varieties and the Canadian varieties Fraser, Garry, the United States varieties Putnam 71 and Froker, the Mexican variety Diamante R-31 and two samples from Equador (K-12028, K-12026). All of these have demonstrated high productivity and increased stability of yield over years.

TABLE 3. Grain yield of 8 H 288 in the final years of strain testing
(tonnes/ha)

Year of testing	Nemtchinovka 8 H 288 : Astor		Ulyanovsk 8 H 288 : Astor	
1981	5.23	4.75	2.95	2.66
1982	5.65	5.23	7.54	6.86
1983	3.13	2.70	6.37	6.00
Average of 3 years	4.67	4.23	5.62	5.17

4. CONCLUSION

The value of oat varieties as parents in a crossing programme is dependent on their genetic differences which determine their different economically useful characters and qualities.

The best results in USSR oat breeding programmes have been obtained by the selection of parental pairs from different ecological and geographical backgrounds. Such hybridisations make it possible to combine the desirable qualities in the necessary combinations and hence the initiation of new improved varieties.

The selection and evaluation of the segregating material under different soil-climatic conditions also contributes to this success.

The promising lines and new varieties developed by this method demonstrate a real genetic advance and improved adaptability which helps to stabilise yields over years.

THE TOLERANCE OF OAT VARIETIES TO HOEGRASS (375 g/l DICLOFOP-METHYL)

A. R. BARR

Department of Agriculture, Adelaide 5001, South Australia

1. INTRODUCTION

The control of weedy grasses in cultivated oats has traditionally been left to cultural methods. Most herbicides developed for wheat and barley show little selectivity between oats and the principal grassy weeds of southern Australia, namely annual ryegrass (Lolium rigidum) and wild oats (Avena fatua, A.sterilis and A.barbata). Currently chlorsulfuron (Glean(R)) is the only herbicide registered for control of annual ryegrass in oats in Australia. While chlorsulfuron represented a major breakthrough in oat husbandry its use is inappropriate where oat/legume mixtures are sown and on soils where pH is greater than 8.5. No product is registered for control of wild oats.

Lack of suitable herbicides is a major factor limiting expansion of the area sown to oats. Firstly, the yield loss due to competition from weeds is often severe. Expectation of weed infestation can often cause severe yield penalties as farmers delay seeding past the optimum time hoping to achieve weed control by cultivation. Oats are promoted to farmers as providing a 'break-crop' for several serious root pathogens of wheat and barley. However, if the oat crop is infested with grassy weeds much of the benefit is lost, leading to disenchantment with this potentially effective biological control method. The control of annual ryegrass in pasture stands has become important in some areas of southern Australia due to the spread of 'annual ryegrass toxicity'. The nematode-bacteria association required to cause toxicity, and livestock deaths, occurs only in ryegrass and control measures centre on eliminating annual ryegrass from pastures. Oats would be a suitable replacement forage if a selective herbicide were available.

Therefore three situations remain unsolved for the oat grower, control of annual ryegrass on high pH soils, in oat/legume mixes and, even more difficult, control of wild oats in cultivated oats.

One herbicide which would be suitable is diclofop-methyl. It is registered in South Australia as Hoegrass(R) (375 g/l diclofop-methyl) and as 'Hoelon' and 'Illoxan' in other countries. Recommended rates for annual ryegrass control are 1.0 litre product per ha and for wild oats, 1.5-2.0 litres product per ha. At these rates, legumes are little affected and Hoegrass is not greatly affected by soil pH. However, oats are moderately susceptible to this chemical(1,2).

The experiments reported here examine the tolerance of domesticated Avena genotypes to Hoegrass.

2. MATERIALS AND METHODS

A preliminary screening experiment undertaken in 1978 demonstrated considerable variation for tolerance to Hoegrass applied at 2.0 l/ha.

Trials conducted in 1980 and 1981 examined genotypes which varied in their response to Hoegrass in 1978 in addition to three selections derived from a bulk population treated with Hoegrass in 1978. Trial design in both years was a split plot with four replications sown at the Turretfield Research Centre. The 1980 trial was sown on 25 July and sprayed with a 1.55 l/ha Hoegrass plus 4 ml/l non-ionic wetter (Agral 60(R)) in 70 l/ha water at the 3-4 leaf stage. The 1981 trial was sown on 16 July and sprayed with 1.5 l/ha Hoegrass plus 4 ml/l wetter in 70 l/ha water at the 3 leaf stage.

The time of application study was undertaken at the Turretfield Research Centre. A 2 metre boomspray swath was applied perpendicular to 9 hoe strips of each of 7 oat cultivars. Treatments were unreplicated. The seeding date was 6 June 1983 and 1.5 l/ha Hoegrass plus Agral 60 wetter at 4 ml/l of spray mixture was applied at the 2 leaf stage (6 July), 4 leaf stage (2 August) and 6 leaf stage (12 August).

The experiment from which 'Savena 1' was selected (Table 4) was conducted at Turretfield Research Centre. The trial was sown on 26 May and 1.5 l/ha Hoegrass plus 4 ml/l Agral 60 wetter mixture was applied at the 6-8 leaf stage (3 August) in 150 l/ha spray mixture. Plot size was 6 hoe widths by 4.5 metres.

3. RESULTS AND DISCUSSION

The most tolerant genotypes in 1980 and 1981 were Avena strigosa cv. Saia and A.sativa cvs New Zealand Cape and Moore (Tables 1 and 2). Saia and A.sativa cv. Sual were the most tolerant cultivars in the 1978 screening experiment. Both New Zealand Cape and Sual are derived from the same parent, that is, Algerian. Cassia is extremely sensitive to Hoegrass application while West is intermediate in its reaction. Since no loss in yield was associated with Hoegrass application in the weed-free trials of 1980 and 1981 using the variety New Zealand Cape, the likelihood of wild oats and annual ryegrass control in oats using 1.5 l/ha Hoegrass seemed high. However, it is apparent from later studies that yield losses of up

TABLE 1. Effect of Hoegrass treatment on plant attributes, Turretfield 1980

Variety	Tiller number +a	Tiller number b	Plants/m +	Plants/m -	Mature height (cm) +	Mature height (cm) -	Yield (kg/ha) +	Yield (kg/ha) -	Yield % control
West	5.4	6.3	10.9	12.8	61.3	68.8	804	1200	67
Moore	6.9	5.6	10.9	11.3	75.0	80.0	1889	2003	94
Cassia	3.7	6.8	2.6	8.6	57.5	81.3	97	1230	8
NZ Cape	7.3	6.9	11.5	10.4	83.8	88.8	714	714	100
Saia	8.6	8.4	9.4	8.1	116.2	120.0	1051	1051	100
H1/2c	8.6	8.3	9.8	9.3	87.5	90.0	1091	1130	96
H2/1c	5.1	6.3	7.8	8.6	70.0	82.5	208	843	25
H2/14c	7.2	8.1	8.6	9.0	88.0	90.0	1091	1150	95
LSD 5%d	1.57		1.98		8.54		456		
LSD 5%e	1.80		1.62		6.42		278		

a:+Hoegrass 1.55 l/ha; b: -Hoegrass; c:Selection from bulk population;
d:between variety levels for same herbicide treatment;
e:between herbicide treatments for same variety

to 40% (Table 4) are associated with Hoegrass application under different climatic conditions. Phytotoxicity seems to be affected by time of seeding and the growth stage at which Hoegrass is applied. The 1980 and 1981 experiments were sown relatively 'late' for the South Australian environment and Hoegrass was applied at the 3 leaf stage. No yield loss resulted in New Zealand Cape. The 1983 trial was sown relatively 'early' and application was not possible until the 6 leaf stage. A 40% yield loss was recorded.

TABLE 2. Effect of Hoegrass treatment on grain yield, Turretfield 1981

Variety	Yield (kg/ha)		% Control
	+Hoegrass 1.5 l/ha	-Hoegrass	
West	661	1805	37
Moore	2327	1962	119
Cassia	1045	2276	46
NZ Cape	1256	1210	104
Saia	1508	1250	121
H1/2	1448	1752	83
H2/2	909	1131	80
H2/14	1521	1772	86
Mean over all varieties	1335	1645	81

LSD 5% Hoegrass x Variety at different levels of Hoegrass - 436
Hoegrass x Variety at same levels of Hoegrass   - 462

The time of application study conducted in 1983 indicated that all cultivars became more sensitive to Hoegrass as application was delayed from the 2- to 4- 6-leaf stages (Table 3). No critical information is available yet on the effect of time of seeding of the crop. Until further studies resolve the interaction between time of seeding, rate and time of application and phytotoxicity, the objective will be lowered to tolerance of 1.0 l/ha Hoegrass, which would allow control of annual ryegrass, but not wild oats.

TABLE 3. Effect of time of application of 1.5 l/ha Hoegrass on oat grain yield, Turretfield 1983

Variety	Yield (g per plot)			Control
	Time of application			
	2 leaf stage	4 leaf stage	6 leaf stage	
West	95	0	10	502
Coolabah	312	161	0	512
Swan	110	0	0	545
Echidna	159	0	0	750
Dolphin	90	5	0	310
Barmah	135	0	0	573
NZ Cape	527	478	433	430
Mean over all varieties	204	92	63	517

While no genetic studies have been undertaken, it is apparent that the heritability of Hoegrass tolerance is high. When 79 F_2-derived F_5 lines from a cross between Hoegrass tolerant breeders lines and West, were sprayed with Hoegrass, 30 had Hoegrass tolerance scores similar to New Zealand Cape.

Hoegrass tolerance has been an objective in the South Australian breeding programme since 1979 and this trait has now been transferred into many backgrounds. One germplasm line was registered by the Registrar of Cereal cultivars in Australia in 1985.

It is named 'Savena 1' and resulted from the cross West/New Zealand Cape//West. Many of the deficiencies of New Zealand Cape have been overcome in Savena 1, although not to the extent that it was considered for variety release. Savena 1 has good tolerance to Hoegrass (Table 4) equal to or perhaps slightly better than New Zealand Cape. It is much higher yielding than New Zealand Cape when not treated with Hoegrass.

Selection for Hoegrass tolerance is simple and rapid in field tests. Treatment of space-planted F_2 rows has resulted in reliable screening and single replicate experiments of lines attaining yield trial status have proved satisfactory.

TABLE 4. Effect of 1.5 l/ha Hoegrass on the yield of 4 oat cultivars, Turretfield 1983

Variety	Grain yield (kg/ha)	
	Hoegrass applied	Control
West	1780	4300
New Zealand Cape	1680	2650
Cooba	177	4000
Savena 1	2440	3780

LSD (5%) Same variety different herbicide means, 604
LSD (5%) Same or different variety or herbicide means, 634

Tolerance to Hoegrass would represent an advance in weed control options. However, if high levels of tolerance to this herbicide exist in Avena sativa, it is possible that it exists in related weedy species. There is little evidence to show that Hoegrass has failed in Australia due to genetic tolerance of wild oats(3), although tolerance in a Lolium rigidum population was recorded in 1982(4). Furthermore it is conceivable that hybridisation between a tolerant, cutivated oat and wild oats would result in the introgression of the tolerance genes into the wild population. This problem must be addressed prior to release of any oat genotype tolerant to a wild oat herbicide, as wild oats are such a serious weed of other crops, one should not prejudice the future use of a herbicide in other crops.

The risk of introgression of tolerance genes will be small but estimating the actual risks would be very complex. An indication of the risk is that Algerian, Sual and New Zealand Cape, varieties with Hoegrass tolerance, have been widely grown in Australia from the early 1900s until the present. If the progeny from chance hybridisation between Algerian types and wild oats have no selective disadvantage against the wild types, they should, by now, have been selected, as Hoegrass has been used in Australia since 1978. This has not yet occurred.

182

Several management and breeding strategies can be suggested to further reduce this risk.

1. Use the herbicide even if low densities of wild oats occur in the field.
2. Do not save seed from wild oat infested crops as this may spread tolerant types.
3. Keep records of herbicide use on each field and rotate them through time.
4. Hoegrass tolerant varieties should be susceptible to other herbicides used for wild oat control.
5. Enforce stringent control measures on the F_1.

The ideal situation would be a herbicide 'safener' system where the safener was so effective that no genetic tolerance was required in the oat plant. Currently safeners have not reached this level of effectiveness, so genetic tolerance still offers the most promise for wild oat control in oats.

4. CONCLUSION

While few chemicals are released for grass control in oats, a good case for breeding for tolerance to 'existing' herbicides can be made. Genetic variability with respect to tolerance to Hoegrass is available. It offers oat breeders a chance to control some annual grasses with genotypes identified thus far and, if further sources of tolerance genes can be identified and recombined with those available, perhaps wild oats can be controlled in domesticated oat crops.

REFERENCES

1. Anderson RN: Weed Sci. 24, 266-269, 1976.
2. Wu C-H and PW Santelman: Weed Sci. 24, 601-604, 1976.
3. Anderson IP: Hoechst Australia Ltd, pers. comm.
4. Heap J and R Knight: J Aust. Inst. Agric. Sci. 48, 156-157, 1982.

THE CHEMICAL CONTROL OF WILD OATS IN OATS - A PROGRESS REPORT

H. F. TAYLOR and T. M. CODD
Long Ashton Research Station, Weed Research Division, Yarnton, Oxford, UK

1. INTRODUCTION

The increasing cultivation of cereals in Britain since the war has resulted in the emergence of several grasses as serious weeds, notably wild oats. Chemical industry has responded with the introduction of a number of herbicides to control this weed in wheat and barley but not, unfortunately, in cultivated oats. Chlorfenprop-methyl was possibly the exception for this chemical did show specificity within the genus Avena(1). Six oats were reported to show tolerance and its use as 'Bidisin' (Bayer) was consequently recommended in the late 1970s for the control of wild oats in specified spring oats. A second herbicide, barban (Carbyne) was also said to be tolerated by certain oat cultivars. Unfortunately, like chlorfenprop-methyl some biotypes of A.fatua were also less sensitive to barban(2), Both of these herbicides have now been largely superseded in Britain. Recently, however, marked differences in the sensitivities of cultivated oats to a third herbicide, diclofop-methyl, (Hoegrass, Hoechst) have been recognised independently by two groups of research workers(3).

An alternative to the conventional use of selective herbicides was suggested by the successful research in the USA with safeners (antidotes). These are chemicals which, when applied to a sensitive crop plant, confer protection against the damage from subsequent herbicide treatments(4).

Research in Canada(5) had already indicated that cultivated oats could be protected against barban damage by giving a seed treatment with 1,8- naphthalic anhydride(NA). Now good protection of oat cultivars by NA seed dressings has been reported for diclofop-methyl(6). This work, currently funded by the Home Grown Cereals Authority, would seem to offer a potential solution to the problem of wild oat control in cultivated oats.

2. MATERIALS AND METHODS

2.1 Assessment of the response of oat seedlings to diclofop-methyl

Seeds of the different cultivars and breeding lines of oats were sown in plastic pots (9 cm diameter) containing a local sandy loam to which a balanced fertiliser had been added. Seeds were 'safened' by shaking with finely powdered 1,8-napthalic anhydride (1 g/100 g of seed) immediately prior to planting. The seeds were covered with soil (1 cm unless otherwise stated), watered, and grown in a glasshouse with minimum temperature of 16°C (day) and 5°C (night). When necessary, artificial light was given, both to increase light intensity and extend daylength to 16 h. Pots, each containing five seedlings at the 1.5 leaf stage, were sprayed with diclofop-methyl at the levels indicated, using a spray volume equivalent to 200 l ha^{-1}.

Herbicide damage was seen as a slight scorching and a cessation of growth within two to three days. At the recommended application rate (about 1 kg ha^{-1}) normal oat plants died within 14 days. Plants were cut at soil level 14 to 21 days after treatment and the total fresh weight of foliage of all five plants in each pot was recorded.

2.2 Diclofop-methyl metabolism in the tissues of normal and safened oat seedlings

For this investigation radioactive (RS)-diclofop-methyl ([2,4-dichlorphenyl-U-^{14}C], 19.8 mCig^{-1}) was used.

Seedlings of the diclofop-methyl sensitive spring oat Margam, were grown to the 1.5 leaf stage from normal and safened seed. 5 µl of a dispersion of herbicide in dilute aqueous surfactant was then dispensed into the opened leaf sheath of each seedling(7). After the specified treatment period, a 20 mm long basal section was removed from each seedling, frozen in liquid nitrogen, powdered and extracted with aqueous methanol at -20°C. Aliquots of these extracts were chromatographed on silica gel plates to separate the unchanged herbicide from the phytotoxic diclofop and the detoxification products (conjugates). These three components were eluted from the silica gel and the ^{14}C-label present in each was determined by liquid scintillation counting.

3. EXPERIMENTS AND RESULTS

3.1 Tolerance to diclofop-methyl

During glasshouse experiments in Spring 1983 with normal and safened oat cultivars it was noted that one of the ten oats tested, Elen, showed tolerance to diclofop-methyl. This occurred whether or not the naphthalic anhydride safening treatment had been given. The result was repeated with further samples of the seed obtained from WPBS, and the observation was then reported(3). Very many more oat cultivars and breeding lines have now been screened by spraying and also by dispensing herbicide into the leaf sheath(7), and further evidence of tolerance obtained. Such susceptibility/tolerance is demonstrated in Figure 1 which shows seven oat cultivars selected from a total of 43 which had been sprayed with diclofop-methyl at 1.5 times the recommended field rate. Four of these oats (E, G, N and P) showed the greatest tolerance, approaching that of Elen(B). Dula(Du), Trafalgar(T), Margam, Rollo, Avalanche and Leander (not shown) were susceptible. It is of interest that of the 43 oats supplied by Twyford Seeds Ltd (Kings Sutton) and included in this test, three (E, G and N) had a common pollen parent.

FIGURE 1. Differing sensitivities of cultivated oats to diclofop-methyl (1.5 kg ha^{-1})

A further experiment was carried out in late April 1985 to compare the tolerance of some of the most promising oats selected from the 220 supplied by WPBS and Twyford Seeds. The results are summarised in Figure 2. In addition to the sister seedlings G and N, Figure 2 also includes a markedly tolerant winter oat(D). Elen(B) gave a disappointing result as it has done frequently when tested in late spring, and its seedling(J) showed the greater tolerance in this test. The Australian oat, CC 6189/3(M) appeared to be exceptionally good, but this result was obtained with an older seed sample and should be repeated. The susceptible spring oat Margam(A) was included.

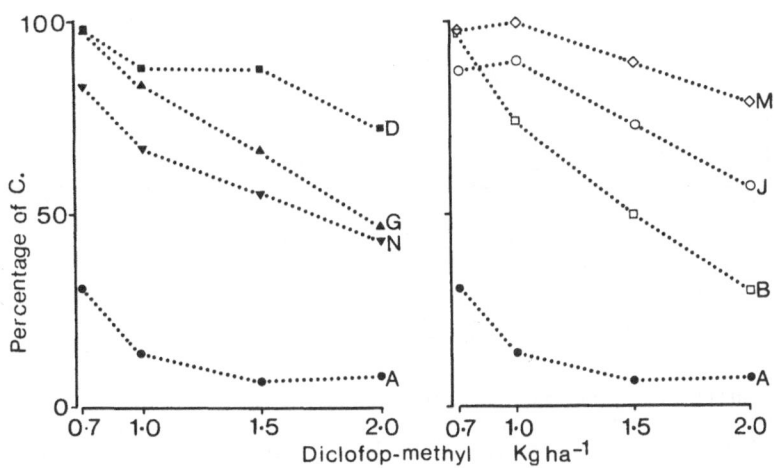

FIGURE 2. Dose-response curves for seven oats obtained in glasshouse tests

3.2 Safening of oats with 1,8 naphthalic anhydride(NA)

Oat seeds dusted with NA at the rate described germinated normally and the seedlings were not visibly different from controls. When sprayed with formulated diclofop-methyl both normal and safened plants retained similar amounts of the herbicide and showed equally, the initial yellowing and scorch. However, safened plants only ceased growth briefly and, provided the dose was not too severe, the damage remained localised and the plants recovered. This has been demonstrated with a number of oat cultivars and all appear to be equally safened, in glasshouse tests. Using diclofop-methyl at the normally recommended dose NA can be expected to give sufficient protection to ensure complete recovery of the oat seedlings under these conditions.

Experiments in the field, however, have been far from encouraging. Some safening is produced, but it is both uneven and inadequate. These failures have necessitated further laboratory investigations, both into the factors affecting the magnitude of the safening response, and also to the mechanisms through which it is mediated.

The results of a typical experiment, in this case to establish the effect of depth of planting upon the safening response, are presented here. Both NA treated and normal seeds of Trafalgar, were either pressed, coleorhiza down, into the soil surface or sown and covered with 1, 2, 3 or 4 cm of soil. When the seedlings had reached the 1.5 leaf stage the pots were sprayed with herbicide at 1, 2 or 3 kg ha^{-1} or left unsprayed.

186

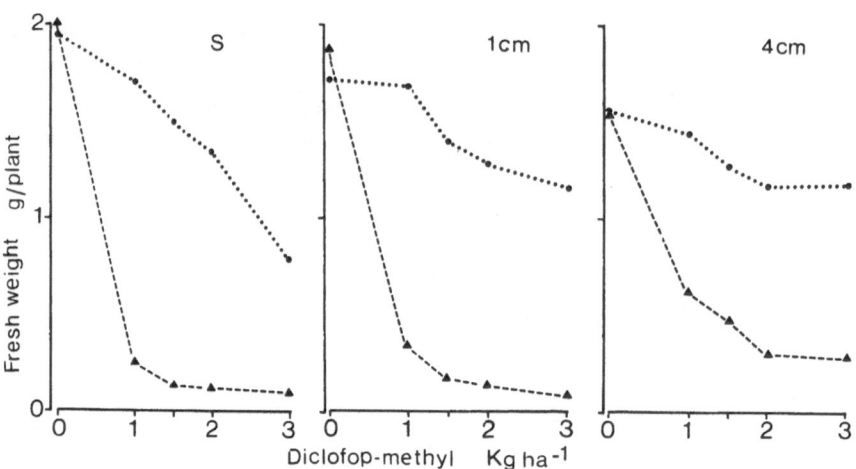

FIGURE 3. The effects of diclofop-methyl treatments on safened (•) and unsafened (▲) Trafalgar oat seedlings grown from seed sown at different depths: surface (S), 1 cm and 4 cm (left to right)

The results for the five planting depths established distinct trends which are well illustrated by the three examples ('surface' 1 cm and 4 cm cover) shown. Increasing the planting depth reduced the vigour of the seedlings but reduced even more markedly, the severity of the herbicidal damage. As the observed safening effect of the NA was greater with the deeply sown plants, it could be suggested that this had given more efficient protection. But equally well, it could have been merely a reflection of the more 'gentle' dose response curve obtained when diclofop-methyl is sprayed onto deeply planted oats. However, these results do indicate that good soil cover is advantageous, particularly if the normal field rate of application may inadvertently be exceeded.

3.3 The effect of NA on diclofop-methyl detoxification

Detoxification of diclofop-methyl is brought about, in oats, by a sequence of biochemical conversions. Firstly, it is hydrolysed to the equally phytotoxic diclofop and this is, in turn, conjugated to comparatively non-phytotoxic products. The relative rates of these reactions may well determine whether an oat survives a treatment with the herbicide. Research here has already indicated that chemicals which reduce the effectiveness of diclofop-methyl probably do so by increasing the rate of its detoxification. This possible explanation for the safening effect of NA has been investigated.

Using the procedures described in 2.2 the formation of conjugates (non-toxic metabolites) from radiolabelled herbicides has been followed in normal and safened oat seedlings for the first 48 h after application. Results obtained for the 5, 10 and 48 h times are shown (Table 1).

These results show clearly that treatment with NA favoured more rapid conjugation of diclofop, and lends support to the theory that the safener operates through accelerating herbicide detoxification. This experiment has been repeated and these differences are very reproducible, but are they too small to explain the observed safening?

187

TABLE 1. Conjugates in oat seedlings after treatment with diclofop-methyl
          (expressed as percentages of extracted radioactivity)

Time after treatment (h)	Control	+ NA	± S.E.
5	34.1	38.9	± 1.1
10	59.8	66.5	± 1.5
48	72.8	77.3	± 2.34

4. CONCLUSIONS
   Tolerance to diclofop-methyl is not widespread amongst cultivated oats
but it has been recognised in a number of cases including three oats which
have a common pollen parent. The conditions under which tests are carried
out may be critical and field trials have not been attempted.   A
diclofop-methyl tolerant spring oat, Elen, has shown no equivalent
tolerance to other wild oat herbicides.
   Safening with NA has been demonstrated with all the oat cultivars used,
but field tests have been disappointing. NA has been shown to increase
herbicide detoxification suggesting a probable mode of action. Although
the changes in conjugation rate for the whole tissue are small it is
possible that greater effects may be located at specific sites of action
to explain the large safening response.
   Progress has been made in both of these aspects of weed control in oats
but much remains to be achieved before field recommendations can be made.
However, the responses of new oat cultivars to safeners and to herbicides
must now be considered seriously before their introduction into commerce.

5. ACKNOWLEDGEMENTS
   The authors thank Twyford Seeds Ltd, Hoechst AG and the staff of WPBS
for their cooperation. Mr Marc Loader is especially thanked for his
contribution to this research, and support from the Home Grown Cereals
Authority is gratefully acknowledged.

REFERENCES

1. Strykers J and M van Himme: Meded Rijksuniversiteit-Gent 17, 32-35,
    1972.
2. Strykers J and M van Himme: Meded Rijksuniversiteit-Gent 15, 32-33,
    1971.
3. Codd TM, AP Day, WG Richardson and HF Taylor: Progress Reports
    1982-1983, Home Grown Cereals Authority, pp.20-23, 1983.
4. Hatzios KK: Advances in Agronomy 36, 265-316, 1984.
5. Chang FY, GR Stephenson, GW Anderson and JD Bandeen: Weed Sci. 22,
    546-548, 1974.
6. Taylor HF and MPC Loader: Outlook on Agriculture 13, 58-68, 1984.
7. Taylor HF, MPC Loader and SJ Norris: Ann. appl. Biol. 103, 311-320,
    1983.

EFFECTS OF HERBICIDES ON OATS

D.L. REEVES and J. LAMMERS
South Dakota State University, Brookings, SD 57007, USA

1. INTRODUCTION
South Dakota is part of the major oat producing region in the United States. Farmers in the state plant about 1,000,000 hectares of oats annually. It is estimated that 40% of the oats are sprayed with a herbicide. Most years we have some farmers complaining of injury to their oat crop. Many of these claim to have done everything properly. Our dilemma is not being able to prove whether the cause is mistake in formulation, rate or timing of application or some other reason such as environmental interaction or varietal sensitivity.

2. MATERIALS AND METHODS
Ten cultivars used in this study were: Burnett, Chief, Clintland 64, Garland, Lancer, Moore, Noble, Nodaway 70, Ogle and Porter. These have all been grown in this region and represent a range of maturities. Five treatments were used: the check 0.56 kg/ha MCPA [(4-chloro-2-methylphenoxy) acetic acid], and 2,4-D amine, [(2,4-dichlorophenoxy) acetic acid], at rates of 0.28, 0.56 and 0.84 kg/ha. A randomized complete block design was used with four replications. Each plot consisted of two rows spaced 30 cm apart and 4 m long. Plants were sprayed in the three to four leaf stages.

3. RESULTS
Visible differences were quite evident with the increasing rates of 2,4-D. The higher rates caused the lateral panicle branches to be reduced in length thus making the panicle more compact. This effect was somewhat dependent upon the cultivar.
Our interest in this type of study was greatly stimulated when one cultivar had a 12% grain yield reduction with the low rate of 2,4-D while four others had none. However, later tests generally showed no yield

TABLE 1.Effects of herbicides on ten oat cultivars
at two locations in 1984

Treatment	Centerville		Watertown	
	kg/ha	kg/hl	kg/ha	kg/hl
Check	2667	41.6	3745	48.6
0.56 MCPA	2921	39.0	3644	47.7
0.28 2,4-D	2482	40.7	3722	48.1
0.56 2,4-D	2613	39.5	3761	47.0
0.84 2,4-D	2530	39.1	3595	46.1

reduction due to various treatments (Table 1). Test weights were significantly affected in 1984 with the largest reductions occurring at the lower yielding site.

Two 30 cm subsamples were taken from each plot to use for yield component analysis. From each subsample 10 panicles were picked at random and individually threshed, grain weighed and seeds counted. There were significant differences between cultivars and treatments in both years (Table 2). The differences between cultivars were expected because the

TABLE 2. Mean panicle weight and number of seeds per panicle of ten oat cultivars treated with herbicides

Treatment	1983		1984	
	g	Seeds	g	Seeds
Check	82.7	40.8	137	45.7
0.56 MCPA	71.9	38.4	132	44.3
0.28 2,4-D	75.2	37.7	130	42.9
0.56 2,4-D	70.4	33.1	128	42.2
0.84 2,4-D	63.2	30.7	120	39.5

ones used represent a range of panicle and seed sizes. There was also a significant cultivar x treatment interaction in 1983. Results of selected varieties are given in Table 3 to indicate the cultivar treatment interaction. Both MCPA and 2,4-D decreased the mean number of seeds per panicle, although the smallest reduction was in the highest yielding environment.

Average seed weight responded differently in the two years. There was a significant cultivar difference each year, but the treatment effect was only significant in 1983. Average seed weight at Brookings was 203, 213 and 188 mg respectively for the check, 0.56 kg/ha of 2,4-D and MCPA.

TABLE 3. Mean panicle weight and number of seeds per panicle of five oat cultivars when treated with 0.56 kg/ha of two herbicides at Brookings, SD in 1983.

Cultivar	Treatment	Means/panicle	
		g	Seeds
Burnett	Check	77.8	38.8
	MCPA	40.8	24.8
	2,4-D	61.6	29.0
Chief	Check	103.2	57.0
	MCPA	95.1	53.7
	2,4-D	72.1	36.5
Lancer	Check	98.6	44.3
	MCPA	81.1	37.4
	2,4-D	92.9	40.0
Noble	Check	69.8	36.4
	MCPA	69.2	38.0
	2,4-D	65.8	32.1

190

Lodging differences were very evident in 1984. The means for two locations were 6, 8, 15, 27 and 34% respectively for the check, MCPA and increasing rates of 2,4-D.

Overall, there has been little effect on yield when recommended rates were applied at the recommended stage of growth. These herbicides have affected other characters measured, however, it appears that plants may compensate if some factor is adversely affected so that yield did not show a similar effect. A considerable environmental effect seems likely for some traits.

IMPROVING WINTER HARDINESS IN WINTER OATS BY SEED TREATMENT WITH PGRS

H.M. ANDERSON, N.D.S. HUBAND, P.J. MURPHY and R.D. CHILD
Long Ashton Research Station, University of Bristol,
Long Ashton, Bristol, UK

1. INTRODUCTION

Winter hardiness is one of the main factors affecting harvestable yields in winter oats. As breeding for high yields is generally accompanied by reduced winter hardiness, the possible beneficial effects of applying an experimental plant growth regulator (pgr) as a seed dressing were investigated in a pot experiment. The cultivars Bulwark and Peniarth, sown at four depths, were used.

2. RESULTS

Neither sowing depth nor pgr concentration affected germination. However, increased pgr concentration and/or increased sowing depth delayed emergence, although eventually acceptable rates were attained by most treatments.

At all sowing depths there was a reduction in mesocotyl length related to pgr concentration; the higher the concentration the shorter the mesocotyl. With the deeper sowings some extension of the lower stem internodes occurred but at all sowing depths, treatment with the pgr resulted in the crown of the plants overwintering deeper in the soil.

Measurements of soil temperatures were made during a freeze-thaw cycle which lasted 9 days. Significant changes with depth occurred both in the total number of hours that soil temperature was below -2°C and in the cumulated degree-hours below -2°C. Thus, for the seedlings, relatively small changes in the depth of sensitive organs allowed exposure to markedly different temperature regimes.

Bulwark was more affected by winter damage than Peniarth. At all sowing depths, plants arising from seed treated with the high concentration of pgr showed little damage as did plants arising from deep sown seed treated with the low concentration of the pgr. Plants from the other treatment showed significantly more damage. Dissections showed that apices of severely damaged plants were usually dead.

WINTER OAT HUSBANDRY - RESULTS FROM ADAS TRIALS 1982-84

R. GREEN*, J.H. BALDWIN** and H.W. ROBERTS***
  *ADAS, Wye Sub-Centre, Wye
 **ADAS, Government Buildings, Cambridge
***WOAD, Trawsgoed, Aberystwyth

1. INTRODUCTION
There has been a recent increase of interest in the general husbandry aspects of winter oats. The ADAS trial programme has concentrated on five main topics, namely varieties, nitrogen use, lodging control, fungicide use, and an examination of the components of yield.

2. RESULTS
2.1 Varieties
A series of Recommended List trials are undertaken by ADAS in conjunction with the National Institute of Agricultural Botany, and the results are presented in the NIAB Farmers Leaflet No. 8.

2.2 Nitrogen
Results from 7 trials from 1982-84, using the variety Peniarth, confirmed that current recommendations (100 kg/ha) were satisfactory, particularly where lodging effects and combining time were taken into consideration. All nitrogen applications were split between 40 kg/ha in late February or early March and the balance applied at GS 31. Different responses could well be obtained with better standing varieties.

2.3 Lodging
Apart from 1982 there was a trend for 5C Cycocel to reduce lodging and increase yields. The 2 x half rate (applied at GS 30 and 32) or the 1 x full rate (applied at GS 32) gave fairly similar results. Split application at one site in 1983, when weather conditions at GS 32 were wet and cool, provided particularly good results.

All treatments were applied with Agral wetter at the rate of 25 ml per 100 l spray.

Split applications, because of weather uncertainties, give a reasonably practical approach.

2.4 Fungicides
Response to fungicide treatments was assessed in both variety trials and in separate fungicide trials and the results indicate that Peniarth and Pennal are less responsive to fungicide treatment than Bulwark and Rosette. It is concluded that the oat crop should be treated according to observed disease levels, and not regarded as a cheap disease free break.

2.5 Components of yield
Data are presented for the mean components of yield, obtained from 17 treatments, with the variety Peniarth, where yields in excess of 7.5 t/ha were obtained.

SESSION IV. CROP PHYSIOLOGY AND PRODUCTION METHODS OF OATS
CHAIRMAN'S COMMENTS AND SUMMING UP

C. M. BROWN
Department of Agronomy, University of Illinois, Urbana, IL 61801, USA

The first two papers in this session dealt with identification and elucidation of physiological parameters that impact oat productivity. Both authors emphasized the need to utilise these physiological parameters as selection criteria in plant breeding programmes to improve oat productivity. Brouwer suggested that growth and development patterns of oats might be manipulated to improve productivity of oat genotypes. The utilisation of variation for such physiological response should lead to the development of cultivars which are more tolerant of environmental stresses or have a developmental pattern that permits them to escape some of the critical stress periods. He encouraged the use of physiological parameters such as root:shoot ratio, seedling vigour, harvest index, photosynthetic activity, leaf area duration and growth rates as selection criteria to be used in oat breeding programmes.

Dr Peterson discussed some of the metabolic limitations to increased yields under more or less optimum growing conditions. He emphasized the need to identify rate-limiting processes in the plant so that these processes might be manipulated by plant breeding methods to improve oat productivity. Once these rate limiting processes are identified, genetic diversity for the processes can be exploited in breeding programmes to increase production potential. Peterson believes that this approach will become even more useful when it becomes possible to use recombinant DNA techniques to alter specific genes without affecting the other desirable genes that are present. He suggested that growth rates, uptake and transport of nutrients, partitioning of dry matter to the grain, and leaf senescence might be some of the physiological processes limiting plant growth and grain filling.

While it is extremely important to identify and characterise these rate limiting physiological processes and enzymes, it should be emphasized that to be useful as selection criteria in practical oat breeding programmes, methods must be developed to adapt such measurements to large populations of plants. Measurements need to be relatively simple, inexpensive, rapid, and adaptable to individual plant analyses of large populations in early generations.

The paper by Tibelius and Klinck examined the effect of genotype, seeding rate, and environment on the primary:secondary seed weight ratio. They found that the ratio was influenced by genotype and environment and that secondary grains were more influenced by environmental variation.

The paper by Lyslov and Küüts provided an excellent account of recent oat variety development in the Soviet Union. They stressed the importance of developing varieties with a combination of useful characteristics and with wide adaptation to the wide variety of soil and climatic conditions of the Soviet Union. Their best results were achieved by crossing the best and most highly productive local varieties and most lodging resistant west European varieties with the more drought- and disease-resistant, but often less productive, types from Canada, USA and Mexico.

The final three papers of the session dealt with the use of chemicals for weed control with special emphasis on control of wild oats in cultivated oats. Barr reported that the herbicide Hoegrass(R) (Diclofop-methyl) is effective for controlling annual ryegrass and wild oats but that most cultivated oats were also moderately susceptible to the chemical. However, the author has observed some useful levels of tolerance in several oat genotypes. This tolerance appears to be highly heritable and should be possible to use in breeding programmes for the development of oat cultivars adapted to the use of Hoegrass for wild oat control. The author cautioned that tolerance to this and other herbicides likely occurs in wild oat populations and he suggested a strategy of using wild oat herbicides in rotation over time to avoid selection of tolerant wild oat biotypes.

Taylor and Codd reported that they have examined two possibilities for controlling wild oats in cultivated oats. The first approach has involved the use of 'herbicide safeners.' to protect the cultivated seedlings against damage from herbicides applied to kill the wild oats. They also observed that some oat genotypes exhibited sufficient tolerance to wild oat herbicides to encourage their use in the development of herbicide tolerant cultivars.

Reeves and Lammers treated ten oat cultivars with several rates of MCPA and 2,4-D and found significant differences among rates for yield, test weight, height and lodging. There was a significant cultivar x treatment interaction for test weight but not for yield.

Weed control is certainly an oat research area that should receive increased attention. Since the oat is not a major crop, chemical companies are not likely to find it economically viable to produce herbicides for specific use in oat production. Therefore, it will be left to oat researchers to devise innovative methods to successfully utilise existing herbicides and those developed for use in other crops. The information presented in the three weed control papers suggests that this objective can be accomplished. The development and identification of herbicide resistant oat cultivars offers excellent potential in this important area.

SESSION V

Quality, use and marketing requirements for oats

Chairman: J.E. Lea

WORLD OATS USE AND MARKETING

DONALD J. SCHRICKEL
Director - Grain Research & Development, The Quaker Oats Company,
Chicago, IL 60654, USA

1. INTRODUCTION

The degree of competitiveness among crops throughout the world is extremely high today. The producer is attempting to get maximum returns from his investment in land and equipment. This in turn is placing demands on the oats breeder to improve varieties and to make oats a more competitive crop.

In this presentation I will discuss the current oats production in the world as well as give an overall view of the uses and markets for oats. I will also project what we may expect in the future with respect to production and needs of oats.

2. WORLD OATS PRODUCTION

Oats are primarily a cool season crop. A large proportion of world production occurs in the northern hemisphere, between latitudes of 35 and 50 degrees.

World oats production, area and yields for major producing countries are shown in Tables 1, 2 and 3.

TABLE 1. World oats production

	1975	1980	1984
	(Thousand metric tonnes)		
USSR	12495	15544	15200
USA	9275	6659	6858
W. Germany	3445	3249	3054
Poland	2920	2245	2750
Canada	4480	3028	2700
Sweden	1321	1567	1990
France	1948	1927	1752
PR China	1549	1800	1680
Finland	1450	1258	1350
Australia	1141	1128	1300
Spain	609	680	800
E. Germany	780	582	608
UK	795	601	550
Argentina	433	433	530
Others	4457	3850	4208
World	47098	44551	45330

198

TABLE 2. World oats area

	1975	1980	1984
		(Thousand hectares)	
USSR	12017	11770	11700
USA	5261	3503	3273
Canada	2414	1515	1426
PR China	1464	1500	1400
Australia	988	1093	1150
Poland	1291	997	1000
W. Germany	920	856	666
Spain	457	458	479
France	655	534	437
Finland	572	448	434
Sweden	464	452	425
Argentina	338	350	400
E. Germany	243	155	160
UK	232	148	110
Others	2864	2327	2405
World	30270	26106	25465

The USSR is the largest producer and has the greatest area devoted to oats, while the highest yields are in the United Kingdom. The quality of soil on which oats are grown differs from country to country, and even within countries. In the United States of America, for example, oats have generally been forced to the lower quality soils for economic reasons. In defence of oats, however, they often prove to be the best crop available for areas of marginal quality soil and reduced rainfall. Another fact to consider in the economic value of oats is the straw, used primarily for livestock bedding. In the livestock regions where demand for oats straw is high, the net return to producers is often equal to 25-30% of the grain value.

TABLE 3. World oats yields

	1975	1980	1984
		(Metric tonnes per hectare)	
UK	3.43	4.06	5.00
Sweden	2.85	3.47	4.68
W. Germany	3.74	3.80	4.59
France	2.97	3.61	4.01
E. Germany	3.21	3.75	3.80
Finland	2.53	2.81	3.11
Poland	2.26	2.25	2.75
USA	1.76	1.90	2.10
Canada	1.86	2.00	1.89
Spain	1.33	1.48	1.67
Argentina	1.28	1.24	1.33
USSR	1.03	1.32	1.30
PR China	1.06	1.20	1.13
Australia	1.15	1.03	1.13
Others	1.56	1.65	1.75
World	1.56	1.71	1.78

3. OATS UTILIZATION
Most of the oats produced in the world are used for livestock feed. However, use for human food is also important. Interesting changes are occurring in the amount of oats used for human food (Table 4).

TABLE 4. Percent of oats used for human food

	1975	1984
World	16.3	17.9
USSR	31.1	22.5
USA	6.9	12.0

The Quaker Oats Company uses oats for human food in 20 different countries, the USA being the largest followed by Canada, then UK.

4. FUTURE TRENDS
Studies made by Phillip M Sisson of the Quaker Oats Company project a decline in world oats production into 1990, although USSR will continue to increase her production (Table 5).

TABLE 5. Oats production - projections to 1990

	1975	1984	1990
	(Thousand metric tonnes)		
World	47098	45330	42500
USSR	12495	15200	16000
USA	9275	6858	5300

The use of oats for human food will increase in the USA, but will decline in the USSR and in the world in total (Table 6).

TABLE 6. Oats used for human food - projections to 1990

	1975	1984	1990
	(Thousand metric tonnes)		
World	7750	8000	7400
USSR	4000	3400	2700
USA	650	900	1130

5. NUTRITIONAL VALUE OF OATS FOR HUMANS
According to Quaker's Director of Nutrition, Dr David Hurt, there have been developments in nutrition which have focussed the attention of the medical world on oats. It has been found that the fibre of oats has unique physiological properties. It is general knowledge that food fibre has laxative properties but the fibre of oats has other effects as well. Research on both test animals and humans has demonstrated that fibre in oats will reduce serum cholesterol and help to promote a more normal blood glucose level in diabetics. These effects are believed to be related to the soluble fibre portion that makes up about one-half of the total dietary fibre in oats. Additional research is underway in these areas. Favourable results could enhance the value and marketing of oats products in the future.

OAT QUALITY - PRESENT STATUS AND FUTURE PROSPECTS

ROBERT W. WELCH
Welsh Plant Breeding Station, Aberystwyth, Dyfed, Wales, UK

1. INTRODUCTION

The previous paper has indicated that although the bulk of the world oat crop is utilised in animal feeding, there is a sustained and substantial demand for milling oats for human consumption. The present paper will describe the value of oats in livestock production and the quality requirements of the oatmeal millers. Recent research into the therapeutic value of oats will be summarised. Variations in nutritionally important grain constituents will be reviewed and biochemical and physiological factors which may constrain the improvement of oat grain quality will be evaluated. The potential of oats as a raw material will be described.

2. FEED QUALITY

Table 1 shows that oats have the highest oil content and consequently the highest gross energy value of the temperate feed cereals. However, as a result of the high fibre content, which is derived primarily from the husk, oats have the lowest metabolizable energy (ME) content. The protein

TABLE 1. Comparative composition and feeding values
of oats, barley, wheat and maize grain

	Oats	Barley	Wheat	Maize
Protein (% N x 6.25)	10.5	11.0[a]	12.5	10.4
Oil (%)	5[a]	2[a]	2	4
Crude fibre (%)	10	5	2	2
Gross energy (MJ/kg DM)	19.6	18.5	18.7	18.9
Metabolizable energy (MJ/kg DM)				
Ruminants	12.0	12.9	13.5	13.8

a = WPBS data; all other data from (1)

content of oats is similar to that of barley and maize and oat protein has a relatively well balanced amino acid composition (Table 2). Although grinding may be required to improve palatability, oats may be successfully fed to cattle, sheep, poultry and horses(2). Nevertheless its lower ME reduces the energy density of the diet, feed conversion efficiency is decreased and thus, compared to other cereals, more oats will be needed to achieve the same level of animal production.

TABLE 2. Essential amino acid content of oats compared
with the requirements of pullets

Amino acids (% of protein)	Pullet requirements[a]	Oats[b]
Phenylalanine + tyrosine	8.6	8.0
Arginine	5.6	6.0
Leucine	5.6	8.1
Glycine + serine	3.9	10.2
Lysine	3.8	4.2
Valine	3.4	5.7
Iso-leucine	3.3	4.3
Threonine	3.1	3.7
Methionine + cystine	2.2	5.3
Histidine	1.4	2.6
Tryptophan	0.9	1.3

a = from (3); b = from (4)

Straw represents a substantial and often underexploited part of the oat crop. Oat straw has a similar proximate analysis to other cereal straws but it has a higher ME which confirms the traditional view of its superior feeding value (Table 3).

TABLE 3. Comparative composition and feed quality
of oat, barley and wheat straw

	Oats	Barley	Wheat
Protein (% N x 6.25)	4.6	4.7	3.8
Oil (%)	2.1	1.6	1.2
Crude fibre (%)	41	42	42
Gross energy (MJ/kg DM)	18.5	18.5	18.2
Metabolisable energy (MJ/kg DM)	7.4	6.4	6.1

All data from (1)

3. MILLING QUALITY

The oat millers aims are to obtain a high yield of groats from whole oats. This is achieved by purchasing clean batches of oats with plump even sized grains and a high specific weight (> 50 kg/hl or 40 lb/bu). As Table 4 indicates, both grain size and specific weight correlate well with groat percentage.

Some samples of oats offered for milling may contain high levels of free fatty acids (FFA) in the oil which will render the finished products unsuitable for certain food uses. The cause of high FFA is not understood and its incidence can vary erratically both between seasons and growth location. However FFA is negatively correlated with both specific weight and grain size (Table 4) and thus the use of these two criteria will tend to ensure both a good extraction rate and low FFA levels. On the other hand these two criteria will also tend to select samples with a lower groat protein content than that desired by the millers (Table 4).

Table 4. Correlations between specific weight, mean grain size
and groat characteristics

	Percentage groat	Free fatty acid (% in oil)	Groat protein (%)
Specific wt	0.78***	-0.79***	-0.72***
Grain size	0.69***	-0.72***	-0.76***

WPBS data; n = 98; *** = significant at P < 0.001

4. THERAPEUTIC EFFECTS

The value of oats in the prevention and treatment of disease has been recognised since ancient times(5). However only comparatively recently have specific therapeutic effects been ascribed to oats(6) and much of the current nutritional interest in oats lies in their reported effects on physiological factors which are related to diseases of the vascular and digestive systems and to the relief of diabetes(7). Oat bran has generally been found to be more effective in this respect than rolled oats. Table 5 shows that dietary oat bran reduces the risk of atherosclerosis in humans by lowering the serum total cholesterol and low-density lipoprotein cholesterol levels. Oat bran can also increase faecal bulk (Table 5), improving intestinal function and thus reducing the risk of diseases such as diverticulitis or bowel cancer.

Table 5. Therapeutic effects of dietary oat bran in humans

	Initial	After 21 days on oat bran diet
Serum analyses (mg/dl)		
Total cholesterol	280	226[a]
Low-density lipoprotein cholesterol	190	149[a]
High-density lipoprotein cholesterol	31	29
Faecal mass (g/d)		
Total	137	191[a]
Dry	27	42[a]

a = Significantly different from initial (P < 0.05)
Data from (8)

Earlier work has attributed the beneficial effects of oats to various grain fractions including the hulls and the oil, protein, gum or phenolic components. Recent collaborative studies between the Cereal Crop Research Unit, Madison, Wis., USA and the Welsh Plant Breeding Station, UK have used the chick as an experimental model to investigate the effects of oat hulls, oat bran and isolated oat bran fractions on plasma and intestinal factors. The results of two experiments are shown in Tables 6 and 7. Significant reductions in plasma total cholesterol were obtained with diets containing oat hulls or oat bran. Feeding oat bran fractions at an equivalent level to that found in the whole bran showed that consistent reductions in plasma cholesterol were obtained with the protein and gum fractions and that these reductions were achieved by improving the ratio of total to high-density lipoprotein cholesterol (Tables 6 and 7). Both

TABLE 6. Effects of oat hulls, oat bran and bran fractions on plasma cholesterol and intestinal function in chicks

Diet supplement	% in diet	Plasma total cholesterol (mg/dl)	Intestinal transit time (mins)
Nil (Control)	-	297	185
Hulls	30	197[a]	118[a]
Oat bran	40	161[a]	164
Insoluble residue	23	275	166
Protein	6.6	207[a]	191
Soluble residue	4.1	250	192
Gum	3.4	163[a]	208
Oil	3.0	250[a]	156

a = Significantly different from control (P < 0.05)
n = 6; 20 days feeding

the hull and oat bran diets reduced intestinal transit times and the insoluble residue and the oil fractions appeared to be responsible for this effect in the oat bran (Table 6). A total of six experiments were carried out with the oat bran fractions and the results for plasma total

TABLE 7. Effects of oat bran fractions on plasma total cholesterol and the ratio of plasma total to high-density lipoprotein (HDL) cholesterol in chicks

Diet supplement	% in diet	Plasma total cholesterol (mg/dl)	Total: HDL cholesterol
Nil (Control)	-	341	2.9
Oat bran	40	168[a]	1.6[a]
Insoluble residue	17.6	347	3.1
Protein	4.5	269[a]	2.5
Soluble residue	3.0	431	2.6
Gum	2.6	215[a]	2.1[a]
Oil	2.3	400	3.0

a = Significantly different from control (P < 0.05)
n = 7; 14 days feeding

cholesterol are summarised in Table 8. This shows that consistent reductions were found only with the gum and protein fractions and that the gum fraction was by far the most effective in this respect.

The level of plasma glucose following a meal is important in the control of diabetes. Table 9 shows that the addition to a diet of only 2.6% oat gum produces large reductions in post-prandial glucose levels. It thus appears that the gum fraction is the prime factor in the observed effects of oat bran on both cholesterol and glucose metabolism.

TABLE 8.   Plasma cholesterol lowering activity
of oat bran and its fractions

Diet component	Range	Median
Gum	10 - 74	43
Protein	4 - 36	15
Oat bran	1.8 - 8.2	4.6
Oil	0 - 45	12
Soluble residue	0 - 36	10
Insoluble resiue	0 - 8.2	1.2

Units - reduction in plasma total cholesterol (mg/dl)
per 1% of diet component

TABLE 9. Effects of oat gum on plasma glucose (mg/dl)
in chicks after meal feeding

Fasted	23 minutes after feeding	
	Control diet	Control diet + 2.6% oat gum
210	485	389

All differences are significant (P < 0.05)

5.   VARIATIONS IN CHEMICAL COMPOSITION

Oat composition has recently been comprehensively reviewed(9) and only three constituents will be considered here: (i) β-glucan, which is the main component of the water-soluble gum fraction(10), (ii) oil and (iii) protein.

5.1 β-glucan

A range of 1.3 - 3.5% β-glucan has been reported from five oat varieties(11) and ranges of 2.8 - 5.6% and 1.0 - 5.0% have been found in 100 oat varieties (Table 10) and eight wild Avena species (Table 11).   The

TABLE 10.   Range of water-soluble β-glucan contents found in
100 diverse oat varieties - field grown

Variety	β-glucan (% in groat)	
Exeter	2.8	
Mapua 70	2.9	
S.84	3.0	
Pusa hybrid G	3.4	Means of 4 replicates
Peniarth	3.5	LSD (P < 0.05) = 0.05
Condor	4.2	
Sun II, Leanda	4.6	
Maris Oberon	4.8	Overall mean = 4.2 ± 0.60
Hinoat	5.1	
Vedette, Greta	5.2	
Dyfed	5.5	
Rigida, Karin	5.6	

Data from WPBS

TABLE 11. Water-soluble β-glucan contents of eight <u>Avena</u> species -
glasshouse grown (means of 2 replicates x 2 N levels)

	β-glucan (% groat)	Groat weight (mg)	Groat protein (%)
A.<u>ventricosa</u>	1.0	3	34
A.<u>canariensis</u>	2.0	10	23
A.<u>maroccana</u>	2.1	25	20
A.<u>longiglumis</u>	2.6	13	18
A.<u>murphyi</u>	2.8	19	20
A.<u>sterilis</u>	3.7	8	17
A.<u>barbata</u>	5.0	4	25
A.<u>hirtula</u>	5.0	5	26
LSD (P < 0.05)	0.4	2.5	0.5

Data from WPBS

data for the wild species was obtained with material grown at low nitrogen
fertility levels and this may have led to the generally lower β-glucan
levels in this experiment(11). However, comparisons of A.sativa genotypes
which were included in both experiments showed similar rankings for
β-glucan contents indicating that this factor is under genetic control.
The reported high β-glucan content of the high protein variety Hinoat(11)
was confirmed. However there was no evidence of a relationship between
β-glucan content and groat protein content or groat weight in either the
wild species or the A.sativa varieties. While high β-glucan oat varieties
may be of value in human nutrition, the β-glucan gums of barley have been
associated with decreases in animal production. If a similar effect can
be demonstrated for oats then low β-glucan varieties may be preferred for
animal feeding.

5.2 Oil
The oil content of oats is higher than that of other cereals and a
range of 3.1-11.6% has been found in the World Collection(12). Winter oat
varieties have a higher oil content than spring varieties and, although
winter varieties are lower in protein(13,14) there is no consistent
relationship between oil and protein in oats(15,16,17). Similarly there
is a positive relationship between oil content and grain yield if spring
and winter varieties are compared, but this relationship is inconsistent
within spring varieties(15).
Oat oil is of good quality being relatively high in linoleic acid and
low in linolenic acid. The range of fatty acid composition in oats is
illustrated in Table 12. Fatty acid composition tends to vary with oil

TABLE 12. Range for oil content and fatty acid composition
found in 105 diverse A.sativa genotypes

Oil (%)	3.8 - 11.5
Fatty acids (% of oil)	
Palmitic (16:0)	16 - 24
Stearic (18:0)	0.7 - 2.1
Oleic (18:1)	22 - 46
Linoleic (18:2)	35 - 50
Linolenic (18:3)	1.7 - 3.7

Data from WPBS

content with the result that oil content is negatively correlated with palmitic and linoleic acid and positively correlated with oleic acid contents(15,18,19). Similar range and interrelationships in oil content and composition have been found in eight wild Avena species(20).

5.3 Protein

The protein of oats is unique among the cereals in having a relatively good balance of essential amino acids (Table 2) which is maintained over a wide range of protein contents(21,22). This results from the predominance of the globulin fraction in oat storage protein(23). Although the protein content of whole oats is similar to that of other cereals, groat protein levels are much higher and can range from 14-24%(22). Of all the quality factors, protein content is the most susceptible to environmental influence and twofold variations may occur within the same variety. Although protein content is heritable and it is possible to breed for increased grain protein(21), negative correlations between yield and grain protein are frequently encountered. These may result from experimental limitations or to the use of inadequate selection criteria. The possibilities for increasing grain protein without yield loss are briefly discussed below.

6. BIOCHEMICAL AND PHYSIOLOGICAL CONSIDERATIONS
6.1 Energetics

The energy content and the biosynthetic glucose requirement of protein and oil are considerably greater than for starch and furthermore the efficiency of energy conversion in protein and oil synthesis is lower than that in starch synthesis (Table 13). Thus the improvement of grain

TABLE 13. The relative energy yields of starch,
protein and oil biosynthesis

	Starch	Protein	Oil
Energy content of product (cals/g)	4000	5700	9500
Glucose requirement for biosynthesis (g/g)	1.2	2.5	3.3
Energy yield (cals/g glucose)	3320	2280	2840
Energy yield (as % of starch)	100	69	86

Calculated from data in (24)

protein or oil content may be expected to impose a penalty on total grain yield. Table 14 which has been calculated from the data in Table 13 indicates that yield losses will exceed 1% for each extra 1% of oil or protein that is synthesised. These losses are based on the assumption that the total energy requirements for biosynthesis are derived from glucose. However many of the biosynthetic steps can be powered directly by co-factors generated in light-dependent reactions(25) and thus these yield penalties must be considered theoretical maxima.

6.2 Nitrogen

Maximising the uptake of available soil N will increase the growth, photosynthetic capacity and yield potential of the crop. However, grain protein production depends both on N uptake and on the partitioning of assimilated N between grain and straw at harvest (nitrogen harvest index, NHI). In oats, varietal differences in grain protein are associated with

TABLE 14. A theoretical evaluation of the effect of changes in
grain protein or oil content on oat grain yields

	Grain composition (%) Starch	Protein	Oil	Glucose equivalents[a]	Relative grain yield
Standard cultivar	45	10	5	95.5	100
Protein raised by 1%	42.9	11	5	95.5	98.9
Oil raised by 1%	42.3	10	6	95.5	98.3

a = grams glucose needed for the starch, protein and oil biosynthesis
in 100 g of the standard cultivar (calculated from data in Table 12).
It is assumed that the same amount of glucose is available for starch,
protein and oil biosynthesis in all cases.

variations in NHI that are attributable to differences in the degree of
remobilisation of N from the vegetative parts during grain
filling(26,27,28). Table 15 shows that the lower grain protein of oats,
compared to wheat, is associated with reduced partitioning of N to the
grain at harvest. Thus significant increases in grain protein without
concomitant yield loss may be obtained in oats by selecting for more
efficient N partitioning.

TABLE 15. A comparison of the nitrogen economies of oats and wheat

	Grain protein content (%)	Total straw protein content (%)	NHI	HI	Plant height (cm)
Oats					
Leanda	9.3	2.2	0.69	0.34	99
	(10.7)	(1.4)	(0.80)	(0.34)	
Trafalgar	9.1	1.9	0.74	0.38	87
	(9.9)	(1.5)	(0.80)	(0.38)	
Wheat					
Sicco	10.0	1.4	0.81	0.38	97
Broom	9.6	1.3	0.82	0.39	87
Timmo	9.8	1.5	0.80	0.38	95
LSD (P < 0.05)	0.6	0.06	0.03	0.04	5.5

Data from WPBS
Figures in parentheses show effects of raising NHI to 0.8

7. OATS AS A RAW MATERIAL

Oats contain three major constituents which are potentially valuable in
the food industry. These are (i) protein which has a high nutritional
value and good functional properties, (ii) gum which has therapeutic
properties and the potential for use in processed foods and (iii) oil of
good quality. Although techniques have been developed for oil
extraction(29), the separation of protein isolates(30,31,32,33) and the
preparation of a gum fraction(33) from oats, none of these are currently
being employed commercially.

Oats have a wide range of other industrial applications(34). However,
the potential of the crop in these areas remains largely unexploited.

8. CONCLUSIONS

Compared to other cereals, oats possess a number of unique quality features (e.g. high oil, high water-soluble gums, nutritionally well balanced protein) and the oat groat is of exceptional value in both animal and human nutrition. The development of naked oat varieties (described in a subsequent paper) is a major step in the improvement of oat quality.

Considerable genetic variation for quality characteristics exists in oats and the full exploitation of this variation by plant breeding would further enhance the value of oats. Future oat quality research should aim at providing more precise definitions of those constituents of importance in nutrition and processing and at determining the full potential for improving the quality of the crop by plant breeding.

REFERENCES

1. Feedingstuffs Evaluation Unit, Rowett Research Institute: 1st Report, 1975; 2nd Report, 1978; 4th Report, 1984.
2. Western DE and WR Graham: In, FA Coffman (ed.), Oats and Oat Improvement. Am. Soc. Agron., Madison, Wis., pp. 552-579, 1961.
3. National Research Council: Nutrient Requirements of Poultry, NRC, Washington, DC, 1977.
4. Zarkadas CG, HW Hulan and FG Proudfoot: Cereal Chemistry 59, 323-327, 1982.
5. Coffman FA: In, FA Coffman (ed.), Oats and Oat Improvement. Am. Soc. Agron., Madison, Wis., pp. 15-40, 1961.
6. De Groot AP, R Luyken and NA Pikaar: Lancet (ii), 303-304, 1963.
7. Gould MR, JW Anderson and S O'Mahony: In, GE Inglett and L. Munck (ed.), Cereals for Food and Beverages. Academic Press, New York, pp. 447-460, 1980.
8. Anderson JW, L Story, B Sieling, WJL Chen, MS Petro and J Story: Am. J. Clin. Nut. 40, 1146-1155, 1984.
9. Youngs VL, DM Peterson and CM Brown: In, Advances in Cereal Science and Technology, Vol. V, AACC, St. Paul, Minn., pp. 49-105, 1982.
10. Wood PJ, IR Siddiqui and D Paton: Cereal Chemistry 55, 1038-1049, 1978.
11. Wood PJ, D Paton and IR Siddiqui: Cereal Chemistry 54, 524-533, 1977.
12. Brown CM and JC Craddock: Crop Sci. 12, 514-515, 1972.
13. Hutchinson JB and HF Martin: J. Agric. Sci., Camb. 45, 411-418, 1955.
14. Brown CM, DE Alexander and SG Carmer: Crop Sci. 6, 190-191, 1966.
15. Forsberg RA, VL Youngs and HL Shands: Crop Sci. 14, 221-224, 1974.
16. Youngs VL and RA Forsberg: Crop Sci. 19, 798-802, 1979.
17. Gullord M: Acta Agric. Scand. 30, 216-218, 1980.
18. Welch RW: J. Sci. Food Agric. 26, 429-435, 1975.
19. de la Roche IA, VD Burrows and RIH McKenzie: Crop Sci. 17, 145-148, 1977.
20. Frey KJ and EG Hammond: J. Am. Oil Chem. Soc. 52, 358-362, 1975.
21. Frey KJ: Z. Pflanzenzuchtg 78, 185-215, 1977.
22. Robbins GS, Y Pomeranz and LW Briggle: J. Agric. Food Chem. 19,536-539, 1971.
23. Peterson DM: Crop Sci. 16, 663-666, 1976.
24. Mitra R and CR Bhatia: Theor. Appl. Genet. 54, 41-47, 1979.
25. Miflin BJ: In, PS Carlson (ed.), The Biology of Crop Productivity.Academic Press, London, pp. 255-296, 1980.
26. Peterson DM, LE Schrader, DA Cataldo, VL Youngs and D Smith: Can. J. Pl. Sci. 55, 19-28, 1975.

27. Welch RW, YY Yong and MV Hayward: J. Exp. Bot. 31, 1131-1137, 1980.
28. Wych RD and DD Stuthman: Crop Sci. 23, 878-881, 1983.
29. Kalbasi-Ashtari A and EG Hammond: J. Am. Oil Chem. Soc. 54, 305-307, 1977.
30. Wu YV, KR Sexson, JF Cavins and GE Inglett: J. Agric. Food Chem. 20, 757-761, 1972.
31. Youngs VL: J. Food Sci. 39, 1045-1046, 1974.
32. Cluskey JE, YV Wu, JS Wall and GE Inglett: Cereal Chem. 50, 475-481, 1973.
33. Hohner GA and RG Hyldon: US Patent 4,028,468, 1977.
34. Caldwell EF and Y Pomeranz: In, Industrial Uses of Cereals, AACC, St. Paul, Minn., pp. 393-411, 1973.

OIL AND PROTEIN CONTENT IN OATS (<u>AVENA</u> <u>SATIVA</u> L.)

MAGNE GULLORD
Apelsvoll Agricultural Experimental Station, 2858 Kapp, Norway.

1.INTRODUCTION

Spring oats are an important break crop in the cereal growing areas of Norway. More than one third of the total acreage of small grains or about 110,000 ha are sown with spring oats annually. More than 90% of the oat crop is used for feeding purposes.

Oats, because of their high fibre hulls are lower in energy value than other cereals. A reduction in the proportion of hull or an increase in oil content would increase energy concentration and make oats more valuable for livestock feed.

In this paper I will deal only with the possibility of increasing oil content by breeding. Studies to date have shown that both the range and heritability of oil content are sufficiently high to suggest that we can expect good progress from breeding and selection(1,2,3,4,5). The relationships between oil content and other important characters have been examined in several studies(2,4,6,7,8). No strong negative correlations have been reported between oil content and grain yield or protein content. In maize, however, it was reported(9) that there was a 70% reduction in grain yield when oil content was raised from 4% to about 18% after selection for 70 generations. It was also found that a smaller increase in oil content, from 4-5% units to 7-8% units, was achieved without a reduction in grain yield.

The main objectives of the present study have been to examine the stability of oil content over years and locations in Norwegian spring oat material and to confirm the relationships between oil content and yield and protein content under Norwegian conditions.

2. MATERIALS AND METHODS

Twenty-two oat lines, selected from crosses between adapted North-European cultivars for high grain yield, lodging resistance and earliness, and three commercial varieties were grown and compared in field plots in six locations (Apelsvoll, Staur, Elverum, Brandval, Aas and Ramnes) in South-Eastern Norway and in one location (Særheim) in South-Western Norway, over the period 1981-84.

The experimental plot layouts were five by five lattice designs replicated two or three times. Each plot consisted of ten 8 m rows, spaced 13.5 cm apart, with a seeding rate of 200 kg per hectare. The lengths of the ten row plots harvested varied between 5.4 and 7.0 m in the different locations.

After harvest and drying, 30 g of seed were sampled from each line/cultivar in each location and ground in a Falling Number hammer mill. Protein content was determined by the Kjeldahl method and oil content by the Soxhlets method. Dehulling percent was calculated on 5 g samples and used to adjust protein and oil values, assuming that the hulls contain no protein or oil.

Analyses of variances were calculated by using a model in which years and locations are random variables and lines/cultivars a fixed variable. Phenotypic correlation coefficients were calculated in the standard way.

3. RESULTS

There were significant differences ($P < 0.01$) between genotypes with respect to grain yield, protein content and oil content (Table 1). Grain yields varied from 5,330 kg per ha for Titus to 6,280 kg per ha for

TABLE 1. Grain yield, protein content and oil content of 25 oat lines/cultivars. (Means of 7 locations in Norway, 1981-1984)

Line/ cultivar	kg grain/ha (15% moisture content)	Protein (%)	Oil (%)
A 9001	5890	10.1	5.58
A 9004	5950	10.0	4.99
A 0001	6110	9.6	4.68
A 9018	5930	10.4	4.56
Titus	5330	10.6	5.09
A 9028	5540	10.8	5.18
A 0002	5610	10.7	4.69
A 9052	5540	10.8	5.14
A 9055	6100	10.0	5.71
A 0005	5750	10.2	5.66
A 9058	5810	10.9	5.48
A 0006	5840	10.7	5.42
A 9060	5820	10.5	5.55
A 0015	5980	10.2	5.09
A 0018	5940	10.4	4.99
A 0022	6070	10.4	6.53
Svea	5990	10.1	5.65
A 0033	5760	10.9	5.41
A 0036	6130	9.8	4.83
A 0040	6040	10.1	5.05
A 0010	6280	10.1	5.81
A 0059	6210	9.7	4.61
Puhti	5990	10.1	5.33
A 0072	6050	10.3	4.91
A 9065	5840	9.7	4.80
Mean	5900	10.3	5.23
SE	72	0.09	0.06

A 0010.   Protein contents showed little variation, 9.6-10.9%, while oil contents varied between 4.56% and 6.53%.   The interaction years x varieties was significant for both grain yield and oil content, while the interaction locations x varieties was significant for grain yield and protein content (Table 2).   The low locations x varieties mean square for oil content relative to the mean square for varieties indicates that oil contents of the lines/cultivars studied were largely determined by genetic effects and the environment modified oil content of oats to a lesser extent that it modified protein content and grain yield.   These results are in agreement with those of experiments carried out in Iowa(8) and indicate that high oil content can be very reliably selected for by measuring oil content in early generations in only one environment.

TABLE 2. Analyses of variance for grain yield, protein content and oil content of 25 oat lines/cultivars grown at 7 locations in Norway, 1981-1984

Source of variation	Degrees of freedom	Mean squares		
		Grain yield	Protein (%)	Oil (%)
Years	3	1,118,160.54**	2,680.20ns	32,539.78ns
Locations	6	680,464.36**	2,547.23ns	14,877.73ns
Years x locations	18	103,648.34**	1,802.93**	13,954.31**
Varieties	24	14,274.15**	419.16**	58,675.55**
Years x varieties	72	1,448.91**	24.34ns	886.86**
Locations x varieties	114	1,467.71**	30.62**	498.78ns
Error	432	1,015.00	18.87	532.00

$**P < 0.01$

The phenotypic correlation coefficients between grain yield and oil content within environments and years were, with one exception, not significantly different from zero (Table 3). Independent or positive

TABLE 3. Correlations of oil content with grain yield for 25 oat lines/cultivars grown at 7 locations in Norway, 1981-1984

	Correlation coefficients				
	1981	1982	1983	1984	Over years
Apelsvoll	0.30ns	-0.18ns	-0.12ns	0.30ns	0.16ns
Staur	0.23ns	-0.08ns	0.09ns	0.36ns	0.15ns
Elverum	-0.14ns	-0.18ns	0.27ns	0.24ns	0.38**
Brandval	-0.00ns	0.18ns	0.15ns	0.01ns	0.30**
Aas	0.24ns	0.09ns	-0.16ns	0.17ns	0.40**
Ramnes	-0.08ns	0.07ns	0.15ns	-0.09ns	-0.15ns
Særheim	0.19ns	0.33ns	-0.59**	-0.05ns	0.17ns
Over locations	0.03ns	0.15*	0.02ns	-0.07ns	0.06ns

$*0.01 < P < 0.05; **P < 0.01$

relationships between oil content and protein content were, with one exception, found within environments and years (Table 4). The results presented along with those referred to in the literature indicate the high heritability of oil content without any strong negative correlation with either grain yield or protein content. This suggests that it should be quite possible to obtain an increase in oil content by breeding without sacrificing either grain yield or protein content.

TABLE 4. Correlations of protein content with oil content for 25 oat lines/cultivars grown at 7 locations in Norway, 1981-1984

	Correlation coefficients				
	1981	1982	1983	1984	Over years
Apelsvoll	0.10ns	0.20ns	0.27ns	-0.46*	0.03ns
Staur	-0.06ns	0.14ns	0.19ns	-0.06ns	0.02ns
Elverum	0.39ns	-0.03ns	0.48*	0.01ns	0.05ns
Brandval	0.04ns	0.10ns	0.45*	0.20ns	-0.42**
Aas	-0.19ns	0.18ns	0.29ns	-0.10ns	0.23*
Ramnes	0.10ns	0.09ns	0.12ns	0.13ns	-0.16ns
Særheim	0.17ns	-0.19ns	0.13ns	-0.07ns	-0.13ns
Over locations	-0.16*	-0.04ns	0.07ns	-0.16*	-0.09*

*$0.01 < P < 0.05$; **$P < 0.01$

REFERENCES

1. Anonymous: Sortssamlingen - Varhavre. Duplikert, 1973.
2. Baker RJ and RJH McKenzie: Crop Sci. 12, 201-202, 1972.
3. Brown CM, AN Aryeetey and SN Dubey: Crop Sci. 14, p.17, 1974.
4. Brown CM and JC Craddock: Crop Sci. 12, 514-515, 1972.
5. Frey KJ and EG Hammond: J. Am. Oil Chem. Soc. 52, 358-362, 1975.
6. Brown CM, DE Alexander and SG Carmer: Crop Sci. 6, 190-191, 1966.
7. Gullord M: Acta Agric. Scand. 30, 216-218, 1980.
8. Thro AM: Personal communication, 1983.
9. Weber EJ and DE Alexander: J. Am. Oil Chem. Soc. 52, 370-375, 1975.

INFLUENCE OF FERTILIZER USE ON THE QUANTITY AND COMPOSITION OF OIL IN OATS

R. LAMPINEN
Länsi - Hahkiala, SF-14700 Hauho, Finland.

1. INTRODUCTION

Oats are the second biggest cereal crop in Finland, barley being the largest. The annual oat crop is around 450,000 ha and the yields per hectare are comparable to those of barley.

The energy content of oats is lower than that of barley and the grain with its high fibre husk is not so suitable for feeding to monogastric animals. Therefore there has been research and breeding work, which aims to increase the amount of oil in the oat kernel in order to increase the energy value. A high proportion of oil with unsaturated fatty acids in the diet leads to a softening of adipose tissue(1) which is a negative character, especially in pigs. This phenomenon can, however, be of use when feeding dairy cows in cases where milk fat tends to be hard. Feeding dairy cows with fodder with a high oil content decreases the digestibility of fibre and lowers the fat content of the milk. This can be avoided by using protected fat. Feeding dairy cows with protected fat, which contains polyunsaturated fatty acids, increases the concentration of those acids in the milk and the milk fat becomes softer(1,2,3). This information also indicates the special features in cereals which should be developed to meet the special requirements of different classes of animals(4).

2. MATERIALS AND METHODS

The effect of fertilizer application was examined in a cultivation trial which included the two oat varieties: Svea [(Voll x Blixt) x Titus] and Stil [Sörbo x Bento].

Fertilization was applied in five treatments: 50 N, 100 N, 150 N, 50 N + 100 N and 100 N + growth regulator. The single applications, and the basic application in the split treatments were applied in between every second seed drill. Seed sowing was by a special combined precision sowing drill. N-P-K (16.0 - 7.0 - 13.3) fertilizer was used.

The additional application of 100 N, in the 50 N + 100 N treatments, was given in the form of calcium nitrate (15.5% N) before heading (Feekes 8-10).

Chlormequat growth regulator was applied in 1981-82, spraying being carried out between tillering and heading (Feekes 7-10) with 1.5 l/ha chlormequat with 0.3 l/ha wetting agent and about 300 l/ha water. In 1983-84, etephon growth regulator was used. Spraying was carried out before heading (Feekes 9-10.1) with 0.24 l/ha in 300 l water/ha.

The laboratory analyses of oil content and fatty acid composition were carried out by Svalöf AB in Sweden. Total oil is given as hexane + hexane-ethanol (1:3) extract and free oil as hexane extract.

Protein analyses were carried out by Viljavuuspalvelu at Helsinki using the Kjeldahl-method.

The work described was carried out over the period 1981-1984.

3. RESULTS
3.1 Total oil content
The mean total oil content over the period 1981-84 was normal (Table 1). Both varieties demonstrated a slight tendency to decrease oil content with increasing nitrogen applied as a single dose. Applying 150 N in two doses, 50 N + 100 N, did not affect oil content in either of the varieties. At the 100 N nitrogen level the use of a growth regulating chemical slightly increased the oil content. Chlormequat increased oil content more under the conditions of 1981-82 than etephon did during the years 1983-84. Etephon slightly increased content of oil in both varieties in both years. Chlormequat increased the oil content of Stil in both years but decreased the oil content of Svea in 1982.

TABLE 1. Total oil and protein contents of the oat varieties Svea and Stil, grown at five fertilizer levels. (Hahkiala 1981-84)

Variety	Fertilizer (kg/ha N)		Total oil content (% dm)	Protein content (% dm)
Svea	50 N		6.9	12.9
	100 N		6.8	14.0
	150 N		6.6	14.4
	50 N + 100 N		6.6	15.0
	100 N + G.r.		6.8	14.4
		Mean	6.7	14.1
Stil	50 N		6.5	11.5
	100 N		6.2	13.0
	150 N		6.0	14.0
	50 N + 100 N		6.1	13.5
	100 N + G.r.		6.5	13.0
		Mean	6.3	13.0
Mean			6.5	13.5

3.2 Free oil content
Free oil was determined as hexane extract in 1983-84 (Table 2). A slight decrease in oil contents can be seen, in both varieties, with increasing single nitrogen doses. Splitting the nitrogen applications, in the way it was carried out in the trial, did not affect the oil content when compared with a single dose of the same amount. However, when compared with the basic dose of 50 N, the addition of 100 N decreased the oil content. The use of etephon did not affect oil content at the 100 N level.

3.3 Fatty acid composition of free oil
The most interesting fatty acids were determined in the free fat (Table 2). At the single fertilizer applications, a slight increase was found in the proportions of palmitic and linoleic acids with increasing fertilizer nitrogen. At the same time the combined proportions of stearic and oleic acids was decreasing. The splitting of the 150 N dose to 50 N + 100 N did not significantly affect any of the fatty acids. Neither did etephon have any affect. The small proportion of linolenic acid also remained unchanged.

TABLE 2. Free oil content and fatty acids of C 16:0 and C 18 series
of the oat varieties Svea and Stil (Hahkiala 1983-84)

Variety	Nitrogen (kg/ha)	Oil content (% dm)	Fatty acids (%)				Protein content (% dm)
			C 16:0	C 18:0+1	C 18:2	C 18:3	
Svea	50 N	5.9	15.7	42.6	39.6	1.7	12.3
	100 N	5.5	16.5	41.2	40.2	1.8	14.5
	150 N	5.3	16.8	40.8	40.2	1.8	15.7
	50 N + 100 N	5.5	16.5	40.6	40.8	1.8	15.9
	100 N + etephon	5.6	16.3	41.2	40.6	1.6	15.4
	Mean	5.6	16.4	41.3	40.3	1.7	14.8
Stil	50 N	5.1	15.1	42.5	40.5	1.6	11.6
	100 N	5.1	14.7	41.5	41.7	1.7	13.6
	150 N	4.9	15.3	40.8	41.7	1.8	14.3
	50 N + 100 N	4.8	15.3	40.9	41.5	1.8	14.3
	100 N + etephon	5.1	15.0	41.4	41.7	1.7	13.3
	Mean	5.0	15.1	41.4	41.4	1.7	13.4
Mean		5.3	15.7	41.4	40.9	1.7	14.1

3.4 Protein content

Both varieties reacted to increasing single doses of nitrogen with
increased protein content. Splitting 150 N to 50 N + 100 N did not
significantly affect the protein content during 1983-84 but caused some
variation in 1981-82. The use of a growth regulator at the 100 N level
slightly decreased the protein content in Stil and increased it in Svea.
The increase in protein content as a result of etephon or chlormequat
treatment is unusual but occurred in Svea in both 1982 (CCC) and 1983
(etephon).

3.5 Oil content in relation to protein content

When the correlation coefficient for the whole material was calculated,
no statistical dependence between the two characteristics was observed,
r = 0.075. When the correlation coefficient was calculated for the single
applications alone, a small negative correlation was observed, r = 0.141.

There seemed to be a clear difference between the varieties. In the
Svea material a faint negative correlation was observed, r = 0.345. The
correlation for comparable applications was again stronger, r = 0.520.

The comparable values for Stil were r = -0.096 and r = 0.089
respectively. Thus Stil, with its lower protein content than Svea, showed
practically no correlation between oil and protein.

4. DISCUSSION

The observed changes in oil content and fatty acid composition were
small. Svea exhibited a slightly higher protein content than Stil, and
this is in agreement with other results with these varieties. Despite its
lower protein content Stil also had lower oil content than Svea. Oil
content and the relationship between oil and protein was dependent on the
variety. This has been reported by several researchers(5,6,7,8). Most
often there is a small negative correlation between these characters and
this was what was found in this material. Increased levels of nitrogen

application slightly decreased oil content while at the same time it increased protein content.

Increased levels of nitrogen slightly decreased the proportions of oleic and stearic acids. A slight increase was found in the contents of palmitic and linoleic acids. The proportion of linolenic acid remained unchanged and this confirms earlier results(8).

5. SUMMARY

The two oat varieties Svea and Stil were examined in a cultivation experiment with five treatments. Applications of nitrogen fertilizer ranged from 50 kg/ha to 150 kg/ha and one of the treatments included spraying with a growth regulating chemical.

Increased levels of nitrogen fertilizer decreased oil content (both total and free). At the same time protein content was increased. Practically no correlation between oil and protein contents was observed in total material nor in the results with variety Stil. Svea exhibited some negative dependence between oil and protein.

Fatty acid composition was also slightly influenced by nitrogen level. The contents of palmitic and linoleic acids were increased and the proportions of stearic and oleic acids were decreased with increasing levels of nitrogen fertilizer.

REFERENCES

1. Gurr MI: Role of Fats in Food and Nutrition, London, 1984.
2. Rudel LL, JS Parks and RE Carroll: Effects of Polyunsaturated versus Saturated Dietary Fats on Nonhuman Primate HDL. In, Perkins and Visek (eds). Dietary Fats and Health. Am. Oil. Chem. Soc. 1983.
3. Ganssmann W: Die Mühle und Mischfuttertechnik 31, 411-414, 1980.
4. Lampinen R: Proc. 22. Nordic Cereal Congr., Lillehammer, pp.263-268, 1984.
5. Johansson H: SUF Tidskrift 5, 279-289, 1976.
6. Youngs VL and RA Forsberg: Crop Sci. 19, 798-802, 1979.
7. Youngs VL and H Püskülcü: Crop Sci. 16, 881-883, 1976.
8. Lampinen R: Proc. 22. Nordic Cereal Congr., Lillehammer, pp.353-362, 1984.
9. Gullord M: Acta Agr. Scand. 2, 216-218, 1980.

218

QUALITY IN OAT BREEDING - A QUESTION OF PRIORITIES

BENGT MATTSSON
Svalöf AB, S-268 00 Svalöv, Sweden

1. INTRODUCTION
In Sweden, we take quality increasingly into consideration when evaluating oats. For many years, hectolitre weight was the only character of economic interest but now we pay attention to protein, oil, husk and fibre contents.

2. GRAIN QUALITY FACTORS
2.1 Hectolitre weight
There is an official scale which regulates price. The zero is set at 55.1 to 56.1 kg and for every kg above this value a 0.25% premium is paid. For every kg below, however, 0.5% less is paid. Considering the average variation in oats from farmer's fields, about 40% of samples may fall within the zero group ± 1 kg (54.1-57.1) but there is large seasonal variation. The total variation, which may span more than 12 kg, is mainly due to variation in climatic and edaphic factors and this is much greater than the variation in hectolitre weights between cultivars grown at any one site, which is at most about 3 kg.
Breeding for high hectolitre weight is not an easy task. Since this character is dependent on the shape and size of the grain and the evenness and consistency of its surface, as well as its measurement being influenced by the frequency of germinated seeds and various contaminations.

2.2 Husk
The oat grain consists of about 25% husk. Variation between samples is largely due to environmental factors, but there are true differences between cultivars. Breeding for low husk content is possible but screening of large populations is laborious. To date there has been no price regulation of this character by the grain trade, but the milling industry producing oat groats are willing to pay slightly more for oats which are easier to hull. In this respsect Sang is an interesting cultivar.

2.3 Fibre
There is a good correlation between husk and grain fibre content. The old Weende-method is still the official method of assessment but newer, modern methods are being discussed. The feeding value of oats is dependent on fibre content and roughly a 1% increase in fibre content decreases the ME feeding value by 2%. In breeding and selection work the near infrared-technique might be a useful tool for screening.

2.4 Protein
The quality of oat protein is high with a good balance of the essential amino acids. The lysine content is higher than in any of the other

cereals cultivated in Sweden. The proportion of lysine also remains about the same at different total protein contents, which is in contrast to barley and wheat where the proportion of lysine decreases with increasing protein content.

In order to encourage the growing of high protein cultivars in Sweden a scale of premium prices is being discussed. We already have such a premium for wheat and both barley and oats will probably follow. The zero for both will be 12% protein content. For each percentage point higher or lower there will be a premium or reduction of price of 1.2% for barley and 2% for oats. The improved feeding value of oats is at least 2% for each 1% higher protein content. In consideration of the fact that most of the oats are fed to cattle, the improved value is closer to 3%.

At Svalöf, over the last ten years, we have been studying protein in oats. A collection of high protein lines has been built up, consisting of foreign lines and induced mutations from different sources. The difference in protein contents between the 'high-lines' and the Swedish standards proved to be about 5% (Table 1). Most of the 'high-lines' have

TABLE 1. Protein content of some oat lines.
Mean of seven years at Svalöf AB

Oat line	Protein content of dry matter (%)
Otee	16.5
Hinoat	15.6
C.I. 6566	15.6
C.I. 4164	15.4
C.I. 6208	15.4
C.I. 1156	14.9
C.I. 6801	14.8
C.I. 2380	14.5
Titus	13.1
Sol II	12.2
Selma	11.5

since been crossed to Swedish cultivars in backcrossing programmes. Usually selection has been made in the F_3 following the first backcross. Selection was made for both protein content and agronomic characters. It has proved to be possible to transfer genes for high protein content. However, when selection pressure was applied to raise yield it became difficult to maintain protein at a high level. Table 2 presents some data as an example. Although it may entail a considerable amount of effort, it would seem to be possible to develop cultivars with higher protein contents than those now cultivated.

Using the HPLC-method at Svalöf we have found a weak negative correlation between proportion lysine in the protein and total protein content. In this investigation protein contents varied between 9.5 and 14.5% and contents of lysine in the protein between 3.3 and 4.3%. Thus we must monitor the quality of the protein in new cultivars before we can accurately assess their contribution to oat crop improvement.

TABLE 2. Lines from crosses with high-protein oats compared to Sang.
Result of trials at three localities during 1984.

Oat line	Relative yield (%)	Straw-strength (0-100)	Hectolitre weight (kg)	Thousand grain weight (g)	Husk content (%)	Protein content of dry matter (%)
Sang	100	91	59.3	38.4	23.6	14.7
SV 83402	101	+1	+0.4	-0.3	+3.9	+0.9
Sv 83403	101	+5	+1.1	+0.1	+2.6	+1.2
Sv 83415	99	+2	+0.3	+0.8	+0.4	+0.2
Sv 83417	101	+1	+0.2	+0.3	+0.2	+0.1
Sv 83419	98	+4	+1.4	-0.2	0.0	+0.2

2.5 Oil

As with protein, the oil in oats is of good quality, and it has a favourable fatty acid composition. At Savlöf, a collection of lines with different oil contents has been built up. The differences range over about 5% units. The proportions of the different fatty acids do not differ so much among the different lines. The average levels of the fatty acids in the lines investigated are given in Table 3.

TABLE 3. Oil content and composition of some oat lines at Svalöf AB

Oat lines	Oil content of dry matter (%)	% of fatty acids				
		Palmitic	Oleic + stearic	Linoleic	Linolenic	C-20 acids
Tarahumara	8.8	16.8	42.3	38.5	1.3	1.1
Chihuauhua	8.3	14.1	44.9	38.8	1.2	1.0
C.I. 3387	8.5	16.0	45.2	37.0	0.8	0.9
C.I. 7447	8.2	13.3	42.7	42.3	1.0	0.7
C.I. 2875	8.1	11.6	44.8	41.5	1.0	1.1
C.I. 4492	7.5	15.4	44.3	37.9	1.2	1.1
Lyon	7.4	14.6	42.5	41.2	0.9	0.8
Avon	6.7	15.7	42.0	39.8	1.4	1.0
Svea	5.8	16.1	41.5	39.6	1.3	1.1
Titus	5.1	16.4	40.5	40.5	1.4	1.1
Sang	5.0	18.3	44.3	34.8	0.8	1.4
Selma	3.9	17.6	40.5	39.2	1.2	1.5

The energy of oil is about three times that of starch. When calculating the feeding values of different oats therefore, it seems reasonable to compensate by 2% for a difference of 1% in oil content. Thus breeding for high oil content should give cultivars with a higher feeding value. At Svalöf we have just started such a programme.

3. MARKET VERSUS FEEDING VALUE

If the market and feeding values of the three most widely cultivated oats in Sweden are calculated, different cultivars will be favoured depending upon their intended use. With a high yield and high hectolitre weight Selma may earn the highest income when sold on the market. If, however, a farmer wishes to use his oats for feeding to his own cattle or pigs then Sang and Svea will indirectly pay him better (Table 4).

TABLE 4. Calculated 'market' and 'feeding' values of three oat cultivars

	Selma	Sang	Svea
Relative yield	100	98.0	98.3
Hectolitre weight	55.0	-0.8	-0.8
Price reduction	-	-0.4	-0.4
Market value	100	97.6	97.9
Fibre content	10.6	-0.7	-0.1
Corrected feeding value	-	+1.4	+0.2
Protein content	11.8	+0.8	+0.8
Corrected feeding value	-	+1.6	+1.6
Oil content	4.3	+0.9	+1.3
Corrected feeding value	-	+1.8	+2.6
Feeding value	100	102.8	102.7

Surveying the breeding material, some lines of the backcross programme
have high protein contents but yield is fairly low. On the other hand,
among the lines with slightly improved protein contents, some lines with
improved yield are to be selected and progressed into varieties. The
problem is summarized in Table 5 in which it can be seen that a difference
between two lines of 10% in yield is not enough to ensure that the best
selection has been made.

TABLE 5. Calculated values of two lines in the programme
for improved quality

	Sang	Line A	Line B
Relative yield	100	105	95
Hectolitre weight	54.0	-0.5	+0.5
Price correction	-	-0.3	+0.3
Fibre content	10.0	+0.5	-0.5
Corrected feeding value	-	-1.0	+1.0
Protein content	12.0	-1.0	+2.0
Corrected feeding value	-	-2.0	+4.0
Oil content	5.5	-0.5	+0.5
Corrected feeding value	-	-1.0	+1.0
Total value	100	100.7	101.3

4. THE IDEAL OAT PLANT
How do we find the best ideotype? We know that it will need twice as
much energy to produce 1 gram of protein as it will to produce 1 gram of
starch and three times as much to produce 1 gram of oil. Thus we have to
search for genotypes which have a greater propensity for transferring
glucose to protein and oil and yet still have enough energy to produce
starch to keep yield at a high level.

EFFECT OF LEMMA COLOUR ON GRAIN QUALITY IN OATS, AVENA SATIVA L.

A. PLOURDE, R.I.H. McKENZIE and P.D. BROWN
Agriculture Canada Research Station, Winnipeg, Manitoba, R3T 2M9, Canada

1. INTRODUCTION

In Canada, oat varieties have white lemmas because of traditional preferences expressed by growers and users. However, it has been observed that red oats may be of higher quality than white oats suggesting an association between lemma colour and grain quality. Such an association would have important implications for selection of higher quality oats since this easily classifiable, visual trait could serve as a selection criteria.

This study is the first attempt to assess a possible association between lemma colour and grain quality in the cultivated oat, Avena sativa, with near-isogenic lines. It encompassed grain quality characteristics, percent hull, percent protein, percent oil, test weight, 1000 kernel weight as well as the possible association of lemma colour with heading date, plant height and yield.

2. MATERIALS AND METHODS

Pairs of near-isogenic F_6 lines of contrasting lemma colour i.e. white versus red, yellow versus red, yellow versus white and black versus non-black were selected from the segregating generations of crosses as described by Atkins and Mangelsdorf(1). For the comparison of white versus red, the white varieties Portmore, Etive, Levin, Dula and Calibre were each crossed to the white line OT233 and the F_1 crossed to the red line OT224. After growing out the F_2 the selection of near-isogenic lines was begun with 56 F_3 lines from each cross. Ten panicles were taken from each line segregating for colour and grown out as F_4 lines in the glasshouse. Ten panicles were again taken from one segregating F_4 line derived from each F_3 line and grown out as F_5 lines in the glasshouse. The final selection of lines homozygous for lemma colour was done on the F_6 generation grown in the field in 1983. One homozygous pair of near-isogenic F_6 lines per F_3 derived line was randomly chosen. The whole one-metre row of each selected line was harvested at maturity. In this manner 36 near-isogenic pairs were developed to compare the agronomic and grain quality traits of white with red lemma colours oats.

Fifteen near-isogenic pairs comparing yellow versus red and ten near-isogenic pairs comparing yellow versus white lemma colour were developed in same manner from 200 F_3 lines, from the cross (Lamar x OT233) x OT224. The ten pairs used to compare black with non-black were selected from the segregating generations of black seeded panicles found in the F_2 populations of a number of different crosses of similar origin, believed to result from outcrossing with black seeded varieties grown nearby.

Percent hull of the primary kernels was determined by manually de-hulling a one gram sample. Whole grain percent oil was determined by nuclear magnetic resonance(NMR)(2). Nitrogen analyses were performed (Kjeldahl methd: AACC 46-12) using whole primary kernels of 22 and 11 pairs of the white-red and yellow-red lemma colour comparisons, respectively. A randomised complete block data analysis and a paired t-test were used.

A large plot (6 row, 3.7 m by 1.25 m) trial was conducted at Glenlea, Manitoba in the summer of 1984 utilising the same entries compared in 1983. A split plot design with the near isogenic lines as the main plot and lemma colour as sub-plots was used. Three replicates were used except for the black-non-black comparison (two replicates). Heading date(days), plant height(cm), yield (kg/ha), percent hull of the primary kernels (using 1 g sample), test weight (kg/hl) and 1000 kernel weight(g) were recorded and analysed using a paired t-test and Anova.

3. RESULTS
3.1 Associations between lemma colour and grain quality characteristics
white versus red

In 1983, the white and red lines did not differ significantly in percent protein or oil (Table 1). However, the red lines were significantly lower (0.85%) in hull content. A similar significant difference (1.17%) was obtained in 1984. The red lines were significantly lower in percent hull in 16 out of the 36 pairs in 1983 and in 30 out of the 36 pairs in 1984. The lines of eight pairs in 1983 and one pair in 1984 showed the inverse significant relationship.

The test weight of the red lines was significantly lower (0.73 kg/hl) than that of the white lines in 1984. The red lines were significantly lower in test wight in 16 out of the 36 pairs while three pairs showed a significant inverse relationship.

TABLE 1. Grain quality measurements in 1983 and 1984 on
near-isogenic lines of oats differing in lemma colour

Lemma colour	Grain quality characteristics					
	1983 data			1984 data		
	% hull	% Protein (grain)	% Oil (grain)	Test wt (kg/hl)	1000 kernel weight (g)	% hull
White	27.60*	14.08	5.01	45.67**	30.94	26.66**
Red	26.75	14.24	5.03	44.93	31.22	25.49
Yellow	29.54	14.18	5.09	44.71*	29.85	28.29**
Red	29.27	13.96	5.15	43.97	30.21	27.18
Yellow	29.03	13.53	5.54	43.01*	30.97*	27.44*
Red	27.56	13.52	5.68	43.55	29.62	25.74
Black	27.44**	15.56	5.73**	44.82**	31.96	27.81
Non-black	25.32	15.48	5.42	47.11	33.03	26.68

*, ** Significant difference at 0.05 and 0.01 level of probability, respectively

224

The near-isogenic white and red lines were derived from different crosses. The paired t-test on percent hull for individual crosses indicate that in 1983 the difference of mean percent hull between lemma colours was not consistent from cross to cross. In 1984, however, all crosses but one, (Calibre x OT233) x OT224, showed a significant difference with the red lines having a lower percent hull (Table 2). Although the number of pairs was large enough to be able to detect a difference of 1.0% hull, a non-significant difference was obtained for the cross because one of the pairs gave a negative difference (-1.54%). When the data were reanalysed without this pair, the mean difference of 1.40% hull was significant at 1% level of probability.

The results of the paired t-test on test weight performed by cross indicate that for three of the five crosses, white lines had a significantly higher test weight. The differences for the other two crosses, although favouring the white lines, were not significant (Table 2).

No significant differences were found in 1000 kernel weight.

TABLE 2. Test weight and percent hull of white versus red lemma colour of primary kernels in near-isogenic lines derived from five classes

Cross	Number of pairs	Colour	% Hull (1983)	% Hull (1984)	Test weight (1984)
					(kg/hl)
1[+]	8	White	27.92*	26.70**	44.42*
		Red	25.90	25.30	43.64
2	9	White	29.02	27.38**	45.87*
		Red	28.00	26.24	45.14
3	7	White	27.38	26.26**	45.99*
		Red	28.22	25.31	44.76
4	6	White	26.75	27.23*	45.71
		Red	26.40	25.85	45.33
5	6	White	26.15*	25.40	46.64
		Red	24.66	24.49	46.18

[+] 1, (Portmore x OT233) x OT224; 2, (Etive x OT233) x OT224; 3, (Levin x OT233) x OT224; 4, (Dula x OT233) x OT224; 5, (Calibre x OT233) x OT224

*, **: significantly different at 0.05 and 0.01 level of probability, respectively

3.2 Yellow versus red
On an average, the yellow and red lines did not significantly differ in percent protein, percent oil and 1000 kernel weight. No significant differnces in percent hull were found in 1983 but in 1984 the red lines exhibited a significantly lower percent hull (1.11%). The red lines showed a significantly lower percent hull in 11 out of the 15 pairs. The red lines also had a lower test weight (0.74 kg/hl) and were significantly lower in six of the 15 pairs and significantly higher in one pair.

3.3 Yellow versus white

There were no significant differences in percent hull, percent protein and percent oil between the yellow and white lines in 1983. However, in 1984, the white lines were significantly higher in test weight, had smaller kernels as shown by a significantly lower 1000 kernel weight and had a significantly lower percent hull (1.7%).

In 1984, the white lines had a significantly higher test weight in five out of the ten pairs; yellow was significantly higher in one of the pairs. The white lines of nine pairs had a significantly lower percent hull; the difference between the lines of the tenth pair was non-significant.

3.4 Black versus non-black

In 1983, the black lines were significantly higher in percent hull (2.12%) than the non-black lines. The difference was not significant in 1984. The black lines had a significantly lower test weight in 1984 with the difference being 2.29 kg/hl.

No significant differences in percent protein were found but the black lines had a significantly higher percent oil (0.32%). The lines did not differ in 1000 kernel weight.

3.5 Association of lemma colour and agronomic characters (1984)

No significant differences in either heading date or yield between lemma colours were found in the four lemma colour comparisons. No associations between lemma colour and plant height were obtained except in the yellow versus white comparison where the yellow lines were 1.59 cm taller than the white lines.

4. DISCUSSION

The agronomic data and observations in the field indicate that there was little difference in appearance between the lines making up each near-isogenic pair. However, inter-pair difference in plant appearance, plant height and heading date were evident.

The growing conditions differed in the two years of the experiment. Results for 1984 are considered more reliable because the conditions were more favourable for a good expression of the genotype than in 1983 where the small plots were affected by hot and dry growing conditions and a serious disease infection caused by stem rust, crown rust and BYDV. The larger plots used in 1984 also provided more uniform and representative samples.

High test weight and low percent hull have been cited frequently as important criteria of grain quality in oats. Because oat hulls have no nutritional value, reduce the energy content of the whole oat kernel and must be removed for human food, a major breeding objective is to reduce hull content in the oat grain without negatively affecting other grain quality characteristics and yield. Red lemma colour was found to be associated with low percent hull and low test weight in the white-red and yellow-red lemma colour comparisons. In comparison to yellow, white oats were of higher quality with a significantly lower percent hull in 1984 and a significantly higher test weight. However, they had smaller kernels. Results of the black versus non-black lemma colour comparison suggest that the black lemma colour was associated with a higher percent oil but the difference was too small to be important in quality evaluation; moreover, the black lines had a significantly lower test weight and a significantly higher percent hull in 1983.

226

The usefulness of associations between lemma colour and kernel quality traits as a tool for selection depends on their inheritance. The nature of the associations found in this study is not clear. The two most likely possibilities are pleiotropism and linkage between the genes responsible for the characters. The presence of recombinants (i.e. pairs showing the inverse relationship) among the near-isogenic pairs of lines supports the hypothesis of linkage. Their frequency gives an indication of the linkage intensity. A smaller number of pairs showing a significant inverse relationship was found for the characteristic percent hull than for the characteristic test weight. This suggests that the genes conditioning low percent hull and red lemma colour are more closely linked than the genes responsible for low test weight and red lemma colour. This gives the breeder the possibility of selecting red lines of low percent hull and high test weight.

However, the presence in 1984 of a single pair showing a significant inverse relationship for the characteristic percent hull among the 36 pairs of white versus red lemma colour lines leads to the possibility that an unfortunate admixture occurred sometime during the process of selecting the near-isogenic lines. If this is true, i.e. the lines are not true recombinants, then pleiotropism would become a plausible explanation. Lyrene and Shands(3) suggested that the gene responsible for shattering which appeared to be closely linked with dark lemma colours in A.sterilis may, in fact, have a pleiotropic effect on the characters associated with it.

The results of this study indicate that in the process of selecting good quality oat lines, one should select among red progenies in preference to other colour classes or among white progenies in preference to yellow. No associations were found between lemma colours and the agronomic characters studied permitting selection for yield, short straw and maturity within any hull colour class.

5. ACKNOWLEDGEMENT
The authors wish to gratefully acknowledge the Quaker Oats Company of Canada for their financial support throughout this study.

REFERENCES

1. Atkins IM and PC Mangelsdorf: J. Amer. Soc. Agron. 34, 667-668, 1942.
2. Robertson JA and WH Morrison: J. Amer. Oil Chem. 56, 961-964, 1980.
3. Lyrene PM and HL Shands: Crop Sci. 15, 361-363, 1975.

PROSPECTS FOR THE NAKED OAT CROP IN THE UK AND ITS AGRONOMY,
WITH PARTICULAR REFERENCE TO THE SPRING OAT CULTIVAR RHIANNON

J. E. JONES, E. CHORLTON and D. A. LEWIS
Welsh Plant Breeding Station, Aberystwyth, Dyfed, Wales, UK

1. INTRODUCTION
   Naked oats are not new to Europe and several attempts have been made to
introduce them commercially with such varieties as Brighton (UK), Caesar
(Germany), and Nuprime (France), more recently with Conradin (Germany),
Rhea (France) and Terra (Canada) and presently with Rhiannon in the UK.
It is only now, however, with over production of some crops into
surpluses that quality crops like naked oats may find a niche. Naked oats
thresh free from the husk, leaving the groats which have an extremely high
nutritive value. Typical chemical compositions of some spring cereals,
are given in Table 1.

TABLE 1. Chemical composition of spring cereals (12% moisture)

	Naked oats	Wheat	Barley	Oats
Crude protein	13.5	12.0	10.0	11.0
Oil	7.5	2.0	2.0	5.0
NFE	63.0	69.0	68.0	58.5
Fibre	2.0	3.0	5.5	10.5
Ash	2.0	2.0	2.5	3.0
ME (MJ/kg)	14.0	12.5	12.0	11.0

   Removal of the husk from oats has a dramatic effect: naked oats have a)
the highest protein content; in addition the amino acids are well balanced
as has been reviewed by Frey(1); b) the highest oil content, giving high
metabolisable energy.

2. THE PRESENT SITUATION
2.1 Evaluation
   Rhiannon is a spring naked oat which was bred at the Welsh Plant
Breeding Station and included on the UK National List of spring oat
varieties in 1983. An average grain yield of 4.3 tonnes/hectare was
achieved from 30 trials in 1982 and 1983, this being 79% of the mean yield
of the National Institute of Agricultural Botany (NIAB) husked controls.
As can be seen from Table 2 the yield of Rhiannon is a significant
improvement on previously available naked oat varieties, and compares
quite favourably with the groat yields of the husked controls (allowing
22-25% for husk). Rhiannon also has good straw strength and mildew
resistance.
   With grain yields averaging between 4 and 5 tonnes/hectare in farm
trials in 1984, and at an estimated market value of approximately
£140/tonne, naked oats could compete economically with many other crops.

TABLE 2.   Relative grain yields of some recently released naked oat varieties in a trial at the Welsh Plant Breeding Station during 1984

Variety	Yield as % of mean of husked NIAB controls	
Rhiannon	81	Naked oat from UK
Conradin	67	Naked oat from Germany
Terra	53	Naked oat from Canada
Rhea	68	Naked oat from France
Leanda	100	Husked oat NIAB control
Saladin	110	Husked oat NIAB control
Dula	96	Husked oat NIAB control
Trafalgar	94	Husked oat NIAB control
SED	3.6	
Mean of controls	5.5 tonnes per hectare	
C.V.	6.0%	

(Leanda, Saladin, Dula, Trafalgar) ⊢ 100

The particularly high oil content (Rhiannon 8.5%) relative to other cereals, could be a key marketing factor. It is hoped that these encouraging statistics will persuade the relevant authorities in the UK that naked oats should be recommended as a crop for general use.

2.2 Husbandry

The general climatic and agronomic requirements for growing naked oats are the same as for covered oats, and no special equipment is necessary for sowing and harvesting the crop nor for handling or storing the grain. However, the grower of naked spring oats should give particular attention to the following points:

1) As with husked spring oats, early sowing is important to ensure good yields and to avoid the danger of frit fly attack. If early sowing is not possible, protection from frit fly attack can be achieved by timely insecticide application.

2) Seed rate can be reduced to 140 kg/hectare. This will still give sufficient seeds per unit area whilst allowing for the lower germination expected of naked oats in comparison with covered oats.

3) Husbandry requirements are as for covered oats. Nitrogen should not exceed 60-90 units per hectare. Although Rhiannon is one of the stiffer strawed spring oat varieties, too much nitrogen may cause lodging.

4) When combining or threshing, the drum speed should be reduced to no more than 1,000 rpm, to avoid seed damage(2,3), although this will depend on the type of combine being used.

5) As the grain remains soft even when dry, rough handling in store can result in damage, particularly to the embryo.

6) When dry (12% moisture), the evidence is that undamaged grain stores quite satisfactorily(4,5).

2.3 Physiology

The essential difference between naked and husked oats is that the husk (lemma and palea) is thin, light and free-threshing instead of being thick, heavily lignified and fused to the grain. This dramatically changes the characteristics of the maturing crop in the field. Some points of interest are:

1) The husk (lemma and palea) makes up 22-25% of the total weight of the groat plus husk in husked oats, whereas in naked oats this is reduced to 12-15% of the total weight. Therefore, if seed number per panicle and straw strength are equal, the naked oat panicle is lighter, and thus less likely to suffer lodging. Consequently, straw strength is not likely to be a limiting factor to higher yields in naked oats. Rhiannon, the first naked oat produced from our spring programme, has standing power equal to that of the best commercially available husked varieties in the UK.

2) The loose thin husk may act as less of a physical constraint to endosperm expansion, resulting in larger, better filled grains. This could be particularly important in crops affected by water stress where, in husked oats, a thin poorly filled sample may result.

3) Following the period of grain filling, ripening is accelerated. The loosely fitting, thin husk allows quicker drying of the seed. Rhiannon, which is classified as a medium maturing spring oat(6), has the same date of maturity as Trafalgar and Leanda although its date of ear emergence is later than both these varieties.

4) In Rhiannon seed retention is good and grain shattering rare. This compares favourably with husked varieties and is perhaps surprising as the seed is loosely held. It may be due to the absence of hard contact when panicles are blown together in the wind, the soft husk absorbing the shock. The morphology of the husk also makes the spikelet more elusive and less likely to entangle in the branches of neighbouring panicles. With husked oats entanglement is often followed by the ripping off of whole spikelets, and is a main cause of grain loss.

5) The loose fitting husk allows easier access for pests and diseases to reach the kernels. This is a potential problem which will need to be monitored if the crop becomes widely grown so that control measures can be employed if necessary.

3. FUTURE PROSPECTS
3.1 Feeding

Naked oats are similar and nutritively the same as the oat groats which are rolled to produce oat flakes, and have been beneficially consumed by humans for centuries. The full value of oats to the human diet is still being discovered, for example, the soluble bran fraction has recently been found to be beneficial in treating some modern diseases(7) e.g. lowering blood cholesterol levels and lowering blood pressure.

From chemical analysis and, in particular, because of the balance of the protein fraction and oil content, naked oats should be valuable for a wide range of farm stock. Non-ruminant animals and poultry which are dependent on a well-balanced protein diet and young ruminant animals, before the rumen is fully functional, are particularly suitable. However, when naked oats are used as a main ingredient of ruminant feed, care should be taken to balance the ration with insoluble fibre. Results of early evaluations include a) favourable comments from farmers who have fed naked oats to their livestock; b) good results from pig-feeding trials at the Rowett Research Institute where naked oats replaced maize and wheat in rations(8); c) establishment of the high ME value of 14.5 MJ/kg for poultry, following laboratory analysis at the Poultry Research Centre, Edinburgh(9).

3.2 Breeding

Figure 1 represents the grain yields of various spring cereals derived from NIAB data, and our estimates of the proportion of the free threshing lemma and palea present in wheat and naked oats. The potential for increasing groat yield in naked oats by plant breeding is enormous. In the short term it is likely that if the weight of each panicle were increased by manipulating groat numbers and groat size to equal the weight of a husked panicle, average crop yields of nakeds could increase by a half a tonne per hectare, from the present 4.1 tonnes/hectare to 4.6. This may be achieved in a relatively short time, in plant breeding terms, by exploiting the variation for grain number and grain size readily available, without having to modify straw strength. Further increases in yield however, would require adjustments to the morphology and physiology of the oat crop, in particular the harvest index. Such improvements are being achieved, for example, in the new covered oat line 07048 Cn III/1 (pedigree Elen x Milo) which is high yielding, short, stiff-strawed and in National List trials this year. These improved types are being hybridised with naked genotypes to raise the grain yield. As yield improvements are made, it is envisaged that grain size will also increase and the proportion of husk will be decreased towards the 7.5% of wheat; this being the long term aim. Figure 1 gives only average yields, much higher yields are achieved by the better crops of all species. The potential for improvement in yield of naked oats is suggested by the harvest index of Rhiannon which is only 0.40.

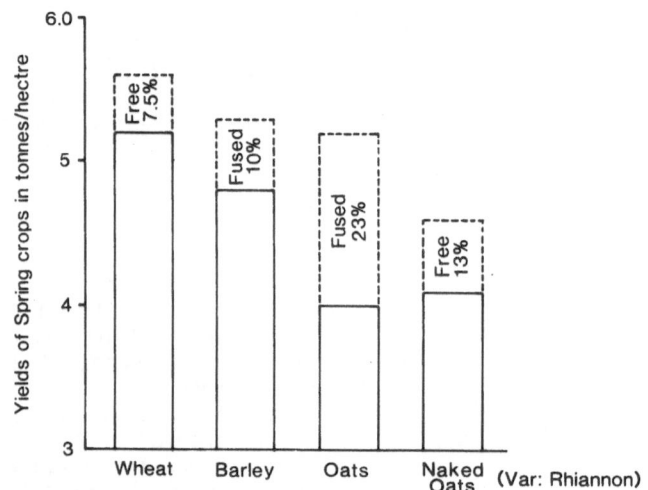

FIGURE 1. Crop yields derived from NIAB Recommended List trial control means (1979-1984). ——— Groat yields; - - - Lemma and palea (husk) added.

3.3 Exploitation

Naked oats are versatile and can be used in several situations:-

i) For on farm feeding, with pigs, poultry and young stock being most suitable.

ii) Whole crops may be ensiled, or dried, with the groats being easier for stock to digest than covered oats. Cutting for silage can be anytime after the cheesy ripe stage, therefore cutting is easy to synchronize with other crops intended for ensilage for which timing may be more critical.

iii) Naked oats could be sold for compounding, with the high oil content (Rhiannon averaging 8.5%) being the most valuable constituent (1984 was a high oil season, and Rhiannon averaged an oil content in excess of 9% on 8 farms in Wales). Potentially the oil might be extracted for human consumption, as the quality is high (approx. 40% linoleic and 40% oleic).

iv) The oat groat is one of the most complete foods known. Naked oats could be used to produce oat-flakes for human consumption without the expensive de-husking process being necessary.

Some seed of the spring variety Rhiannon is now available for farm trials and the winter varieties Branwen and Kynon, also bred at WPBS, are undergoing National Trials.

REFERENCES

1. Frey KJ: Z. Pflanzenzuchtg 78, 185-215, 1977.
2. Thornton MS and DA Lawes: Progress Reports on research and development 1979/1980, Home-Grown Cereals Authority, pp. 58-60, 1980.
3. Kittlitz von E. and M Uhling: Z. Pflanzenzuchtg 72, 305-316, 1974.
4. Welch RW: J. Sci. Food Agric. 28, 269-274, 1977.
5. Sinha RN, HAH Wallace, JT Mills and RIH McKenzie: Can. J. Pl. Sci. 59, 949-957, 1979.
6. National Institute of Agricultural Botany: Classified list of cereal varieties 1984/85, 1985/86.
7. Gould MR, JW Anderson and S O'Mahony: In, GE Inglett and L Munck (eds), Cereals for Food and Beverages. Academic Press, New York, pp.447-460, 1980.
8. Evans JB: A comparison of cereals in diets for early weaned pigs with special reference to naked oats. MSc Thesis, University of Aberdeen, 1984.
9. Fisher C: ARC Poultry Research Centre, Edinburgh, (personal communication), 1983.

HULL-LESS OATS FOR THE MILLING INDUSTRY

S. H. WEAVER
The Quaker Oats Company, Chicago, IL 60654, USA

Hull-less oats offer many interesting opportunities for commercial oats millers. Some obvious advantages include improved milling yield, reduced energy and storage space requirements, elimination of certain pieces of milling equipment and reduced transportation costs.

Milling yield improvement is probably the most significant cost saving afforded by hull-less oats. Milling yield is measured in units of raw material required to manufacture certain units of finished product. Theoretically, a yield improvement of 30% is attainable. Economically, this would result in the purchase of approximately 30% fewer oats in order to yield a given amount of product.

With all of the cost savings and advantages of hull-less oats, why are they not used to any large extent by millers in the free-world? Basically, because there are no commercial areas seeded to hull-less oats. Producers probably would grow hull-less oats if they could select high yielding and adapted varieties when making seed purchases. The lack of hull-less varieties to choose from could change in the very near future because of advanced breeding programmes located at various breeding locations in Canada, USA, Mexico and Chile. If farmers find new hull-less varieties advantageous to their farm operation and livestock feeding programmes, oats millers should then be able to enjoy the many benefits exhibited by hull-less oats at reasonable prices.

Comparisons of the utilization of hull-less and hulled oats indicate very minor and insignificant differences in storability, free fatty acid content[1], protein content, total oil content, product flavour and product acceptability. Additionally, microbial counts and pesticide contamination have been shown to be at comparably low levels[2].

Consequently, the challenge for the plant breeder is to develop high yielding, agronomically acceptable hull-less varieties. Subsequently, farmers must find them advantageous to their operations and be willing to sell their surplus production at reasonable prices. If millers were able to purchase hull-less oats at a price comparable to conventional oat groats, then the many advantages of hull-less oats could be realised by the milling industry.

REFERENCES

1. Welch RW: J. Sci. Fd Agric. 28, 269-274, 1977.
2. Lenz, Marvin: Interoffice memo, 1983.

OATS - THE HEALTH STORY

G. A. SMITH
Quaker Oats Ltd, Southall, Middlesex, UK.

In the years that followed the founding of Quaker Oats Ltd in 1899, porridge sales expanded regularly until the cereal was as regular a part of the English breakfast as it was in Scotland. This trend continued until the start of the massive growth of the ready-to-eat cereals in the 1940s and 50s. As speed and convenience took over from the more traditional virtues of natural nutrition and unprocessed sustenance, the porridge habit started to decline. That movement has become a steady slump over the last two decades - until two years ago.

When the opportunity arose for Quaker to buy Scott's Porage Oats, its main branded rival, Quaker clinched the deal to secure one main target - to rebuild the oats and hot cereals market with the money freed from costly and counter-productive branded battles for supermarket shelf space. With that aim in mind, Quaker committed itself to a long-term campaign at the end of 1982 to get porridge back to the British breakfast table, using the good news from medical scientists and researchers about the valuable contribution of oats to a healthy diet. A public relations programme was developed around two main activities - the sponsorship by the Quaker and Scott's brands of the Oats Information Bureau, which works to increase the popularity of oats generically, and events such as National Porridge Week in 1983 that included publicity more geared to the specific brands.

In the two years since its launch, the Oats Information Bureau has established itself as a highly credible authoritative source of information for the consumer media, with features, news stories, recipes and other information going to television, radio and newspapers on a national and regional level, as well as to specialists in the medical world. The main message promoted by the Oats Information Bureau has been the soluble fibre story, based on years of research work in America and Britain that shows oats to be one of the best, most readily-available and versatile sources of soluble fibre. This range of plant fibres differs from water-insoluble fibre, of which the best example is wheat bran, in that soluble fibre is partially digested and has been shown to have metabolic effects, as well as providing some roughage. Some respected researchers believe these benefits include a reduction in cholesterol levels in the blood, and a levelling-out of blood sugar levels, which is obviously of importance to diabetics.

Over the last two years, the Oats Information Bureau has kept the country's 30,000 family Doctors informed, by direct post, of the developments in the oats story and hundreds of their requests for further information have been dealt with. Hospital and health authority staff including health visitors and midwives are now far more aware of the benefits of oats, and nearly 100,000 leaflets have been distributed via doctors, hospitals and magazines directly from the Oats Information Bureau.

Many consumer magazines and newspapers have printed stories from the Oats Information Bureau on various aspects of the oats story. Most recently a page feature in the Sunday Times Magazine has dealt with the beneficial effects of the soluble fibre in oats, especially for slimmers - the fibre helping slow digestion and reducing hunger pangs. Recent TV coverage includes a piece on soluble fibre and oats on the popular BBC Food and Drink programme, which has a viewing audience of 4 million.

The nett effect to date of the Oats Information Bureau plus major publicity events like National Porridge Week, has been to turn the tide for oats, from steady decline in the hot cereals market to an increase over the last two years. Last year, the market increased by about 8% compared to the previous 12 months, and this year looks like showing even stronger growth.

SESSION V. QUALITY, USE AND MARKETING REQUIREMENTS FOR OATS
CHAIRMAN'S COMMENTS AND SUMMING UP

J. E. LEA
Morning Foods Limited, Crewe, UK

The facts and figures presented by Shrickel left no doubt that the present world production of oats of fifty million metric tonnes, of which the Soviet Union is by far the highest producer, is very much on the decline. The bulk of oats grown are for animal feed and only 17% of world production is used as human food.

It is interesting to note that whereas in most of the developed countries of the world, the area sown to oats is declining very rapidly, in the less developed countries it is declining more slowly. The more agriculturally developed parts of the world do not use oats for livestock feed because of the emphasis on competitive crops which have a higher energy value, and they also have highly developed animal feed industries.

Considerable improvements are being made in the yields of oats per hectare and the Plant Breeder has contributed to this increased yield. By far the best yields are obtained in the United Kingdom, where averages of 5.5 tonnes per hectare are achieved.

In the USA the usage of Oats for Human Consumption has increased enormously with the introduction of new products such as Granola Crunchy Bars, Muesli and different types of instant oatmeal. However, it should be pointed out that the figures are small in relation to the size of the crop and there would have to be a massive increase in consumption of oats for human food to make any noticeable difference to oat usage overall.

The prospects for the oat crop and how it could be developed, were dealt with by Welch. Naked oats and their future prospects were discussed. Many of the problems previously encountered were being ironed out and shedding, which had been a problem with the earlier naked varieties, has been largely overcome. The nutritional value of oats compared to other cereals was described. The oat possesses a high oil content, good quality protein and very high levels of soluble and insoluble fibre. Variations in the determinants of oat quality are caused by environmental effects and by genetic differences. There is the possibility that some of the valuable wild oat genes can be introduced into the Plant Breeders improvement programmes.

Gullord gave a detailed analysis of the oat crop in Norway and explained that a breeding programme had fairly recently been initiated with the objective of increasing both oil and protein content in oats.

Nothing conclusive had yet come out of studies of the interaction between high oil and protein content. It was generally considered that a high protein content meant low oil and vice versa but this has not been proved. However, it was apparent that the interaction between varieties and locations was small. Lampinen discussed the work in Finland investigating the effects of N fertilizer on oil and protein contents. N decreased oil content whilst increasing protein.

Mattson explained that the Swedish law imposed certain controls on the cereal production industry. This year the Swedish Government were introducing a payment by results system; this payment was called 'A Quality Premium' and was set according to bushel weight. It was understood that a difference of 2% hectolitre weight increased the value of the yield by 1%. The other parameters for grain quality were 1000 grain weight, and the amounts of husk, fibre, protein and oil. All these factors will have varying effects on the price that the farmer obtained.

Much of the work in Sweden is to develop varieties of oats that have higher protein and oil contents. They have not found any high yielding oat with a higher protein content. As other workers have shown, variation can be found among cultivars for both oil and protein contents but much of the variation is due to environment and growing conditions.

The leading question left however is: 'Is it easier to improve lines by 1% for both protein and oil or to improve yield by 5%?' Some disagreement between researchers was apparent as to which was the most important aspect of the oat crop requiring attention. Was higher oil content desirable as against other aspects of grain yield and quality. It would of course depend on which market one is aiming for; whether it was animal or human food and if for on farm feeding, the type of animals.

The effect of lemma colour on grain quality in oats was discussed by McKenzie. The inter-relationships between black, red, yellow, grey and white oats had been studied with the aim of producing an oat with a lower percentage hull weight than was available from any individual colour oat on its own. Lemma colour was found to have no significant effects on 1000 kernel weight, post harvest dormancy, protein content or oil content. Furthermore, no association was found between lemma colour and the agronomic characters that had been studied.

The new naked oat variety Rhiannon was described by Jones. This variety has been placed on the United Kingdom National List and the reported yield is four to five tonnes per hectare. The differences between a maturing naked oat crop compared with a covered crop were discussed as was standing power, grain filling, ripening and shedding, and disease resistance; all were modified by the free threshing lemma and palea. It was pointed out that the high oil (8.5%) and protein contents of the naked oat could be well received in pig and poultry feeds and suggested that naked oats could be a replacement for maize.

Weaver was of the opinion that there were plenty of opportunities for 'hull-less oats' with commercial oat millers, and producers would probably grow hull-less oats if they were high yielding and if the price was attractive enough. Mill yield improvement within the oat industry is probably the most significant area for cost saving and this should be afforded by the hull-less oat. However, it should be remembered that milling yield is measured in units of raw material required to manufacture a given unit finished product. Therefore, if a yield improvement of 30% is obtainable then economically this would result in the purchase of approximately 30% fewer oats to yield the same amount of finished product. There was also an economic benefit achieved from the hull which must be borne in mind when making a direct comparison.

Turner, of Quaker Oats Ltd, discussed the health giving qualities of the oat cereal. His Company had done a lot of public relations work to make sure the general public understood the medical benefits to be derived from the oat cereal. He also discussed the reduction in the consumption of oat products since the War, which had largely been brought about by the increased consumption of 'ready to eat' cereals. Originally porridge was almost the only cereal available but immediately after the end of the war its place was taken by Corn Flakes and other cereals of this nature. However, two years ago this trend was reversed and it was suggested that the Quaker Oats purchase of the Scottish Porridge Oats business from R.H.M. had enabled the brand competition to disappear.

The higher priced branded products on the Supermarket shelves had enabled more money to be spent on advertising and promotion and this had made a contribution to halting the decline in oat product consumption. This was indeed fortunate at a time when the world's attention was centred on the value of fibre in the Human Diet particularly with reference to diverticular disease and cholesterol levels. Indeed some researchers have shown quite conclusively that oats are one of the best and most readily available and versatile sources of solube fibre. The range of plant fibres includes both water insoluble fibre which is only partially digested and provides roughage, and the soluble fibre which has an excellent metabolic effect and could lead to the reduction in cholesterol levels in the blood and a levelling out of blood sugar levels, which in itself is of great importance to diabetics.

The Chairman pointed out that his own Company, Mornflake Oats Ltd, had also sponsored a great deal of work with Dr J.W. Anderson, Professor of Medicine and Clinical Nutrition at the University of Kentucky, Lexington, USA, and it had been instrumental in bringing Dr Anderson to the United Kingdom twice in the last three years.

All speakers spoke very highly of oats as a nutritive source for the human diet and applauded the supporting research work being undertaken to develop both the oat crop and oat food products. Indeed it was suggested from the audience that, as far as the UK was concerned, a great debt of gratitude was due to the oat workers at the Welsh Plant Breeding Station for the development of the variety Peniarth which was just one variety that had proved its worth over the last few years.

It should also be noted that all members of the British Oatmeal Association have worked tremendously hard to promote the sale of oat products within the United Kingdom for the benefit of the industry and oat growers alike.

It is very encouraging to see so much work and effort being undertaken in the oat cereal industry and it is to be sincerely hoped that this will continue and be encouraged in the years to come. This Conference was a great step in this direction and everyone involved in its organisation and operation should be congratulated.

SESSION VI

World status of oats and biological constraints
to increased production

Chairman: B. Mattsson

WORLD STATUS OF OATS AND BIOLOGICAL CONSTRAINTS TO INCREASED PRODUCTION

R. A. FORSBERG
Department of Agronomy, University of Wisconsin-Madison,
Madison, WI 53706, USA

1. INTRODUCTION

The title of this paper tends to convey a negative feeling about oat production, and this scepticism, in a historical sense, is probably justified. After all, oat hectareage world wide has decreased by half during the past 30 years (Fig. 1), from 53,671,000 hectares in 1954 to 26,588,000 hectares in 1983(1). Furthermore, it is readily apparent that biological and environmental constraints to production abound in many oat-growing regions of the world based on the fact that 'high-yielding' countries have average grain yields as high as 4,000 to 4,600 kg/ha. As conscientious oat workers, I suggest that we approach this topic in a positive sense, and I will proceed as though the title of this paper was 'Genetic and Production Improvements in Oats, and Biological Opportunities for Increased Production'.

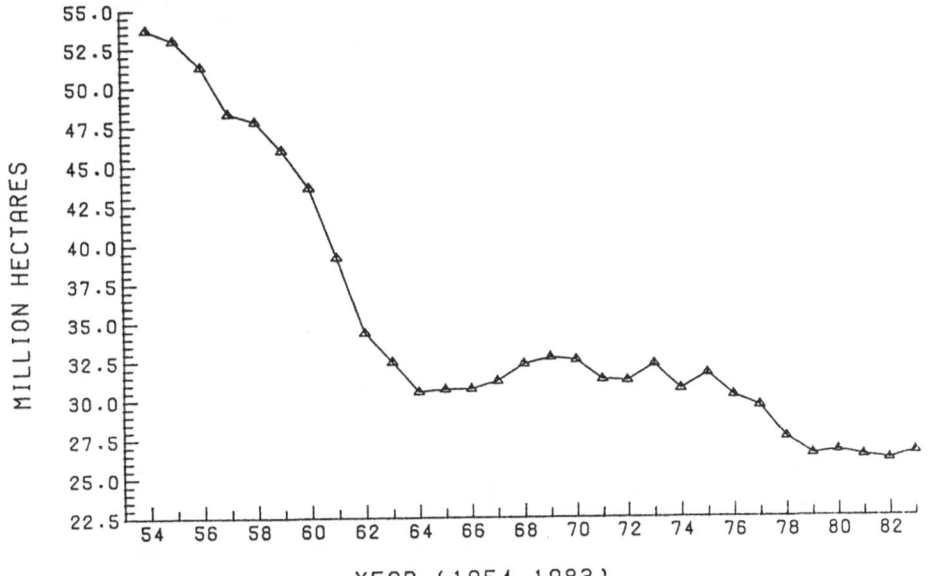

FIGURE 1. Hectares sown to oats world wide during the period 1954 to 1983.

2. FACTORS INFLUENCING OAT PRODUCTION
2.1 Market demands and prices

Many factors have contributed to the decline in oat hectareage. In many countries, grain yield fluctuates considerably from year to year due to

242

uncontrollable climatic fluctuations. High on-farm use, low export volume, and the fact that only a small portion of the oat crop (about 18%) is prepared industrially for human food all combine to cause oat prices in the market place to be influenced primarily by the prices of other cereals, especially corn and wheat. Continental or geographical needs generally have been met by production in the same or in a relatively near geographical region which minimizes the oat export market. The decrease in the area seeded to oats has been partly compensated for by an increase in grain production per unit area. Thus the total world oat production in 1983 of 43,101,000 metric tonnes is 69% of the total oat production of 62,900,000 metric tonnes in 1954 on only half as many hectares.

2.2 Geographical considerations

Oat areas harvested in 1983 and grain yields in 1953 and 1983 for the leading countries on an area basis are given in Table 1. The USSR harvested by far the greatest area (12,516,000 ha) followed by the USA, Australia, Canada and Poland. Average grain yields in these 11 countries in 1983 varied widely, from 1,009 kg/ha in Spain to 3,443 kg/ha in the Federal Republic of Germany. The increases in grain yield per unit area for the 30-year period 1953 to 1983 ranged from 25.3 to 106.0% for individual countries, with a 10-country mean increase of 59.8%.

TABLE 1. Oat areas harvested in 1983 and grain yields in 1953 and 1983 for the 11 countries top-ranked on an area basis

| | | Grain production | | |
| Country | Area 1983 (1000 ha) | 1953 | 1983 | Increase (%) |
		(kg/ha)		
USSR	12,516	940[a]	1,278	36.0
USA	3,682	1,110	1,882	69.5
Australia	1,995	620	1,183	90.8
Canada	1,400	1,580	1,980	25.3
Poland	1,042	1,420	2,281	60.6
Germany FR	601	2,420	3,443	42.3
China	450	-	1,778	-
Spain	466	720	1,009	40.1
Finland	449	1,890	3,130	65.6
France	443	1,610	3,316	106.0
Sweden	404	1,940	3,139	61.8
Mean	2,131.6	1,425.0	2,219.9	59.8

[a]Yield value for USSR for 1959 crop

This same information for countries ranked on a grain yield/unit area basis is presented in Table 2. Four countries appear in both lists, the Federal Republic of Germany, France, Sweden and Finland, a worthy achievement. The 1983 average grain yield of the countries in Table 2 (high unit yields) exceeded the 1983 average yield of the countries in Table 1 (high oat areas) by 73.2% (3,844.7 versus 2,219.9 kg/ha). Improvements in grain yields per unit area were relatively similar, 59.8% for 'area' countries versus 65.7% for 'yield' countries.

TABLE 2. Oat areas harvested in 1983 and grain yields in 1953 and 1983 for the 11 countries top-ranked on a <u>yield</u> basis

Country	Area 1983 (1000 ha)	Grain production		
		1953 (kg/ha)	1983 (kg/ha)	Increase (%)
Ireland	22	2,490	4,680	88.0
Switzerland	11	2,890	4,677	61.8
Netherlands	14	3,100	4,521	45.8
UK	108	2,490	4,306	72.9
Belgium - Luxembourg	34	2,870	4,000	39.4
Norway	116	2,480	3,553	43.3
Austria	83	1,800	3,527	95.7
Germany FR	601	2,420	3,443	42.3
France	443	2,610	3,316	106.0
Sweden	404	1,940	3,139	61.8
Finland	449	1,890	3,130	65.6
Mean	207.7	2,361.8	3,844.7	65.7

2.3 Environmental factors

2.3.1 General considerations. We all would like to know the precise reasons for high grain yields in some countries (Table 2) as compared to lower yields in other countries. Many causes exist and many cause-effect relationships deserve further study and elucidation. Daily maximum and minimum air temperatures, night temperatures, soil temperatures, photoperiod, precipitation frequency and amounts, soil nutrient levels, and altitude, all play key roles in (a) plant morphological development, including tiller and inflorescence formation and development, and (b) physiological processes such as photosynthate assimilation and dry matter accumulation during the grain-filling stage. Two additional geographical/environmental factors, latitude and length of growing season, merit special attention.

2.3.2 Latitude. The latitudes of the principal oat growing regions in the countries listed in Table 1 (area basis) and Table 2 (yield basis) are tabulated in Table 3(2). Considering the fact that four countries are common to both lists, the southern production border chosen for the USSR, USA, Canada, Poland, China and Spain in the high-area group caused a 4° lower southerly average than did those for the seven non-common countries in the high-yield group (47 versus 51°). The mean northerly latitudes for the two sets of countries were quite similar, 54 versus 55°. The influence of maritime temperatures, in combination with latitude, undoubtedly has an impact on oat production in many of the European countries.

2.3.3. Length of growing season. Lengths of growing seasons for spring oats in eight European countries are compared with those at Madison, Wisconsin (USA) and at Winnipeg, Manitoba (Canada) in Table 4(3). Differences among genotypes for maturity, and variation among planting, heading, and harvest dates from region to region within countries cause the growth periods listed in Table 4 to be somewhat arbitrary.

TABLE 3. Latitudes of principle oat growing regions in countries
top-ranked for oat grain yield (left) or for oat area (right)

Country (yield order)	Latitude, primary oat regions (°North)	Country (area order)	Latitude, primary oat regions (°North)
Ireland	52-54	USSR	50-60
Switzerland	46-48	USA	40-49
Netherlands	51-53	Australia	23-38 (South)
UK	51-58	Canada	49-53
Belgium - Luxembourg	50-51	Poland	49-54
Norway	58-62	Germany FR	48-54
Austria	47-49	China	40-50
Germany FR	48-54	Spain	36-43
France	43-50	Finland	60-65
Sweden	55-60	France	43-50
Finland	60-65	Sweden	53-60
Mean boundary (South and North)	51-55		47-54[a]

[a]Australia (south latitude) not included in average

Nevertheless, the contrast in lengths of growth periods is dramatic.
Number of days from sowing to heading ranges from 80 to 95 days in the
European countries compared to 60 days for the North American sites. In
like manner, the number of days from heading to harvest is considerably
higher in the European countries, from 50 to 75 days compared to 31 to 34
days.

TABLE 4. Number of days from sowing to heading, heading to harvest,
and sowing to harvest in eight European countries
and at two North American sites

Country	Sowing to heading	Heading to harvest	Length of oat growing season
Ireland	90	52	142
Switzerland	95	50	145
Netherlands	95	60	155
Wales	80	70	150
UK	90	75	165
Belgium - Luxembourg	95	70	165
Germany FR	80	60	140
France	90	60	150
Wisconsin (USA)	59	34	93
Manitoba (Canada)	60	31	91

The growing season is therefore from 50 to 75 days longer in the high
yielding countries, i.e. 149 to 165 days versus only 91 to 93 days. There
is little doubt that the extended growth periods under favourable

environmental conditions contribute to higher oat grain yields in the top-ranked countries. Early planting of spring oats is beneficial in nearly all countries.

The fact remains that within relatively restricted management options, most of us cannot significantly alter the length of growing period or the climatic environment in our respective countries. We must therefore continue our task of developing improved genotypes which meet farmer needs and acceptance in our respective geographical regions. Let us therefore consider some of the biological options at our disposal.

3. BIOLOGICAL AND MANAGEMENT OPPORTUNITIES
3.1 Biological traits

It is convenient to express oat grain yield in terms of the three primary yield components, number of panicle-producing tillers per unit area, number of florets (seeds) per panicle, and seed weight. While large increases in any one yield component appear highly unlikely, small but consistent improvements in many morphological and physiological traits which influence the three yield components can have a cumulative and significantly favourable impact on crop yield. The list of direct and indirect biological effects to be considered is lengthy and includes disease resistance; morphological traits, stiffness of straw, leaf area, groat percentage, straw yield, harvest index, and the three main yield components; and physiological traits and processes, photosynthetic rate, photosynthetic capacity, growth rate, photosynthate source-sink relationships, nitrogen and carbon assimilation, leaf area duration, leaf senescence, and mineral uptake, transport and deposition.

Improvements in the nutritive and energy levels of oats would enhance their use in human foods and animal feeds. However, farmers are reluctant to sacrifice agronomic performance and disease resistance for improvement in biochemical traits such as protein concentration.

Thus far, new knowledge about important physiological processes has not led to a recipe for breeding improved oat genotypes. The main criteria for selection of parents to be used in crosses primarily still fall in the nonphysiological categories such as grain quality, disease resistance, morphological traits, and actual grain yields, all of which obviously result from physiological processes.

3.2 Improved germplasm

Progress is clearly being made in some geographical regions where oat yields are restricted by shorter growing seasons. An excellent example is the cultivar Ogle developed by CM Brown and H Jedlinski at the University of Illinois. Although this cultivar is midseason in maturity and has a shorter growing cycle than several commonly grown later-maturing cultivars, it continually is among the top-ranked entries for grain yield in North Central US performance tests. Ogle typically yields from 3,600 to 5,000 kg/ha, and abundant tillering, a good number of spikelets per panicle, and above-average groat size are reasons for this superior performance. Fortunately Ogle appears to be an excellent parent, and we have identified stiff-strawed, disease resistant, and higher-yielding progenies from Ogle crosses in preliminary performance tests in Wisconsin. Ogle is a significant departure from early maturity-lower yield, later maturity-higher yield stereotypes.

3.3 Intensive crop management

Management practices required to obtain high oat yields involve high levels of nitrogen. This in turn requires very stiff-strawed genotypes to minimise the probability of severe lodging. Again the future is bright when we consider relatively new breeding stocks at our disposal. The stiff-strawed germplasm developed by USDA scientist HG Marshall at Penn. State University is an excellent example. These lines are very stiff and they have a desirable range in plant height. For example, Pennsylvania line PA 8098-13900 is extremely stiff for its height (100 cm). The challenge will be to retain this degree of stiffness in combination with other required levels of performance including high straw yields where bedding is in demand for use on dairy and livestock farms. The cultivars Noble (from Indiana) and Egdolon (from New York) have been widely used as parental sources of stiffness.

Intensive crop management as a total production system has focussed attention on the need for total integration of management practices to obtain maximum and economically profitable yields. For the small grain cereals, including oats; sowing date, rate, and depth; row spacing; fertilization; pest management in the form of herbicides, fungicides, and insecticides; and use of growth regulators all are management variables which can be patterned to fit individual cultivars. The use of tramlines facilitates the application of sequential chemical treatments throughout the growing season. In many countries, the point of maximum yield is surprisingly close to the point of maximum economic yield.

3.4 Future opportunities

The challenge facing oat breeders in both developed and developing countries is to develop cultivars, i.e. morphologic and physiologic ideotypes, which maximize the probability of producing high yields of high quality grain under specific management systems in specific environments. While we must rely heavily on adapted germplasm in the final stages of our respective breedings programmes, we also must continually infuse specific genes and gene combinations from nonadapted breeding stocks and noncultivated Avena species. The goals are long term and require much patience, but we have excellent germplasm, new knowledge, new techniques, and a great collaborative spirit, and I believe that we should look to the future with considerable optimism and enthusiasm.

REFERENCES

1. FAO Production Yearbooks: FAO Statistics Series No. 55. Vol. 5 to 37, 1951-1983.
2. Agro-ecological Atlas of Cereal Growing in Europe: Vol. II. Atlas of the cereal growing areas in Europe. Elsevier Pub. Co., Amsterdam, London, New York, 1969.
3. Agro-ecological Atlas of Cereal Growing in Europe: Vol. I. Agro-climatic atlas of Europe. Elsevier Pub. Co., Amsterdam, London, New York, 1969.

OAT BREEDING AT THE UNIVERSITY OF PASSO FUNDO, BRAZIL, 1976-1984

E. L. FLOSS
School of Agronomy, University of Passo Fundo, Passo Fundo, RS, Brazil

1. INTRODUCTION

The successive failure of wheat crops has increased farmers' interest in finding alternative winter crops. Although in the South of Brazil in the summer practically all tillable land is planted with soybeans, corn, beans, sorghum, etc., land is often idle during the winter season. In the State of Rio Grande do Sul(RS) alone, about 3,000,000 ha remain follow.

In RS, in 1984, cultivation comprised 700,000 ha of wheat; 220,000 ha oats for grazing and green manuring; 60,557 ha grain oats; 52,000 ha barley; 32,000 ha flax; 30,000 ha lupins and 2,000 ha seed rape. Thus, the cultivation of oats for grain is the second most important crop in the State during the winter months. In realisation of the present situation that the national demand and consumption of grain oats is higher than production, the cultivation of grain oats provides a feasible alternative crop for Southern Brazil.

2. AREAS OF OAT CULTIVATION AND PRODUCTION

At present, the cultivation of both grain and fodder oats is concentrated in the South of Brazil, mainly in the States of Rio Grande do Sul(RS), Santa Catarina(SC) and Paraná(PR). In other regions, such as Mato Grosso do Sul and southern Minas Gerais, oats are cultivated exclusively for fodder production. The increase in grain oat cultivation and production in Southern Brazil over recent years is shown in Table 1.

TABLE 1. Cultivation, production and yield of oats in Brazil over the period 1976-1984

Year	Area (ha)	%	Production (metric tonnes)	%	Yield (kg/ha)
1976	36,200	100	39,958	100	1104
1977	44,735	123	39,052	98	773
1978	56,000	154	53,947	135	971
1979	62,629	173	57,564	144	919
1980	77,000	212	75,609	189	982
1981	83,387	230	89,787	224	1077
1982	94,349	260	61,144	153	648
1983	95,105	262	92,824	232	976
1984[a]	120,147	332	131,041	327	1090

[a]Estimate   (Source: IBGE-FEE-CEPA-CACEX)

In RS, the cultivation of grain oats increased progressively between 1976 and 1984; increasing in total by 163% (Table 2). In 1984 there were 60,557 ha of grain oats while an estimated 220,000 ha were used for fodder, which was also a significant expansion.

This latter increase is due to an increase in dairying and beef fattening during the winter. Many farmers in RS sow oats immediately following summer crops, graze them during the winter and harvest grain from the regrowth.

TABLE 2. Cultivation and production of oats in Rio Grande do Sul, and the states contribution to the overall production of oats in Brazil, 1976-1984

Year	Area (ha)	%	Production (metric tonnes)	%	Yield (kg/ha)	Contribution to national production (%)
1976	23,000	100	22,158	100	963	56.9
1977	29,500	128	24,700	111	837	63.3
1978	41,800	182	39,800	180	952	73.8
1979	45,469	198	40,334	182	880	70.1
1980	51,394	223	47,942	216	933	67.5
1981	57,187	249	58,838	265	1029	60.0
1982	60,798	264	35,027	158	576	57.3
1983	54,154	235	52,947	238	978	57.0
1984	60,557	263	60,543	263	1000	46.2

(Source: IBGE-FEE-CACEX)

Rio Grande do Sul is still the main oat producing area of Brazil, but there has also been recent intensive development elsewhere, particularly in the State of Paraná(PR).

Within RS, the main grain producing area continues to be the region of Cotrijui cooperative, whereas in PR the largest producing area is in the region of the Cooperativa Agrária Mista Entre Rios Ltda.

The increases in oat cultivation are due, in part, to the frequent failures of wheat and thus the need to find alternatives, as well as to the more favourable prices of home grown grain relative to imports and the growing development of dairying.

3. OAT VARIETIES

The cultivars most widely grown for grain production are Coronado (A.byzantina) and Suregrain (A.sativa). Common Black Oats (A.strigosa) are most widely grown for fodder. Since 1981 new varieties, developed by the University of Passo Fundo and the Federal University of Rio Grande do Sul have been sown. The programme for oat improvement at the University of Passo Fundo began in 1977 with the transference of genetic material from the Centro Nacional de Pes quisa de Trigo - EMBRAPA. The same year the programme of exchanges with the Universities of Wisconsin and Texas A & M began. This consists of annual introductions of collections and segregating material in the F_3 generation.

The major objective of the programme is the development of new grain and fodder oat varieties with higher yield potential and resistance to the most important diseases, such as leaf rust (Puccinia coronata), stem rust (Puccinia graminis f.sp. avenae) and Barley Yellow Dwarf Virus.

Current investigations are concerned with various improvement projects which are carried out at the University of Passo Fundo. These include the crossbreeding and selection of material involving introduced germoplasm, preliminary assays of grain and fodder yield and regional tests including grain yield trials in southern Brazil. The programme includes experiments on sowing date, different times and frequencies of clipping, various plant densities and row spacings. Also the evaluation of chemical control of diseases, the resistance of oats to aphids and the identification of physiological races of leaf rust. Genetic stocks of seeds are produced with the purpose of providing new varieties for the farmer in the shortest possible time.

As a result of the breeding programme, the first cultivar, UPF-1, was released in 1981. This variety was selected from Flaab 113, a line sent by Professor H.L. Shands, University of Wisconsin, USA, in 1976. However, in 1983, this cultivar lost its resistance to leaf rust and is no longer recommended.

In 1982, basic seed of UPF-2 was distributed. This variety originated from selection X2505-4, a line also introduced from the University of Wisconsin, USA. This cultivar is dual purpose, for both grain and fodder production.

TABLE 3. Grain yields of new oat cultivars, selected at the University of Passo Fundo, 1978-1984, in different regions

Identification	1978[a]	1979[b]	1980[c]	1981[d]	1982[e]	1983[f]	1984[g]	Average	%
UPF-5	-	-	-	3302[y]	2776	2988	2772	2960	180
UPF-3	-	-	3362	3527	2372	2051	1971	2657	162
UPF-7	-	-	-	-	2444[z]	2508	2558	2503	153
UPF-4	-	1779	2849	2847	2525	2384	1884	2376	145
UPF-8	-	-	-	-	2400[z]	2514	2265	2393	146
UPF-6	-	-	-	3015[y]	1998	2253	2077	2335	142
UPF-1	3376	1580	2766	2301	1499	1523	1504	2078	127
UPF-2	-	1668	2441	2787	1517	1674	1643	1955	119
Suregrain	2408	1078	1953	2451	854	1189	1442	1625	99
Coronado	2263	1032	2070	2364	818	1180	1741	1638	100

[a]Passo Fundo, Vacaria, Cruz Alta, Bagé, Ijuí and Ibirubá.

[b]Passo Fundo, Vacaria, Cruz Alta and Bagé.

[c]Passo Fundo, Vacaria, Cruz Alta, Entre Rios(PR), Guaíba and Ijuí.

[d]Passo Fundo, Vacaria, Cruz Alta, Entre Rios(PR), Guaíba, São Gabriel, Dom Pedrito, Campos Novos(SC), Castro and Viamao.

[e]Passo Fundo, Vacaria, Cruz Alta, Entre Rios(PR), Guaíba, Campos Novos(SC), Ponta Grossa, Dom Pedrito and Ijuí.

[f]Passo Fundo, Vacaria, Cruz Alta, Entre Rios(PR), Guaíba, Sao Gabriel, Campos Novos(SC), Ponta Grossa, Ijuí, Maracajú and Nova Prata.

[g]Passo Fundo, Chiapeta, Ijuí, Campos Novos, Vacaria, Entre Rios(PR), Julio de Castilhos, Guaíba and Ponta Grossa(PR).

[y]Grain yield of Regional Trials at Passo Fundo, Guaíba, Campos Novos(SC), Vacaria & Sao Gabriel

[z]Passo Fundo, Campos Novos(SC), Guaíba, Vacaria and Ijuí

Two newer varieties, UPF-3 and UPF-4, were recommended in 1984. UPF-3 derives from the cross Coronado x X1779-2, made at the University of Wisconsin, and introduced to the University of Passo Fundo in 1977; this variety is also recommended as dual purpose. UPF-4 was selected from a line X2055-1 introduced from the University of Wisconsin in 1977. This is a short, early variety and is recommended for grain production.

This year, UPF-5, selected at the University of Passo Fundo, was officially released. It derives from the cross 2185 x Ill 151, made at the University of Wisconsin, and was introduced as an F_3 in 1977. Experimental results indicate that this new variety is suitable for dual purpose.

During the current season basic seed of cultivars UPF-6 (Coker 1214 x Lang), UPF-7 (TCFP x X2503-1) and UPF-8 (X2505-4 x OTEE) is being produced. All these varieties were selected from F_3s introduced under the project 'Breeding Oat Cultivars Suitable for Production in Developing Countries'.

Table 3 shows the grain yields of the new cultivars, developed by the University of Passo Fundo during the period 1978-1984, compared to the traditional Corondao and Suregrain cultivars, in different regions.

OAT PRODUCTION AND BREEDING IN IJUÍ, RS, BRAZIL

R.B. MEDEIROS, J.E.G. ZAMBRA, L.V.M. VIAU, C. PITTOL and R. CARBONERA
111-Cotrijui, Ijuí,(RS), Brazil

1. INTRODUCTION

The 'Cooperativa Regional Tritícola Serrana Ltda - Cotrijuí' was founded in 1957 and is an association of 22,250 farmers. It is concerned with the production, processing and commercialization of agricultural products in three areas of Brazil. In Mato Grosso do Sul State, its area of influence extends to about 10,000,000 ha of agricultural land, of which only 1,500,000 ha are devoted to arable cropping. The administrative office is in Campo Grande (20°25'S, 54°35'W). In Rio Grande do Sul State its area of influence is in Dom Pedrito (31°S, 54°35'W) where there are 72,000 ha of cereal growing land and in Ijuí (28°20'S, 53°50'W) where 400,000 ha are very intensively and continuously cropped with wheat and soybeans.

The Head Office is in Ijuí, where the Cooperative also has the 'Centro de Treinamento, Cotrijui - CTC' where oat research is carried out. In the Ijuí area, oats, barley and flax are only minor crops in the winter/spring growing season in comparison with wheat, which is still the main crop in spite of its low yields (Table 1).

TABLE 1. Production areas and average grain yields of wheat, oats, barley and flax in the Ijuí area. (Ijuí, RS)

Crops	1981 ha	1981 kg/ha	1982 ha	1982 kg/ha	1983 ha	1983 kg/ha	1984 ha	1984 kg/ha
Wheat	90,400	1,170	166,000	430	81,500	990	70,180	570
Oats	10,340	1,270	14,700	360	8,930	870	5,410	830
Barley	5,100	1,080	5,630	360	5,660	1,130	7,090	670
Flax	4,910	960	1,880	500	2,790	830	7,190	630

The growing season in the Ijuí area is usually rainy with high humidities and air temperatures during the spring. This provides an ideal environment for crown and stem rust attacks.

2. OAT BREEDING

CTC introduced the varieties Coronado, Suregrain and Epecoen, in 1973. From then until 1976, Coronado and Suregrain yielded on average about 1,500 kg per ha, while Epecoen yielded less than 1,000 kg per ha.

These varieties were at first grown as dual purpose crops but more recently most farmers have moved to oats as a grain crop due to the frequent failure of wheat and a good market for oats.

In 1981, the CTC multiplied up seed of the new cultivars UFRGS 1 and UFRGS 2 produced by the Federal University of Rio Grande do Sul (Table 2). Unfortunately these varieties were almost completely wiped out by a new race of crown rust. In 1983, the University of Passo Fundo, RS, released the cultivars UPF 3 and UPF 4. UPF 3 was severely attacked by crown rust last year. CTC B 207 was selected at Ijuí (CTC), out of 35 entries received from the Federal University of Rio Grande do Sul in 1978; these came originally from Madison, Wisconsin. Unfortunately we were unable to identify the CI number of this line. CTC B 207 is not a good cultivar as far as kernel weight and proness to lodging are concerned but we are continuing with it until we are able to replace it with something better.

TABLE 2. Production areas and grain yields of five oat cultivars distributed to farmers in 1982, 1983 and 1984 (Ijuí, RS)

Cultivar	Breeding/selection description	1982		1983		1984	
		ha	kg/ha	ha	kg/ha	ha	kg/ha
UFRGS 1	Dalx Tex 71C 3039-2	30	70	29	245	Withdrawn	
UFRGS 2	Dalx Tex 71C 3039-2	22	20	9	100	Withdrawn	
UPF 3	Coronado XX 11779-2	-	-	8	965	90	700
UPF 4	Pure line selection from Wisconsin	-	-	8	2,230	210	1,550
CTC B 207	Pure line selection from Wisconsin	3	1,800	44	1,720	920	1,450

3. FUTURE PROSPECTS

Although the Ijuí area does not have ideal weather conditions for oat growing, it appears that oats may be one of the main cereal crops in the future. Oats also fit in well in rotation with the main summer crop of this area - soybeans. In recognition of this trend, we started the oat breeding programme at CTC in 1983. This work is carried out with the support of 'Breeding Oat Cultivars Suitable for Production in Developing Countries', sponsored by the Quaker Oat Company. The objectives are to breed oats with better yields and resistance to lodging and crown and stem rust. From the International Quaker Oat Nursery we selected 108 F_3 panicles in 1983 and 624 F_3 panicles in 1984. In 1983 and 1984, we made 190 and 174 crosses, respectively, using as parents genetic stocks we received from the International Quaker Oat Nursery in 1982, 1983 and 1984.

PERFORMANCE OF OATS IN THE QUAKER-SOUTH AMERICAN NURSERY, 1983-84

M.A. BRINKMAN and H.L. SHANDS
Department of Agronomy, University of Wisconsin, Madison, WI 53706, USA

1. INTRODUCTION

In 1974, Dr Hazel Shands of the University of Wisconsin initiated a project entitled 'Breeding Oat Cultivars Suitable for Production in Developing Countries' that was funded by a two year grant from USAID. Quaker Oats International assumed financial responsibilities for the programme in 1976 and has provided excellent funding since that time. In recent years the Quaker project has focussed its testing in Central and South America because Quaker is interested in the development of improved oat cultivars in these areas.

2. THE 1983 REPORT

The 1983 report of 'Breeding Oat Cultivars Suitable for Production in Developing Countries', which was distributed in May of 1985, has the same general format as the previous annual reports. The Introduction discusses historical perspectives, defines some abbreviations and describes the nursery. The 1983 nursery consisted of 314 test lines (pure lines) and 46 segregating populations (F_3 bulks). The Results and Discussion section summarises the performance of the test lines and bulk populations in each of the locations in which they were grown. The South American locations are discussed chronologically according to the order in which they were visited by ME McDaniel, SH Weaver and MA Brinkman in November of 1983.

Most of the 1983 Quaker report consists of tables that describe the performance of test lines and bulk populations in various locations throughout South, Central and North America. The last four tables (Tables 40 through 43) summarise the performance of 42 test lines that were considered to be among the better lines in the nursery for crown rust and stem rust resistance. Oat workers interested in widely adapted germplasm with good crown and stem rust resistance can identify lines with these traits by integrating the information in these four tables.

The performance of a random group of 10 of these 42 test lines is summarised in Tables 1-4. Table 1 describes crown rust (Puccinia coronata) responses in Brazil (Passo Fundo ad Ijuí), Uruguay (La Estanzuela), Argentina (Castelar), Mexico (Chapingo) and the United States (Madison). Test lines 64, 65, 97, 101 and 270 were very resistant to crown rust in all locations.

Reactions to stem rust (Puccinia graminis) and barley yellow dwarf virus (BYDV), are summarised in Table 2. Test lines 43, 44, 64, 65, 97, 101 and 270 had excellent resistance to stem rust in Uruguay (La Estanzuela), Argentina (Castelar), Mexico (Chapingo) and the United States (College Station and Beeville, Texas and Madison, Wisconsin). Test line 97 had good BYDV tolerance at both Davis, California and Chapingo, Mexico.

TABLE 1. Crown rust reaction of Quaker test lines in 1983

1983 entry no.	Passo Fundo	Ijui	La Estanzuela 10 Nov.	5 Dec.	Castelar	Chapingo	Madison
43	0	0	2MR-R	2MR	0	0	VR
44	Tr	0	0	0	0	0	R
64	0	0	0	0	0	0	VR
65	0	0	0	0	0	0	VR
85	0	10	TrMS	10MS	Tr	0	VR
97	0	0	0	0	0	0	VR
101	0	0	0	0	0	0	VR
151	0	0	0	10MR	0	0	MR
169	5S	20	5MR	5MR	Tr	40S	R
270	0	0	0	0	0	0	VR

TABLE 2. Stem rust and BYDV reaction of Quaker test lines in 1983

1983 entry no.	Stem rust						BYDV	
	LE	Castelar	Chapingo	CS	Bee	Msn	Davis (1-9)	Chapingo (0-100)
43	0	0	0	R	VR	VR	7.0	10
44	0	0	0	VR	R-MR	VR	6.0	20
64	0	0	0	R	R-MR	VR	5.0	70
65	0	0	0	R	R	VR	5.5	30
85	10S	Tr	25MR	MR	VR	VR	6.0	20
97	0	0	0	VR	VR-R	VR	3.0	30
101	0	Tr	0	VR	R	R	7.5	50
151	70S	60	40MS	VS	MS	MR	6.0	40
169	10MS	10	40MR	MS	MR-MS	MR	4.0	-
270	0	0	0	VR	VR	VR	7.0	50

The agronomic performance of these test lines is presented in Table 3.
Grain yields are from non-replicated rows and should be interpreted with
caution. The 10 test lines had considerable variability for heading date,
plant height and lodging.

TABLE 3. Agronomic performance of Quaker test lines in 1983

1983 entry no.	Grain yield		Heading date (2 loc) (October)	Plant height (2 loc) (cm)	Lodging	
	Passo Fundo (kg/ha)	Maracaju (g/row)			Chapingo (%)	Madison (1-9)
43	1211	54	19.0	125	90	8
44	2750	175	19.0	110	80	7
64	4378	110	29.5	90	0	3
65	3935	154	29.5	96	0	4
85	3034	145	19.0	95	0	1
97	1576	129	23.0	91	60	1
101	1045	104	23.0	112	10	1
151	2158	139	22.5	93	20	1
169	1957	140	19.0	90	0	1
270	5331	97	Nov./1.0	114	0	3

Comments on agronomic type are presented in Table 4. Several test lines
(85, 151 and 169) received good notes in all locations and are considered
to be widely adapted. Other test lines were attractive in some locations
but not in others.

TABLE 4. Comments on Quaker test lines grown in 1983

1983 entry no.	Brazil		Argentina		United States	
	Passo Fundo	Guaiba	Bordenave	Castelar	Madison	Beeville
43	tall, weak	tall	weak	weak	lodged	early
44	++, tall	+, early	+, sl weak	+, early	lodged	early
64	++++, stiff	unattr	late	++, late	+	nice type
65	++++, stiff	unattr	late	++, late	+	fair +
85	++++, stiff	+++	+++, stiff	+++	++	productive
97	++, stiff	unattr	late	++	+	strong
101	+, het	+++, early	het ht	++	+	good
151	++++, stiff	++++	+++, stiff	++	++	good type
169	+++, stiff	++, early	+++	++	++	good type
270	++	discard	late	late	++	fair

3. THE 1984 AND 1985 NURSERIES

The 1984 Quaker nursery consisted of 245 test lines, 90 F3 bulks and 25
selections from China (Table 5).

TABLE 5. Composition of the Quaker Nursery 1983-85

1983	314 Test lines
	46 F3 populations
1984	245 Test lines
	90 F3 populations
	25 Hull-less oat selections from China
1985	238 Test lines
	13 Illinois BYDV lines from the USDA
	47 BYDV lines from Illinois
	24 BYDV lines from Canada
	110 F3 populations

SH Weaver, ME McDaniel and MA Brinkman visited the South American locations in November of 1984. Their observations and reports submitted by
nursery cooperators indicate that test lines from four crosses were highly
resistant to crown rust and stem rust in all locations in 1984 (Table 6).

The 1985 Quaker nursery consisted of 238 test lines, three sets of lines
included as sources of BYDV tolerance and 110 F3 populations. In Texas,
many of the test lines matured much earlier than the BYDV lines but, in
Wisconsin, many of the test lines were later than the BYDV lines. Most of
the BYDV lines from Illinois and Canada had an attractive agronomic
appearance at Madison, Wisconsin in 1985, but they were susceptible to
stem rust.

TABLE 6. Sources of crown and stem rust resistance

C5-2, 1563 CRcpx/SRcpx
Cortez[4]/C5-2, 1563 CRcpx/T312/SRcpx
79 Bordenave selection/Kenya SR resistance line
78 SA-33/69Ab5245/Unknown

CRcpx = one of several combinations of commercial
        pure line varieties x A.sterilis
SRcpx = one of several combinations of TAM 0-301,
        TAM 0-312, Coker 227, Coker 234 and CI 9221

4. OTHER DEVELOPMENTS

During the past two years, a number of crosses have been made to combine rust resistance and tolerance to BYDV. In addition, testing for BYDV tolerance has been increased. Testing the Quaker nursery for BYDV tolerance was initiated at Urbana, Illinois in 1984 and continued in 1985. In addition to BYDV results from the South and Central American locations, the BYDV tests at Davis, California and Urbana, Illinois will assist in developing widely adapted, rust resistant oat cultivars with tolerance to BYDV.

The Quaker project has made significant progress towards accomplishing its objectives in South America. Dr Elmar Floss has released five cultivars, UPF-1, UPF-2, UPF-3, UPF-4 and UPF-5, that were derived from Quaker nurseries and three more releases are planned in the next few years. These will be named UPF-6, UPF-7 and UPF-8. Most of these cultivars have good resistance to the races of crown and stem rust that are present in the Ijuí area, where many of the oats in Rio Grande do Sul are grown.

Another important trait that is needed in South American oat cultivars is regrowth after grazing. A large percentage of the oat hectarage in South America is grazed, primarily by beef cattle. After the livestock are removed, the oats are grown to maturity and harvested. Some farmers graze their livestock long after the oats have headed and do not harvest the oats for grain. For many years diploid oats (Avena strigosa) have been grown extensively for grazing, particularly in Brazil, Uruguay and Argentina.

Many entries in the Quaker nursery had non-pathogenic leaf discoloration at Temuco, Chile in 1984. The leaf discoloration has been referred to as PLS (physiologic leaf spot) by the coordinators of the Quaker nursery. PLS seems to be associated with a source of stem rust resistance that has been used in a number of crosses. A limited amount of Spermospora avenae and halo blight were also present at Temuco in 1984.

SELECTING OATS FOR NORTH ISLAND CONDITIONS:
COLLABORATION BETWEEN NEW ZEALAND AND CANADA

J. M. MᶜEWAN
Crop Research Division, DSIR, Palmerston North, New Zealand

1. INTRODUCTION

The production of oats has a long history in New Zealand cropping, particularly important when horses were the power source for transport and cultivation. Up to 1915 large quantities of chaff and grain were exported to Australia. Since then the crop has declined in significance, though oats are still produced as livestock feed, either as a whole crop conserved feed (chaff) or as grain or for forage. As well, milled oaten products are produced for human consumption. The main areas of production have traditionally been in the South island, but oats were also cropped successfully in some North Island districts. Over the years there has been a marked decline in North Island production as the diseases barley yellow dwarf virus (BYDV) and crown rust (Puccinia coronata Cda. f.sp. avenae) have increased in incidence. These diseases are also present in the South island, but the problems there are less acute. Total North Island areas sown: 15,000 ha in 1932, 2,900 ha in 1962 and 1,370 ha in 1983.

A sustained oat breeding programme at the Crop Research Division (CRD), DSIR, Lincoln has been successful in the development of new cultivars for all aspects of production in the South island. The standing ability of the crop was improved by using the Welsh cultivar Milford to produce the short strawed cultivars Amuri and Makuru. More recently improved tolerance to diseases, particularly to BYDV, has been attained by using the Australian cultivar Avon and the Canadian cultivar Rodney in the development of Omihi and Ohau.

There is a substantial North Island demand for high quality grain oats for the rations of pedigree racehorses. This is met by freighting grain or chaff from the South Island at a high cost. Local production of oats would be a useful addition to North Island cereal cropping but the cultivars previously available ran a high risk of serious losses to disease. The combined effects ran of BYDV at an early stage of crop development and crown rust at grain filling have a very detrimental effect on the productivity and grain quality of the crop. Though a cereal improvement programme was instituted by CRD in the mid 1960s for the North Island, the principal thrust of the work was on wheat and barley, with oats receiving little attention.

BYDV in New Zealand cereal crops is a serious problem, probably of long standing. The 'red leaf' symptom in oats was described early in the century, although the causal agent was not then identified. The significance of BYDV was not recognised until the 1950s when Dr HC Smith began his studies on the disease. In the North Island it can be particularly severe, since cereal crops tend to be relatively small and isolated amongst large areas of pasture which act as a reservoir for BYDV and as a refuge for the infective aphids, principally Rhopalosiphum padi L.

Flights of viruliferous aphids occur during October, after spring-sown cereals have emerged.

2. COOPERATION IN OAT BREEDING BETWEEN NEW ZEALAND AND CANADA

During the mid 1970s, outbreaks of BYDV in the grain crops of Eastern Canada led to the development of a cooperative cereal breeding project between Agriculture Canada Research Stations in Quebec, Ontario, and Manitoba and CRD stations at Gore, Lincoln and Palmerston North. Early generation oat populations, principally F_2 and F_3 bulk populations for spaced planting, were screened for BYDV reaction in New Zealand.

This programme concentrates on material being exchanged between the Winnipeg programme and Palmerston North, where good natural infections can be guaranteed. Agronomic performance is assessed. A small amount of wheat and barley included. The programme is cooooperative to the extent that CRD has agreed access to the breeding stocks for selection and testing for local adaptation, and this material has formed the basis of an oat improvement programme for the North Island. The Agriculture Canada crossing programme has included some New Zealand derived material, particularly Omihi and its derivatives for BYDV resistance, ensuring a component of local adaptation; and selected lines have shown satisfactory performance at Palmerston North. In addition Agriculture Canada has screened promising New Zealand selections for their genetic resistance to crown rust and oat stem rust (Puccinia graminis f.sp. avenae Pers.).

From samples collected in New Zealand, PAV, MAV and RPV strains of BYDV have all been identified (CW Schaller, pers. comm.). Though some strain differences between the two countries may occur, the responses to BYDV of a number of cultivars have usually been similar. Therefore selection for BYDV tolerance in New Zealand should be effective for Canadian programmes.

3. BYDV YIELD REDUCTION IN NEW ZEALAND TRIALS

An experimental assessment of the amount of damage that BYDV causes to grain crops is not easy, but estimates of yield losses up to 25% have been reported (1). In comparative trials, the aphid control measures applied may considerably influence the level of virus infection in the untreated plots, giving unreliable results. However, a comparison of trials that have been exposed to different levels of natural infection and containing stocks of differing reaction to BYDV should provide valid information. Such a comparison was possible during the 1983-84 season when two local trials of a set of breeding lines developed markedly different levels of BYDV infection. The 10 selections, from Canadian material, had had several seasons of screening in the North Island under conditions where BYDV infection was prevalent, and were thought to have good tolerance to the disease. In one trial (code 83ROMT) however, three of the selections showed severe BYDV damage, while the other selections, the two standard cultivars and the surrounding crop of a resistant cultivar were little affected (see Table 1). Further, the level of infection showed a marked gradient over the four replicates of the trial, from 50% yield loss near the western perimeter of the crop to little damage where the trial was most protected. Such gradients are frequently observed in infected cereal fields, and show the distribution pattern of the aphids. In the second trial (83ROKG) BYDV infection was apparently not a factor in yield reduction, since the three groups of oat lines had similar grain yields (Table 1). These observations illustrate the sensitive interaction displayed by cereal cultivars to different levels of BYDV infection, to the amount of damage observed and to the yield reduction sustained. Such dosage effects may be of critical importance in the reaction of a cultivar or crop(2,3).

TABLE 1. Comparisons of grain yields, in two trials differing
in level of BYDV infection

	Mean yield (t/ha)	Relative yield
83ROMT (high BYDV incidence)		
Standard cultivars(2)	4.68	100
BYDV tolerant selections(7)	4.99	107
BYDV susceptible selections(3)	3.07	66
Standard deviation	0.60	

Yield differences within replicates

	Rep. 1		Rep. 2	
	Mean yield	Rel. yield	Mean yield	Rel. yield
Standard cultivars(2)	4.95	100	4.78	100
BYDV tolerant lines(7)	5.05	102	5.29	109
BYDV susceptible lines(3)	2.50	50	2.39	50

	Rep. 3		Rep. 4	
	Mean yield	Rel. yield	Mean yield	Rel. yield
	4.78	100	4.20	100
	5.07	106	4.62	110
	3.21	67	4.15	99

	Mean yield	Relative yield
83ROKG (low BYDV incidence)		
Standard cultivars(2)	8.00	100
BYDV tolerant selections(7)	7.99	100
BYDV susceptible selections(3)	7.77	97
Standard deviation	0.30	

Cooperative breeding programmes, such as the oat programme described, have proved effective for rapidly establishing cereal breeding projects in new areas. Already one oat selection, Awapuni, of Canadian origin has been approved for commercial production. In addition to having an excellent level of BYDV tolerance (equivalent to Omihi) it has two effective factors for resistance to crown rust (P.c. 38 and P.c. 39) and one for stem rust resistance (P.g. 13) and has complete field immunity to rusts. Awapuni has a good agronomic type and out yields all other cultivars at North Island sites. It also has the large grain size and high test weight demanded by the racehorse feed industry. It is anticipated that Awapuni will find ready acceptance in the North Island.

4. ACKNOWLEDGEMENTS
    The contribution of staff members in the Agriculture Canada oat breeding programmes and the CRD, DSIR, New Zealand cereal groups to this work is gratefully acknowledged.

REFERENCES

1. Smith HC: NZ J. Agric. Res. 6, 229-244, 1963.
2. Smith HC: The Canadian Entomologist 95, 537-547, 1963.
3. Smith HC: NZ J. Agric. Res. 10, 445-466, 1967.

OAT BREEDING RESEARCH AT PANTNAGAR - INDIA

S. N. MISHRA and J. S. VERMA
Department of Plant breeding, G.B. Pant University of Agriculture
and Technology, Pantnagar 263145, UP, India

1. INTRODUCTION

Oat breeding research in India is of recent origin in comparison with approximately a century of advanced breeding research in western countries. Interest in growing oats has increased in recent years because of its high value as green fodder for livestock. The area sown to oats for herbage production is still increasing gradually, but interest in oat grain production and utilisation is not increasing at a similar rate. Breeding research on oats started on a small scale at the G.B. Pant University of Agriculture and Technology, Pantnagar, India, in 1972. Pantnagar is located 29°N latitude, 79.3°E longitude and at an altitude of 243.8 m above mean sea level. The initial oat material to start the breeding programme was obtained from USA by courtesy of Drs H.L. Shands and R.A. Forsberg of University of Wisconsin, Madison; Dr K.J. Frey of Iowa State University, Ames, and Dr J.C. Cradock of USDA, Beltsville, Maryland. Since then, germplasm from other countries has been collected and assessed for its potential use in the current breeding programme. The major objectives of the research programme have been to increase herbage production, enhance regrowth for repeated cuttings, to achieve a satisfactory level of grain production, incorporate resistance to the major diseases, and improve herbage quality. This paper presents the results of some of the more significant studies carried out during the past decade.

2. MATERIALS AND METHODS

The material used in the research programme is comprised of more than 4,000 germplasm lines which have been collected from USA, UK, Canada, Denmark, Australia, New Zealand, Argentina and Columbia. The collection has also been enriched regularly by adding a range of breeding material, such as F_2s, F_3s and other lines, from the ongoing breeding programme. The entire germplasm collection has been grown in 1 m rows for evaluation of the important plant characteristics, including resistance to the major diseases and pests. Potential parents for hybridisation are chosen from this germplasm pool. The hybridisation programme is mainly based on the use of multiple parents, although single cross combinations are also made. In addition to matings among pure line cultivars and land races others, such as biparental (BIP), NC Design-I, II, III and triple test crosses, are also practiced in the F_2 and F_3 generations of single and multiple crosses. These programmes are designed to release additional genetic variability through greater gene reshuffling and breakage of linkage blocks. Segregating progenies are usually handled by the pedigree method, using single plants or single panicles as the units of selection.

261

Green fodder or grain yield tests begin in the F_5 generation and continue up to the F_9 and F_{10} generations. The experimental row length, number of rows and number of replications, differ according to the generation under test. National yield evaluation trials of the most outstanding progenies are conducted at several locations across the country to assess stability of yield performance and other characters. After 3-5 years of testing in National yield evaluation trials, the best lines are recommended for release to farmers.

Apart from intraspecific crosses, crosses between species are also being carried out with the objectives of incorporating genes for disease resistance, greater adaptability, increased protein content and high tillering capacity. The most frequently used species in interspecific hybridisation is Avena sterilis. Genetic upgrading is being achieved by backcross breeding and intermating within segregating populations. These studies only commenced fairly recently but early results are encouraging.

3. RESULTS AND DISCUSSION

A little more than a decade of breeding research at Pantnagar has contributed substantially to the improvement of fodder oats in India (Table 1). Herbage yields from improved new varieties/lines have gone up to 2.5 times those of the long standing check cultivar, Kent. Herbage production continues to increase with the additional characteristic of regrowth for repeated cuttings; these attributes provide a continuous supply of green fodder for an extended period.

TABLE 1. Herbage yield levels of new varieties/lines of oats developed at Pantnagar (India) in comparison with Kent

Variety/line	Herbage yield (t/ha)	Status
Kent	30.0	Old var. (1972-1973)
UPO 94	52.5	New var. (1981-1982)
Ox 12-16-9-1	84.7	New line (1984-1985)
Ox 12-10-10-4	81.9	" " "
Ox 76-6-9-2	80.0	" " "
Ox 105-2-2-1	83.3	" " "
Ox 105-9-4-5	81.4	" " "
Ox 189-12-29-2	82.3	" " "

Collections of exotic, indigenous and breeding material have shown a considerable range of genetic variation for a number of plant traits, including herbage yield. Additional genetic variability for obtaining even further genetic gain has been released by using hybridisation procedures such as multiple crosses, diallel crosses, backcrosses, and interspecific crosses. Intermating between plants within segregating populations has also released additional genetic variability.

Genetic variability for such characters as forage yield, leaf width, number of leaves per plant and number of tillers per plant, is substantial(1). In an 8 x 8 diallel of diverse parents it was observed(2,3) that an array of genetic variation occurred in the F_2 generation for height, tiller number and green and dry matter yields. There was also a high frequency of transgressive segregates both in positive and negative directions due to gene association and dispersion. One of the most important cross for increased fodder production was identified as Orbit x Bingham. However, the varieties Portal, Bingham, Burt and Kent were identified as the best general combiners for high number of tillers, and green and dry weight fodder yields. Various crosses between these have produced some very useful lines with higher productivity.

Intermating among plants of F_2 and F_3 generations was carried out with the objective of maximising genetic variation through reshuffling of the parental genomes and to break linkages. Yield improvements and changes in linkage relationships come about(4) by such matings which can be carried out in several ways. The biparental (BIP) and NC Design-II (D-II) matings showed on average about 15 t/ha of green fodder yield superiority over the direct descent (DD) progenies(5). The BIP and D-II progenies were equal in their performance (Table 2). Advanced generation progenies of such matings are being tested under a range of environmental conditions to assess their stability of performance.

TABLE 2. Performance of different types of oat progenies

Pedigree	Progeny	Plant height (cm)	Days to heading	Green fodder yield (t/ha)
Ox 63-3-1	DD	143	127	55.2
Ox 63 (BIP)-4	BIP	144	121	70.7
Ox 63 (D-II)-3	D-II	148	125	69.4
Ox 68-3-2	DD	157	116	57.4
Ox 68 (BIP)-7	BIP	153	122	70.7
Ox 68 (D-II)-11	D-II	151	114	69.6

When the BIP progenies were compared with the selfed progenies of direct descent, it was observed that the BIPs and the BIP selfs had grain yield advantages of about 4-10 g per plant over the selfed progenies. Improvements also occurred in plant height, number of spikelets per panicle and biomass (Table 3). Variation in the BIPs was also higher than in the selfed progenies and some new correlations developed in both populations indicating breakage of linkages(4).

TABLE 3. Effects of biparental mating and selection in oats

Cross	Progeny	Plant height (cm)	No. of spikelets per panicle	Biomass (g)	Grain yield (g)
Coachman x Swan	F_2 BIP	116	92	60	18
	F_3	113	74	59	14
	F_2 BIP*	114	80	83	17
	F_3 BIP	140	124	106	32
	F_4	130	111	82	22
Goodland x UPO 94	F_2 BIP	94	80	86	17
	F_3	87	60	51	10
	F_2 BIP*	99	74	72	20
	F_3 BIP	126	94	118	30
	F_4	121	88	91	21
Lyon x Montezuma	F_2 BIP	123	80	81	16
	F_3	87	60	51	10
	F_2 BIP*	112	77	81	22
	F_3 BIP	127	102	112	30
	F_4	123	86	95	23

*Selfed F_2 BIP progenies

Interspecific crosses of A.sativa x A.sterilis have shown encouraging results in both productivity and productivity traits. At present, three A.sterilis accessions, CI 8007, PI 292561 and PI 295932, are being used. A.sativa x A.sterilis F$_1$ hybrids have shown remarkable superiority over the higher parent for such vegetative characters as plant height, leaf length, leaf width and number of tillers per plant(6). The heterobeltiosis for tiller number was maximum and varied from cross to cross (Table 4).

TABLE 4. Heterobeltiosis for some plant characters in A.sativa x A.sterilis crosses

| Cross | Heterobeltiosis (%) | | | | |
	Heading date	Plant height	Leaf length	Leaf width	Tiller number
Forward x CI 8077	1.6	15.2	12.8	2.6	78.4
Menominee x PI 292561	-3.2	1.9	13.5	-2.1	45.3
WA 1470 x PI 295932	-2.9	-11.2	33.7	60.8	87.0

Cross WA 1470 x PI 295932 produced maximum heterobeltiosis for leaf length, leaf width, and tiller number. It is, therefore, expected that some lines with increased herbage yield can be derived from this cross.

Programmes are underway to produce high grain yielding varieties by the similar use of interspecific crosses. Improvements in tiller number, number of spikelets per panicle and grain yield have been demonstrated by backcross derived lines. Three sativa varieties are being crossed with one sterilis line, PI 295932.

Differences between crosses have been apparent (Table 5). Progenies have not, however, shown improvements in grain or kernel weights. It is hoped that it may be possible to improve these traits by using other sativa lines.

TABLE 5. Effects of A.sterilis genes on certain traits in oats

Cross	P$_1$	P$_2$	F$_1$	(F$_1$ x P$_1$)	(F$_1$ x P$_1$) P$_1$
No. of tillers per plant					
Kent x PI 295932	9.68	10.95	12.49	11.23	10.86
Nodaway x PI 295932	11.69	11.28	16.59	14.79	12.40
WA 1470 x PI 295932	9.97	10.91	14.93	13.57	11.83
No. of spikelets per panicle					
Kent x PI 295932	71.03	31.40	59.70	73.92	75.55
Nodaway x PI 295932	59.47	31.29	47.27	49.85	58.40
WA 1470 x PI 295932	48.97	30.87	50.54	49.36	47.19
Grain yield/plant (g)					
Kent x PI 295932	36.89	18.06	43.39	37.52	36.82
Nodaway x PI 295932	21.41	17.18	37.87	33.70	22.41
WA 1470 x PI 295932	22.77	17.86	44.24	37.19	29.47

Progeny evaluation and yield testing are two important aspects of the crop improvement programme. Progenies are advanced according to the pedigree method on an individual plant or panicle basis. Yield tests commence at F_5 and are carried out over a range of locations for F_7 and later generations. Assessments are made for herbage production and grain yield in comparison with a popularly grown check variety(7).

It is expected that further improvements in herbage and grain yields will be achieved by applying the appropriate breeding methods and utilising cultivated, indigenous and exotic germplasm, including genes from wild species of different ploidy levels.

4. ACKNOWLEDGEMENTS

The senior author is grateful to G.B. Pant University of Agriculture and Technology, Pantnagar, and the Indian Council of Agricultural Research, New Delhi for financial assistance.

REFERENCES

1. Dhumale DB and SN Mishra: Pantnagar J. Res. 2, 15-18, 1977.
2. Dwivedi S, SN Mishra and JS Verma: Oat Newsletter 34, 37-38, 1983.
3. Dwivedi S, SN Mishra and JS Verma: Forage Res. 10, 103-106, 1984.
4. Prasad R: Ph.D. Thesis, G.B. Pant University of Agriculture and Technology, Pantnagar, India, 1984.
5. Mishra SN, JS Verma, R Prasad and R Rastogi: Oat Newsletter 33, 38-39, 1982.
6. Mishra SN, JS Verma, R Rastogi and R Prasad: Oat Newsletter 33, 35-37, 1982.
7. Mishra SN, JS Verma and R Prasad : Oat Newsletter 34, 40-41, 1983.

OAT BREEDING IN THE USSR AND THE INITIAL BREEDING MATERIAL

V. D. KOBYLYANSKII*, N. A. RODIONOVA* and H. D. KÜÜTS**
*All-Union Research Institute of Plant Industry (WJR), Leningrad, USSR
**The Jõgeva Plant Breeding Station, Estonia, USSR

1. INTRODUCTION

In USSR oats are a basic agricultural crop and the area under oats in the USSR constitutes over one-third of the world total. Oat breeding started at the end of last century and at the beginning it was based on the selection of the best aboriginal forms which represented a mixture of biotypes which had passed through evolution in different soil-climatic zones and were well adapted to local conditions. By selection of this material the varieties Shatilovski, Zolotoi and Khersonski, amongst others, were developed. The relative constancy of these varieties was preserved by purchasing seed from the originating sources where uniformity was maintained.

Later these variety-mixtures were replaced by pure-line varieties which were derived from single elite plants, examples are, Shatilovski 56, Tulunski 86/5 and Moskovski A-315.

2. THE ROLE OF COLLECTIONS

An important role in the development of scientific selection was played by the Bureau of Applied Botany (now named the N.S. Vavilov All-Union Research Institute of Plant Industry) by organising collections and carrying out botanical studies of the plant resources of Russia.

The Oat Collection in the pre-revolutionary period numbered 1,100 samples, and was chiefly of local strains from different provinces of Russia; amongst these was some very valuable material. Some of our early-ripening, drought-resistant oats from the steppe regions of the Ukraine exerted a considerable influence on the development of oats in the USA, appearing in the genealogical registers of many varieties. Local Siberian oats, exported from Russia, formed the basis of several early Canadian varieties.

Also, at the end of last century, coarse-grained oat varieties started to be introduced into Russia from Sweden, England and Germany. Some of these varieties were evaluated in practice, for example Pobeda (Victory) and Zolotoi dozhd (Golden Rain) from Sweden and Lokhovski, Leitevitski and Dippe from Germany, and were also used in oat breeding. The varieties Maganski, Sovetski, Onokhoiski 547 and Yakutski, were developed by selection from local samples. The varieties Oryol, Tayezhnik, Falenski I, Tchernigovski, Yaroslavski 15 and many others were also selected from an assortment of foreign material.

Appreciation of the developing requirements of oat breeding for conditions of intensive cultivation emphasised that attention should be given to ensure that all the best domestic and foreign selections were added to the collection of VIR. Now, a comprehensive range of the genetic variability within both domestic and wild oats, of importance for oat breeding, is represented in the VIR collection.

In the last ten years the Vavilov Institute has collected further valuable material representing large genetic diversity from various ecological zones and of breeding importance. Among these are heat and drought-resistant forms from North Africa and Latin America. The fund of disease resistant sources has been considerably supplemented from North America.

Study of this world oat collection through the network of experimental stations and bases of VIR, which are located in different soil-climatic zones of the USSR, makes it possible to systematically select and recommend to breeders the appropriate starting material for programmes from which to develop varieties to satisfy local requirements.

3. VARIETY DEVELOPMENT

In the period 1976-1983, 30 new varieties were developed in the Soviet Union and 28 of these were developed from the base of the VIR collection. Selectionists are directing their efforts to the development of varieties with a high degree of plasticity, suitable for intensive agriculture and which have a high potential for productivity. New varieties surpass their predecessors in productivity and are characterised by a greater adaptability to local conditions. The level of productivity at some testing sites reaches 7.5-9.3 tonnes per hectare.

Some of the increase in productivity is due to the plasticity of the varieties which have tolerance to unfavourable factors of the environment, particularly to drought, to which the greater part of the Soviet Union is subject. Oat varieties differ from each other in their degree of drought and heat resistance. The most promising are the byzantine oats from the Mediterranean genetic centre. The variety L'govski 1026, bred from byzantine oats, occupies considerable areas in different zones of the USSR. It is characterised by a high level of drought resistance and high plasticity. From L'govski 1026 in combination with varieties from Belgium, France and USA the high-yielding varieties L'govski 78, Mirny, Horizont and Sinelnikovski 21 have been developed.

The local adapted varieties of the steppe regions of the USSR, West and East Siberia, are characterised by high drought resistance. Now new material, which stands out because of even higher levels of drought resistance, has been recommended for cultivation. New breeding material which has increased drought resistance includes the varieties Junon, Anita (Belgium), Fany, Flamo (France), Turra (Austria), Mariner, Niagara and Chief (USA).

High resistance to lodging is one of the basic requirements of any variety. The variety Astor (Netherlands) is characterised by the greatest degree of resistance to lodging and is among the varieties recommended for cultivation. Amongst the Soviet lodging-resistant oat varieties, Isetski and Alo should be mentioned.

Other varieties recommended to be included in breeding programmes include those with short straw as well as resistance to lodging: examples are, Uralski karlik (Ural dwarf), Gnom M (USSR), Palu, Diane (Belgium), Tiger, Permit (FRG), Angus, Mapua, Maris Elf (UK), K-13013 (USA) and D-104 (Peru).

The requirement for high-yielding, early-ripening varieties for Siberia is being successfully solved by Kolpashevski and Tayezhnik; varieties for the non-blackearth zone of the RSFSR - by Kirovski and Falensky 3; varieties for the Southern and South-eastern region by Kubanski and Sinelnikovski 28. In their genealogies are varieties of Belgian, German and Australian origin. For the first time in Soviet selection work varieties have now been developed which are resistant to crown rust: for example, Drug, L'govski I and Omski kormovoi. The fund of varieties resistant to loose smut has been supplemented by Kubanski, Kirovski, Ruslan and Tayezhnik. In their family trees are varieties of American, Canadian, German and Australian origin.

Sources of complex resistance to crown rust and loose smut have been found amongst foreign material: Holden (USA), K-13360 (Mexico), D-20 (Peru), A-3 (Equador) and K-11528 (Algeria). With resistance to powdery mildew: Maris Tabard (UK) and Salem (USA). With resistance to the oat nematode: Krasnodarski 73, L'govski 1026, Minski 17 (USSR), Orient (Australia), Manod (UK) and Markton (USA).

Breeding and research has been intensified on the feeding and nutritional qualities of oat grain. Uspeh, the first variety of hull-less oats to be developed has been recommended for cultivation. The protein content of Uspeh ranges up to 20% and the lysine content in the protein over 4.5%. The varieties K-6060 (USSR) Garland and Clintland (USA) have been used in breeding as donors of increased protein content in the grain.

Specialised varieties have been bred for green fodder production, these include: Zelyony, Uzbek broad-leaved, Omsk kormovoi and Ural. The yielding capacity of these varieties is 40-50 tonnes green mass per hectare.

OAT BREEDING AND PRODUCTION IN HUNGARY

A. PALÁGYI
Cereal Research Institute, H.6701 Szeged, Hungary

1.INTRODUCTION

The area sown to oats in Hungary in the 1950s was well above 200,000 ha. By 1960 it had decreased by half and the acknowledged reason for this has been the diminishing number of draught horses. The area of oats is now 40,000 ha, which is only 0.7 to 0.9% of the total arable land; a slight increase compared to the 1970s. Total oat production has shown a proportionately greater increase, and average yields are now over 2.4 t/ha. This increase is due partly to recently introduced varieties and partly to improvement in cultural practices.

Only spring varieties are now grown commercially. These include Leanda, Perona (from the Netherlands) Solidor (from the GDR) Szegedi korai and GK3 (developed in Hungary).

Oats produced in Hungary are, almost totally, fed to animals. The quantity used for human consumption is insignificant.

Hungary is not a typical oat-producing country. It lies south-east of the main oat-producing regions of Europe and the dry, continental climate of the Carpathian Basin does not normally induce high average yields. Nevertheless, in some years, yields of more than 4 t/ha are attained in western areas of the country, where there is more rain.

2. PRACTICAL BREEDING RESULTS

Oats have been bred at the Cereal Research Institute in Hungary since 1970. The main objectives of our breeding programme are to breed improved, adaptable varieties, resistant to disease and of high nutritional value.

The first significant result of this work was the official registration of Szegedi korai, a variety which ripens 10 to 12 days earlier, has better standing ability and higher grain quality than the control variety Leanda. This variety was selected from an F$_3$ bulk population which originated from crosses of genotypes with 'continental character'. Another variety GK3, from a Nestor x Astor cross, was registered in 1985 and seed is being multiplied up. There has been an encouraging increase in the demand for naked oats for human consumption in the past few years. Our basic material is Táplani csupasz, an early, naked variety which has been crossed with sativa types. In this way we are trying to develop naked lines with higher productivity.

We can report only moderate initial success in breeding winter oats. Few genotypes can tolerate the cold winters of mid-Europe without considerable loss by winter kill, and this is why the two Mediterranean type varieties, Ujszegedi and Karcagi, which were bred in Hungary in the 1960s did not become widely grown in our country. On the basis of several years' observations, the lines with the best winter hardiness seem to originate from crossings of Dubois, Checota, Windsor, C.I. 7300 and the

cross combinations of winter bulk populations from the USA. It is
expected that, by using this basic material, we shall soon have candidate
varieties with acceptable winter hardiness, which will ripen earlier than
spring oat varieties (of basic importance in our summer) and which also
give higher yields of better quality green fodder.

3. MAIN RESEARCH RESULTS
Besides our practical breeding work, we are also studying the
correlations between the components of yield. We are trying to draw some
general conclusions which can help us in our selection work. Kernel
number per spikelet, the palea: caryopsis ratio, percentage lemma and
harvest index are amongst the most important quantitative characteristics
studied. I wish to discuss some of the more important results of these
studies.
It is well known that the spikelets of sativa type oats contain three
florets. The tertiary floret, which develops last, is in most cases
rudimentary or is not fertilised. From the sparse data reported in the
literature, it can be concluded that seed set of spikelets is a
multiallelic trait which is characteristic of a genotype but one that can
be modified by epistatic effects and environmental circumstances, for
example, plant density, nutrition, microclimate, etc.(1,2).
The requirements for grain production and processing are different from
those for forage production(3). According to official Hungarian seed
standards, only seeds retained by a 1.8 mm mesh sieve can be sown. The
grain fraction which goes through such a sieve comprises poorly developed
grains of low 1000 grain weight.
From an experiment examining 35 oat varieties, Takeda and Frey(1)
reported that primary, secondary and tertiary seed setting had
heritability values of 57.2%, 72.0% and 68.3%, respectively. The most
frequent spikelet type (69.4% of total) was that in which only primary and
secondary seeds developed; tertiary seeds being sterile. Only 20% of the
35 genotypes studied in Iowa developed a tertiary floret. The conclusion
was drawn that, under favourable environmental conditions, the tertiary
seed can form an effective supplement for photosynthate and provide the
possibility of increasing productivity in oat varieties which have this
tendency to develop tertiary seeds.
We analysed spikelets of several varieties over several years; 50
panicles per variety per year(4). Grains of spikelets were carefully
separated into primary, secondary and tertiary seeds. The proportions of
the individual fractions and the fractions remaining on sieves with meshes
of 2.2 mm and 1.8 mm were also determined. Thousand grain weights and
lemma percentages were then calculated (for the latter the caryopses were
hand-shelled).
The data for three varieties are given in Tables 1 and 2. The mean
values for the quantitative parameters of grain yield fractionated in two
ways, when weighted according to percentage of total, were very similar.

From the data the following conclusions can be drawn:
- There is considerable difference between varieties tested with regard
  to the individual fractions.
- Leanda produced the highest number of tertiary seeds, but had the
  lowest 1000 grain weight and the highest percentage lemma. More than
  three-quarters (78.13%) of the tertiary seed fraction passed through a
  1.8 mm mesh sieve, and thus these seeds are not suitable for sowing.

- Szegedi korai developed the lowest number of tertiary seeds (0.79%), had a high 1000 grain weight and a low lemma percentage. Less than half of the tertiary seeds passed through a 1.8 mm mesh   sieve.
- 1000 grain weights and lemma percentages decrease markedly in the sequence-primary seed ⟶ secondary seed ⟶ tertiary seed. 1000 grain weight decreases and lemma percentage increases in the 2.2 mm ⟶ 1.8 mm ⟶ less than 1.8 mm sieve size categories. The conclusion to be drawn is that all three sieve fractions (even the 2.2 mm) contain tertiary seeds.
- Szegedi korai and GK3 have similar percentages for the fraction smaller than 1.8 mm (about 10%). Leanda on the other hand, produced more than 16% seeds unsuitable for sowing, had a low 1000 grain weight and a high lemma percentage.
- For GK3 the proportion of the 2.2 mm fraction was the largest in spite of its high content of tertiary seeds.

TABLE 1. Proportions of seed types (% total), 1000 grain weights and percentages lemma, of primary, secondary and tertiary seeds of spikelets of three varieties, 1979-1982

Fraction	Leanda			Szegedi korai			G K 3		
	% total	1000 gr. wt. (g)	% lemma	% total	1000 gr. wt. (g)	% lemma	% total	1000 gr. wt. (g)	% lemma
Primary seed	51.67	30.26	36.17	59.43	33.51	31.27	56.70	31.84	32.93
Secondary seed	38.46	21.83	33.04	39.78	24.41	26.02	38.69	23.41	29.27
Tertiary seed	9.87	14.15	29.42	0.79	16.10	19.74	4.61	14.69	25.16
% Tertiary seed below 1.8 mm	78.13			49.20			63.09		
Weighted mean*		25.44	34.30		29.76	29.09		27.79	31.15

*Weighted according to percent total

In summary: the development of tertiary seeds, is limited and has a negative influence on the development of primary and secondary seeds, as seen even in the variety Leanda. Thus, the favoured ideotype amongst sativa varieties is the genotype which develops two kernels per spikelet of approximately the same value; the optimum being without tertiary seeds. (The majority of genotypes examined by Takeda and Frey were of this type). In the present study, based on the weighted means of the individual fractions, Szegedi korai has the most advantageous characteristics.

The yields of the varieties tested were similar. An endeavour to increase yield by selecting lines which develop tertiary seeds is only a possibility if there are no accompanying decreases in the quantitative and qualitative parameters of the primary and secondary seeds. This proviso is especially valid when growing for seed production.

TABLE 2. Proportions of grain sizes (% total), 1000 grain weights and percentages lemma of three grain size fractions of three varieties, 1979-1982

Fraction	Leanda			Szegedi korai			G K 3		
	% total	1000 gr. wt. (g)	% lemma	% total	1000 gr. wt. (g)	% lemma	% total	1000 gr. wt. (g)	% lemma
2.2 mm	44.81	31.07	32.80	50.23	35.06	27.22	54.77	33.16	29.87
2.2 - 1.8 mm	38.73	23.29	35.41	38.81	26.42	31.25	34.59	23.81	32.27
1.8 mm	16.46	14.21	37.22	10.96	16.60	32.34	10.64	14.58	35.09
Weighted mean*		25.28	34.54		29.68	29.34		27.95	31.25

*Weighted according to percent total

The heritability of harvest index was studied in variety maintenance experiments involving the two early spring oat varieties Szegedi korai and GK2(5). Harvest indices were determined in three experiments over two years; 1979 (two planting dates) and 1980. Then using a method based on the analysis of variance of progeny groups, a broad sense calculation was carried out and genetic progress estimated (Table 3).

TABLE 3. Numerical values of variance components for harvest index (HI), heritability (h^2) and estimated genetic progress (G_s)

	HI (%)	h^2 (%)	G_s (%)	Variance components		
				o_g^2	o_{gy}^2	o_e^2
Szegedi korai	51	93	4.65	7.49**	1.11**	1.99
GK 2	48	13	0.38	0.36	5.73**	4.62

o_g^2 = genotypic variance o_e^2 = error (environmental) variance

o_{gy}^2 = interaction variance for genotype x year and genotype x planting date; ** = Significant at 1% level

Significant differences between the varieties were found and Szegedi korai had the higher H1 heritability. Positive selection for harvest index may be effective in lines of this variety, making genetic progress. On the other hand, in the case of selection in GK2 (and in the population produced from the selected lines), practically no progress can be expected because of the lower genetic variance.

Both to aid the choice of breeding methodology to be used and also to increase the effectiveness of selection aimed at increasing grain yield, an understanding of harvest index and its heritability is required.

Presumably, there will also be some additional differences between genotypes in respect of this important quantitative parameter. We need to find genotypes with high heritability of harvest index and by using these in our breeding programmes achieve increases in grain yield.

Finally, I would like to refer to a series of trials we started in 1977, in which grain quality of our oat varieties was compared to the nutritive value of the grains of the other fodder cereals, barley and sorghum. It was shown that oats are the most valuable feed. We can, with certainty, find oat cultivars which will fully satisfy the demands of both monogastric and polygastric animals(6,7).

REFERENCES

1. Takeda K and KJ Frey: Crop Sci. 20, 771-774, 1980.
2. McBratney BD and KJ Frey: Cereal Research Communications 11, 91-97, 1983.
3. Palágyi A and L Benke: Vetőmaggazdálkodás 7, 54-62, 1980.
4. Palágyi A: Cereal Research Communications 11, 269-274, 1983.
5. Palágyi A: Növénytermelés 23, 393-398, 1983.
6. Palágyi A and I Herold: Acta Agr. Acad. Sci. Hung. 33, 130-141, 1984.
7. Palágyi A and I Herold: Acta Agr, Acad. Sci. Hung. (in press), 1985.

THE OAT CROP IN THE UK

D. A. LAWES
Welsh Plant Breeding Station, Aberystwyth, Dyfed, Wales, UK

1. INTRODUCTION

During the last 40 years there has been a marked reduction in the area of oats sown in the UK and this decrease has been largely by a transfer to barley. The reasons for this change have been variously suggested to include the following:

1. The demise of horses as on-farm power.
2. The comparatively low yield of metabolisable energy per hectare produced by oats.
3. The replacement of oats by barley as a major constituent of animal feed.
4. The advent of new short-strawed barley varieties which were easier to manage and harvest.
5. Lack of availability of improved oat varieties.

2. THE INCREASE IN YIELDS AND THE CONTRIBUTION OF PLANT BREEDING

If the latter factor is true, lack of progress by plant breeders is implied. The general increases in cereal crop yields since 1947 and the contribution of new varieties to these increases have been discussed by Silvey(1) and her data provide some interesting evidence.

Table 1 presents Silvey's figures indicating the improvements in England and Wales national average cereal yields between 1947 and 1978.

TABLE 1. England and Wales national average yields
(5 year moving averages as percent of 1947 figures)

	Wheat	Barley	Oats
1947	2.4 t/ha = 100	2.3 t/ha = 100	2.15 t/ha = 100
1957	135	131	120
1967	162	156	167
1978	205	176	187
Total % increase	105	76	87

Taking the average yields produced in 1947 as a base-line, it is shown that over the ten year period 1957-67 the increases in average yields were: wheat 35%, barley 31% and oats only 20%. Over the next ten year period there were further increases in wheat yields of 27%, barley 25% and oats 47%. The next eleven years up to 1978 showed further additions of 43% for wheat, 20% for barley and 20% for oats. Comparing oats with its major competitor barley shows that, following less progress in the decade 1947-1957, increase in average yields since 1957 has been greater with oats than it has with barley. Silvey(1) carried out further analyses

274

to produce estimates of the contributions of new varieties to these increased cereal yields. Table 2 presents her estimates for the same periods of time. Again, the comparison we are most interested in is between barley and oats.

TABLE 2. Contribution of 'new varieties' to increased cereal yields
(5 year averages as percent increases over 1947)

	Wheat	Barley	Oats
1947	2.4 t/ha = 100	2.3 t/ha = 100	2.15 t/ha = 100
1947-57	13	16	3
1957-67	13	5	12
1967-78	37	11	10
Total (31 years)	63	32	25

In the first decade, 1947-57, there was a 16% increase in barley yields due to the advent of new varieties. During the same period there was little progress, only 3%, in increasing the yield of oats by release of new varieties. However, progress in increasing grain yield of oats by breeding during this period would have been hampered by two constraints.

a) Prior to this, oat varieties had been developed not only for grain but also for the value of the straw.
b) Grain quality, as assessed by grain appearance, had previously been considered to be of paramount importance and had been accorded higher priority than yield.

Breeding programmes had to be reorientated so that yield of grain was the major objective.

From the figures presented for the second decade, it can be seen that the 12% progress achieved in oats was greater than the 5% progress made in barley. It was during this period that new oat varieties emerged which had markedly improved straw character and were thus better able to utilize nitrogen fertiliser and still remain standing for harvest. For the last period, 1967-78, progress in both barley and oats was similar. Overall therefore, and particularly latterly, it can be seen that progress in oat breeding compares quite favourably with the progress made in barley. When consideration is given to the resources and effort expended on oat breeding and research in comparison with barley, then the improvements by oat breeding appear to be quite respectable. The contribution made by the development of new varieties of oats has been of the order of 1% per annum over recent years.

3. THE PRESENT SITUATION

The present situation concerning the UK oat crop can be itemised as follows:

1. Oats accounts for about 3% of the annual UK total of 4 m hectares of cereals (48% in the 1940s)
2. The present 110-140,000 hectares produce 450-600,000 tonnes of oats annually.
3. Of this some 140-160,000 tonnes go for milling (human consumption), and the remainder for animal feed.
4. Of the oats used for milling, up to 40,000 tonnes may be imported in years when there is not an adequate supply of good quality home grown oats.

In the UK both winter and spring sown oats are grown and these figures include both crops. On a national basis the proportion of winter sown oats to spring sown is approximately 50:50 but there is considerable variation by region. Almost no winter oats are grown in northern England or Scotland. In the more southern areas it is the spring sown crop which continues to decrease; there has been less decline in the area sown to winter oats.

It is interesting to examine the more recent trends in the areas sown to cereals in the UK (Table 3).

TABLE 3. UK Cereal crop areas ('000s hectares)

	Wheat	Barley	Oats
1980	1441	2330	148
1981	1491	2327	144
1982	1663	2222	129
1983	1695	2143	108
1984	1939	1978	106
1985	1890	1980	142

Source of data: MAFF(2)

Since 1980 the area sown to wheat has been on the increase. Barley, on the other hand, has now entered a period of decline. The oat area continued to decrease until this year but we now see a sudden upsurge in interest. The 1985 figure is 34% up on last year and takes the crop back to the level that it was in 1981. What is the reason for this recent increase? It has been apparent over the last two or three years that the value of oats in a cropping system has become increasingly appreciated. Oats have been shown to demonstrate real advantages in crop rotations, particularly in the control of pests and diseases of other crops. For instance, it has been shown that oats can reduce take-all and therefore increase subsequent wheat yields. Oats have also been shown to control root-knot nematode and thus increase barley yields in grass/cereal rotations. In a multiple factorial experiment carried out at Rothamsted(3) over recent years, wheat following oats has been shown to yield over 2 tonnes per hectare more than wheat following barley. This advantage has been ascribed to the fact that a preceding oat crop provided a site virtually free of take-all. This reawakened appreciation of the value of crop rotations is also important in the context of potential interest in farming systems which are not so dependent on high inputs.

The gross margin to be obtained from a crop is obviously part dependent upon the market value of the produce. It can be shown that, with the price strutures pertaining over recent years, oats can compare favourably with other cereals. The following data (Table 4), extrapolated from Nix(4) illustrates this point.

These data give the relative costs and gross margins to be obtained from five tonne per hectare crops. The variable costs expended on winter barley are less than those on winter wheat. This difference is due to slightly lower inputs in added fertiliser and chemical sprays. The variable costs expended on winter oats are lower still; this is due to a further decrease in the level of applied fertiliser and an even more

TABLE 4. Winter cereal production (5 tonne per hectare crops)

	Winter wheat	Winter barley	Winter oats
Variable costs (£)	200	170	135
Gross margin (£)			
Feed	351	367	390
Milling/malting	410	429	440

Wheat @ £110/t feed; £120/t milling
Barley @ £107/t feed; £120/t malting
Oats @ £105/t feed; £115/t milling

marked decrease in the number of fungicidal sprays normally applied. With the value of feed oats approximately equivalent to that of feed barley, winter oats can achieve a gross margin slightly in excess of either winter feed wheat or winter feed barley. If the comparison is between oats for milling and wheat for baking or barley for malting, the differential in favour of oats still exists. Figures can be similarly presented for spring cereals but, although spring oats still compare fairly favourably with spring barley, the differential is slightly in favour of spring barley.

A further factor which almost certainly had an influence on the recent increase in the size of the oat crop was the price structure for grain last autumn. For a period oats were attracting a higher price than either wheat or barley and, at a time when farmers are looking for ways to diversify their enterprises, this most certainly attracted interest. Whether the increase in area sown will be maintained remains to be seen. Unfortunately by experience we know that a change in price structure in favour of a particular crop may only be temporary; inevitably as supply grows prices fall.

Nothing encourages interest more than experiences of bumper crops and the more frequent occurrence of high yielding crops has inevitably improved the image of oats. There are now fairly frequent examples of oat crops which yield over 8 tonnes per hectare and a few have approached the 10 tonne mark. This level of yield is of the order achieved by the best wheat and barley crops.

4. FUTURE PROSPECTS

To come to a conclusion about the future role of oats in UK agriculture we need to consider several factors. To make the crop more attractive agriculturally we must continue our efforts to improve some of its agronomic attributes. If there is to be a future for spring oats then the development of earlier maturing varieties is essential. This pertains to the whole of the UK but is particularly true for Scotland and the north where at present harvest is far too late. In winter oats we need increased winter hardiness, so that the crop can be extended into the north of England and to Scotland, as well as to remove the more general possibility of loss by winter kill which is still too liable to occur with present varieties. It is general experience that oat crops are more difficult to harvest than either wheat or barley and with both winter and spring crops there is need to shorten the straw and improve lodging resistance. The whole image of the oat crop must be improved: there are too many memories of difficult to manage crops and high crop losses.

Of pre-eminent importance however is the question of crop use and market
outlets. There is an increasing awareness of the value of oats as human
food and this should lead to some increase in the quantity required by the
millers. However, it is likely that this will only lead to a modest
increase in requirement; perhaps 10% or so has been suggested. Milling
will therefore still remain a limited market outlet. If at all possible,
of course, we must try and meet our total requirement for milling from the
home-grown crop and not have to resort to importing.

To achieve any large expansion in the crop, oats must break into
processing and the animal feed trade. For this purpose the future depends
upon the delicate balance of grain quality (nutritional constituents) and
price structure. As far as the farmer is concerned his receipts are
dependent on a) the price of grain per tonne and b) the yield per hectare
he is able to produce. As it is the intrinsic value of the grain which
will dictate the relative price of oats in comparison with other
commodities, the objective must be to produce oats of the composition that
would interest the processors and compounders at a competitive price.

A point of major thrust of our research must therefore be into grain
quality and recognition of the factors required by processors and the
animal feed trade. This could then lead to the development of varieties
to meet these requirements. Oats of high feeding quality would also find
a niche for 'on-farm' use. Another paper at this conference has reported
the work on naked oats at the WPBS. This is an attempt to produce a crop
of high quality grain which will hopefully attract new market interest.
Rhiannon, the new naked spring oat, has an oil content of 8.5%. In winter
naked oats we are confident that we could achieve 10% and we intend to
pursue this objective.

Only when market outlets are firmly established can we expect to see any
real expansion in the area of oats sown. Although agriculturally oats as
a farm crop may seem a good idea, interest at the farm level cannot
generate impetus if there is no reliable market outlet available. The
recognition of a firm requirement for wheat and barley in the EEC led to
the establishment of intervention price guarantees. Oats have not been
considered eligible for this preferential treatment and their exclusion
from the intervention price system has inevitably contributed to the
present lack of enthusiasm for growing the oat crop.

REFERENCES

1. Silvey Valerie: J. Natl Inst. Agric. Bot. 15, 399-412, 1981.
2. Ministry of Agriculture, Fisheries and Food: Various issues of the
   Animal Review of Agriculture, HMSO:London, 1980-1985.
3. Rothamsted Experimental Station. Rep. of the Rothamsted Exptl Stn
   for 1984, pp.23-38, 1985.
4. Nix John: Farm Management Pocketbook, Wye College, University of
   London, 1985.

SESSION VI. WORLD STATUS OF OATS AND BIOLOGICAL CONSTRAINTS TO
INCREASED PRODUCTION
CHAIRMAN'S COMMENTS AND SUMMING UP

BENGT MATTSSON
Svalöf AB, S 268 00 Svalöv, Sweden

The oat hectarage of the world has decreased by half over the last 40 years. This has been mainly due to decreased areas in USA, Canada and Western Europe; oats having been replaced by other crops such as maize and barley. This downward trend now seems to have declined at least in parts of Europe and, as reported by Lawes, there might even be a slight interest in increasing oat production in UK. In USSR, which accounts for about one third of the world oat area, there has been less variation over recent years and the oat area is now greater than it was during the 1960s. In papers given by Floss, Mishra and Palágyi we have also learned that oat areas are increasing in Brazil, India and Hungary, respectively. Thus, there seems to be reasonable evidence suggesting that the decline in the oat area is easing and may even cease.

Oats are mainly cultivated for grain production. However, in some countries the crop is cut green for direct feeding or for silage. Thus, according to Küüts, special cultivars suitable for green fodder have been released in Russia. From Mishra we learned that 85-90% of oats in India are cut green. In the south of Europe and in the Middle East oats are also mainly cultivated for green fodder. Floss mentioned that in Brazil, oats are sown for grazing and then later harvested for grain.

Whatever type of feed grain-oats are used for, the greatest part of production is used on the producing farms. Some trade takes place within countries but the volume exported is small. As there is no especial interest in oats, the price of oats more or less follows that of barley or maize.

Grain oats are mostly used for feeding animals and to a lesser extent for human consumption. Schrickel (in an earlier session) gave the figure for world use as human food as 17.9% in 1984, which was slightly higher than 16.3% in 1975. There have also been changes within countries. Thus, in the USSR, human consumption accounted for 31.1% in 1975 and 22.5% in 1984. Corresponding percentages for USA were 6.9% in 1975 and 12.0% in 1984. However, these figures are not necessarily indicative of real changes in the volume consumed as the total amount of grain available may also have changed. As overall production decreases the proportion used for human consumption will increase. Thus Lawes reported that, in the UK, about 140,000-160,000 tonnes are used for milling out of a total of 450,000-600,000 tonnes, or over 25% of production. There is a growing interest in oats for human consumption in many countries. This is largely due to recognition of the good quality of oat protein and oil as well as the soluble fibre, which give oats a valuable role in healthy diets.

What are the limits of oat production and what can be done to improve yields further? Forsberg discussed the lengths of the growing seasons in different countries. In USA, for instance, the average period for plant growth is about two thirds of that in Europe. This is of course one reason for the lower yields in USA. Different climatic conditions have a big influence and the effects of stress due to drought and heat, in particular, were discussed by Forsberg and Küüts.

Another major stress factor in many countries is attack by various parasites. The need for good resistances was discussed in several papers. Brinkman gave an interesting report of the excellent Quaker Oats South American Programme. Nurseries are planted in various locations throughout North, Central and South America. Each nursery contains lines with different genes for resistance to rust and tolerance to BYDV. By observations of the severity of disease attack and then the combining of different sources of resistances in breeding programmes new cultivars have been developed. Thus Floss has released five new cultivars in Brazil and three more are on the way. Another good example of cooperation to speed up breeding programmes for the development of resistant varieties was given by McEwan. Canadian breeders are getting an additional generation per year by growing an 'out of season' crop in New Zealand. In addition New Zealand breeders also have the opportunity of making selections within this material. Already one variety has been released in New Zealand from this programme.

To improve the oat plant, a number of morphological and physiological characters need to be considered and Forsberg gave examples. Information on the three yield components - number of panicle-producing tillers per unit area, number of seeds per panicle and seed weight - could be of value, particularly in planning crosses. According to Palágyi the number of grains per panicle, the ratio between palea and caryopsis and the harvest index, are some of the more important characters to consider.

The session ended with a discussion as to how to make oats more popular. As Lawes pointed out, farmers expend less inputs on the oat crop as compared to barley and wheat as they need not apply so much nitrogen and such comprehensive fungicide treatments are not required. Oats are also a valuable crop in rotations and Lawes gave an example indicating 30% higher yield of wheat after oats compared to wheat after barley. This is in agreement with the results of Swedish investigations in which oats have proved to have the same rotational value as a preceding oil crop. The value of oats for human consumption needs further attention and firm evidence of their dietary qualities would increase interest in the crop. In comparison with other cereals, there are fewer breeders and researchers working on oats. If more resources were allocated, there would undoubtedly be more rapid progress with the improvement of the crop and also in the exploitation of oats as a commodity.

AUTHOR INDEX

LIST OF PARTICIPANTS

Altosaar I,
Biochemistry Dept,
Univ. of Ottawa,
Ottawa, Ontario, K1N 9B4,
Canada

Ambrose MJ,
John Innes Institute,
Colney Lane,
Norwich,
Norfolk, UK.

Anderson HM,
Long Ashton Res. Station,
Univ. of Bristol,
Long Ashton,
Bristol, BS18 9AF, UK

Bacon R,
Dept. of Agronomy,
Univ. of Arkansas,
Fayetteville,
AR 72701, USA

Baldwin JH,
Agric. Dev. & Advisory Service,
Government Buildings,
Cambridge, UK.

Barr AR,
Northfield Res. Lab. & Res. Centre,
Dept. of Agric., Adelaide 5001,
South Australia.

Bickelmann U,
Lufa-Augustenberg,
Nesslerstr 23,
Postfach 430230,
D-7500 Karlsruhe 41,
Federal Republic of Germany.

Black RA,
Border Oats Ltd.,
Edington Mill,
Chirnside, Duns,
Berwickshire,
UK.

Brace Jane,
Dept. of Biological Science,
Manchester Polytechnic,
Oxford Road,
Manchester, UK.

Breese EL,
Welsh Plant Breeding Station,
Aberystwyth,
Dyfed, Wales, SY23 3EB,
UK

Brinkman MA,
Dept. of Agronomy,
Univ. of Wisconsin,
Madison, WI 53706,
USA.

Brouwer JB,
Victorian Crops Research Institute,
Horsham,
Victoria,
Australia 3400.

Brown CM,
Dept. of Agronomy,
Univ. of Illinois,
Urbana, IL 61801,
USA.

Brown PD,
Agric. Canada Research Station,
195 Dafoe Road, Winnipeg,
Manitoba, R3T 2M9, Canada.

Bunch RA,
Dept. of Agronomy,
Univ. of Wisconsin, Madison,
WI 53706, USA

Buraas T,
Bjorke Forsokgard,
2344 Ilseng,
Norway.

Burrows VD,
Ottawa Research Station,
Ottawa,
Ontario, K1A OC6,
Canada.

Catherall PL,
Welsh Plant Breeding Station,
Aberystwyth, Dyfed,
Wales, SY23 3EB,
UK.

Chorlton Elizabeth,
Welsh Plant Breeding Station,
Aberystwyth, Dyfed,
Wales, SY23 3EB,
UK.

Clamot G,
Station d'Amélioration des Plantes,
Rue du Bordia 4,
B-5800, Gembloux,
Belgium.

Clifford BC,
Welsh Plant Breeding Station,
Aberystwyth, Dyfed,
Wales, SY23 3EB,
UK.

Clothier RB,
Welsh Plant Breeding Station,
Aberystwyth, Dyfed,
Wales, SY23 3EB,
UK.

Codd TM,
Long Ashton Research Station,
Weed Research Division,
Begbroke Hill, Yarnton,
Oxford, OX5 1PF,
UK.

Cook R,
Welsh Plant Breeding Station,
Aberystwyth, Dyfed,
Wales, SY23 3EB,
UK.

Eskilsson L,
Weibullsholm PBI,
Box 520,
S-261 24 Landskrona,
Sweden.

Evans Rhiannon,
Twyford Seeds Ltd,
Scotts Farm,
Kings Sutton,
Banbury, Oxon,
UK.

Exley Diane,
MAFF,
Great Westminster House,
Horseferry Road, London,
SW1P ZAE, UK.

Flahavan J,
E Flahavan & Sons Ltd,
Kilmacthomas,
Co Waterford,
Ireland.

Floss EL,
School of Agronomy,
Univ. of Passo Fundo,
Passo Fundo, RS,
Brazil.

Forsberg RA,
Dept. of Agronomy,
Univ. of Wisconsin,
Madison, WI 53706,
USA.

Freed R,
1014 Touraine,
East Lansing,
Michigan,
MI 48823, USA.

Frey KJ,
Agronomy Department,
Iowa State University,
Ames,
Iowa 50011,
USA.

Fritz Sue E,
Cornell University,
Ithaca,
NY 14850,
USA.

Ganssmann W,
Peter Kölln,
Mühlenwerke am Hafen,
D-2200 Elmshorn,
Federal Republic of Germany.

Gooding R,
Dept. of Agronomy,
Ohio State University,
Ohio Agric. Res. & Dev. Centre,
Wooster, OH 44691,
USA.

Grant B,
North Eastern Farmers Ltd,
Bannermill,
Aberdeen, AB9 2QT,
UK.

Griffiths TER,
Welsh Plant Breeding Station,
Aberystwyth, Dyfed,
Wales, SY23 3EB,
UK.

Gullord M,
Agricultural Experiment Stn.,
Apelsvoll,
2858 Kapp,
Norway.

Harder DE,
Agric. Canada Res. Station,
195 Dafoe Road,
Winnipeg,
Manitoba, R3T 2M9,
Canada.

Hayward MD,
Welsh Plant Breeding Station,
Aberystwyth, Dyfed,
Wales, SY23 3EB,
UK.

Hees W,
Federation of the Association of
  Oat & Barley Millers in the EEC,
Postfach 190165, D-5300 Bonn 1,
Federal Republic of Germany.

Holden JHW,
IBPGR/ECP/GR,
Crop Genetics Resources Centre,
FAO, Rome,
Italy.

Jedlinski H,
USDA-ARS,
Dept. of Plant Pathology,
Univ. of Illinois,
Urbana, IL 61801,
USA.

Jones DM,
Welsh Plant Breeding Station,
Aberystwyth, Dyfed,
Wales, SY23 3EB,
UK.

Jones ERL,
Welsh Plant Breeding Station,
Aberystwyth, Dyfed,
Wales, SY23 3EB,
UK.

Jones IT,
Welsh Plant Breeding Station,
Aberystwyth, Dyfed,
Wales, SY23 3EB,
UK.

Jones JE,
Welsh Plant Breeding Station,
Aberystwyth, Dyfed,
Wales, SY23 3EB,
UK.

Klinck HR,
Plant Science Department,
Macdonald College,
Ste Anne de Bellevue,
Quebec, H9X 1C0,
Canada.

Kostov KD,
Plant Introduction Institute,
Sadovo,
Plovdiv,
Bulgaria.

Küüts HD,
Jõgeva Plant Breeding Station,
Estonia 202350,
USSR.

Ladizinsky G,
Faculty of Agriculture,
The Hebrew University,
P.O. Box 12, Rehovot,
Israel.

Lafever HN,
Department of Agronomy,
Ohio Agric. Res. & Dev. Centre,
Wooster,
OH 44691,
USA.

Lambert T,
Twyford Seeds Ltd,
Scotts Farm,
Kings Sutton, Banbury,
Oxon, UK.

Lampinen R,
Länsi-Hahkiala,
SF-14700 Hauho,
Finland.

Lawes DA,
Welsh Plant Breeding Station,
Aberystwyth, Dyfed,
Wales, SY23 3EB,
UK.

Lea JE,
Morning Foods Ltd,
North Western Mills,
Crewe, Cheshire, CW2 6HP,
UK.

Lea P,
Morning Foods Ltd,
North Western Mills,
Crewe, Cheshire, CW2 6HP,
UK.

Leggett JM,
Welsh Plant Breeding Station,
Aberystwyth, Dyfed,
Wales, SY23 3EB,
UK.

Leist N,
Lufa-Augustenberg,
Nesslerstr 23,
Postfach 430230,
D-7500 Karlsruhe 41,
Federal Republic of Germany.

Lewis DA,
Welsh Plant Breeding Station,
Aberystwyth, Dyfed,
Wales, SY23 3EB,
UK.

Luke HH,
Plant Pathology Department,
University of Florida,
Gainesville, FL 32611,
USA.

Lyons RD,
Odlum Group Ltd,
Sallins,
Co. Kildare,
Ireland.

Lyslov E,
Res. Inst. of Agric. Management
  of Central District of the
  Nonchernozem Zone,
Nemtchinovka, Moscow Region,
USSR.

Maldonado U,
Inst. Nacional de
  Investigaciones Agricolas,
Chapingo,
Mexico.

Mazaraki Maria,
Plant Breed. & Acclim. Institute,
4 Zawila Str.,
Cracow,
Poland.

Marshall HG,
Department of Agronomy,
Pennsylvania State University,
University Park, PA 16802,
USA.

Mattsson B,
Svalöf AB,
S 26800 Svalöv,
Sweden.

McDaniel ME,
Dept. of Soil & Crop Sciences,
Texas A & M University,
College Station,
Texas 77843,
USA.

McEwan JM,
Crop Research Division,
DSIR, Private Bag,
Palmerston North,
New Zealand.

McKenzie RIH,
Agric. Canada Res. Station,
195 Dafoe Road, Winnipeg,
Manitoba, R3T 2M9,
Canada.

McLean R,
Department of Agriculture,
Jarrah Road,
South Perth 6151,
Western Australia.

McMullen MS,
Agronomy Department,
North Dakota State Univ.,
Fargo,
ND 58105,
USA.

Medeiros RB,
III-Cotrijui,
Ijuí,
RS, 98700,
Brazil.

Middleton BT,
Welsh Plant Breeding Station,
Aberystwyth, Dyfed,
Wales, SY23 3EB,
UK.

Mishra SN,
Dept. of Plant Breeding,
GB Pant Univ. of Agric. & Technol.,
Pantnagar 263145,
Nainital, UP,
India.

Morgan DR,
Brit. Oat & Barley Millers' Assoc.
6, Catherine Street,
London, WC2 B5JJ,
UK.

Morikawa T,
Lab. of Genetics & Plant Breeding,
University of Osaka,
Sakai, Osaka 591,
Japan.

Murai K,
Faculty of Agriculture,
Kyoto University,
Sakyo-ku, Kyoto 606,
Japan.

Murphy CF,
USDA-ARS-NPS,
Beltsville,
MD 20705,
USA.

Murphy P,
University of North Carolina,
Raleigh,
NC 27607,
USA.

Palãgyi A,
Cereal Research Institute,
H.6701 Szeged,
Pf. 391,
Hungary.

Papavoine F,
The Quaker Oats Company,
Spuiboulevard 366,
3311 GR Dordrecht,
The Netherlands.

Parry Anne L,
Welsh Plant Breeding Station,
Aberystwyth, Dyfed,
Wales, SY23 3EB,
UK.

Peterson DM,
Department of Agronomy,
Univ. of Wisconsin,
Madison,
WI 53706,
USA.

Postoyko J,
Dept. of Biological Sciences,
Manchester Polytechnic,
Oxford Road, Manchester,
UK.

Reeves DL,
Plant Science Department,
South Dakota State Univ.,
Brookings, SD 57007,
USA.

Rekunen M,
Hankkija Plant Breed. Inst.,
SF-04300 Hyryla,
Finland.

Richardson WG,
Long Ashton Research Station,
Weed Research Division,
Begbroke Hill, Yarnton,
Oxford, OX5 1PF, UK.

Rines HW,
Dept. of Agron. & Plant Genetics,
University of Minnesota,
St Paul, MN 55108,
USA.

Roderick HW,
Welsh Plant Breeding Station,
Aberystwyth, Dyfed,
Wales, SY23 3EB,
UK.

Rogers WJ,
Welsh Plant Breeding Station,
Aberystwyth, Dyfed,
Wales, SY23 3EB,
UK.

Rothman PG,
Cereal Rust Laboratory,
University of Minnesota,
St Paul, MN 55108,
USA.

Saur L,
Inst. Nat. de la Recherche Agron.,
Station d'Amélioration des Plantes,
BP 29 - 35650 Le Rheu,
France.

Scanlan J,
National Oats Company,
1515 H. Avenue N.E.,
Cedar Rapids, IA 52402,
USA.

Scheffel A,
The Quaker Oats Company,
Spuiboulevard 366,
3311 GR Dordrecht,
The Netherlands.

Schrickel DJ,
The Quaker Oats Company,
Merchandise Mart Plaza,
Chicago, IL 60654,
USA.

Sebesta J,
Res. Inst. for Crop Production,
Plant Protection Division,
Ruzyne 507,
16106 Prague 6,
Czechoslovakia.

Semple Jayne T,
Nickerson RPB Ltd,
J. Nickerson Res. Centre,
Rothwell, Lincoln, LN7 6DT,
UK.

Shands HL,
Agronomy Building,
University of Wisconsin,
Madison, WI 53706,
USA.

Simons MD,
ARS-USDA, Dept. of Plant Path.,
Iowa State University,
Ames, IA 50011,
USA.

Smyth W,
Speedicook Ltd, Scarva Road,
Tandragee,
Co. Armagh, BT62 2BY,
Ireland.

Sorrells ME,
Dept. of Plant Breeding,
Cornell University,
Ithaca, NY 14853,
USA.

Souza E,
Cornell University,
Ithaca,
NY 14853,
USA.

Stähle FC,
Hallebaan,
Zandhoven,
Belgium.

Stuthman DD,
Dept. of Agron. & Plant Genetics,
University of Minnesota,
St Paul, MI 55108,
USA.

Taylor HF,
Long Ashton Research Station,
Weed Research Division,
Yarnton,
Oxford, OX5 1PF,
UK.

Taylor M,
Speedicook Ltd, Scarva Road,
Tandragee,
Co. Armagh, BT62 2BY,
Ireland.

Thomas D,
Quaker Oats Ltd,
Bridge Road, Southall,
Middlesex, UB2 4AG,
UK.

Thomas Hugh,
Welsh Plant Breeding Station,
Aberystwyth, Dyfed,
Wales, SY23 3EB,
UK.

Tibelius Christine,
3 Grantham Bank,
Barcome, Nr Lewes,
E.Sussex, BN8 5DJ,
UK.

Tsunewaki K,
Laboratory of Genetics,
Faculty of Agriculture,
Kyoto University,
Sakyo-ku, Kyoto 606,
Japan.

Turner JC,
Quaker Oats Ltd,
Bridge Road, Southall,
Middlesex, UB2 4AG,
UK.

Twigg PA,
Welsh Plant Breeding Station,
Aberystwyth, Dyfed,
Wales, SY23 3EB,
UK.

Valentine J,
Welsh Plant Breeding Station,
Aberystwyth, Dyfed,
Wales, SY23 3EB,
UK.

Vanselow M,
Inst. für Angewandte Genet.,
Universitat Hannover,
Hannover,
Federal Republic of Germany.

Walter H,
Peter Kölln,
Mühlenwerke am Hafen,
D-2200 Elmshorn,
Federal Republic of Germany.

Weaver SH,
The Quaker Oats Company,
Merchandise Mart Plaza,
Chicago, IL 60654,
USA.

Welch RW,
Welsh Plant Breeding Station,
Aberystwyth, Dyfed,
Wales, SY23 3EB,
UK.

Williams HT,
A & R Scott,
Cupar,
Fife,
UK.

Williams Pam,
Welsh Plant Breeding Station,
Aberystwyth, Dyfed,
Wales, SY23 3EB,
UK.

Wilson DW,
Welsh Plant Breeding Station,
Aberystwyth, Dyfed,
Wales, SY23 3EB,
UK.

Wouda KP,
Semundo BV,
PO Box 2,
9970 AA Ulrum (Gr.),
The Netherlands.

Wright DJ,
Food Research Institute,
Colney Lane, Norwich,
Norfolk, NR47 7UA,
UK.

York PA,
Welsh Plant Breeding Station,
Aberystwyth, Dyfed,
Wales, SY23 3EB,
UK.